Mining and Selling Radium and Uranium

Carnotite, $K_2(UO_2)_2(VO_4)_2$, the American radium ore. From the Happy Jack mine in Utah. Public Domain. Courtesy of the US Geological Survey

Roger F. Robison

Mining and Selling Radium and Uranium

Springer

Roger F. Robison
Bloomington
Indiana
USA

ISBN 978-3-319-11829-1 ISBN 978-3-319-11830-7 (eBook)
DOI 10.1007/978-3-319-11830-7

Library of Congress Control Number: 2014955176

Springer Cham Heidelberg New York Dordrecht London

Printed on acid-free paper

Springer is part of Springer Science+Business Media (www.springer.com)

This book is dedicated to four superb historians, whose writings never fail to educate as well as entertain.

Ed Landa: geologist

Jim Marshall: chemist

Joel Lubenau: health physicist

Richard Mould: radiation physicist

Foreword

Human progress was dramatically and forever altered after the Curie's discovery of radium near the end of the nineteenth century. Fascinating lore revolving around that auspicious event and the subsequent use and misuse of radium as a medical treatment, and for other less noble purposes, is relatively well known. However, the complete radium story is much larger, antedating the isolation of radioactive element itself, and with ramifications that are felt even today.

We are grateful to the author for exposing this lesson-rich history in a single accessible volume. The scholarship involved is remarkable for its depth and completeness. Many scientific and commercial pursuits have been cross-cut as shown in this complete chronicle of radium. Anyone with professional or personal interest in radium, its origins and use, radioactivity and related topics, or a general interest in the history of science will enjoy exploring this book. Some may even find it difficult to put down without reading it through.

14 July 2014

J. Frank Wilson
MD, FACR, FASTRO

Preface

This book brings together reports from multiple scientific disciplines to produce a comprehensive history of the radium and uranium industries of the twentieth century and their effect on world events. Whereas much has been written about the utilization of uranium to produce atomic bombs, this narrative is concerned with the mining of uranium and its byproducts during the centuries before and the decades after August of 1945. It begins with the discovery of pitchblende and the appearance of Miner's Plague in the mid-sixteenth century and ends with the political and diplomatic events of 1990. The initial status of uranium as a periodic table curiosity and its minor commercial role in ceramics is described. The discovery of ionizing roentgen rays and radioactivity is seen to pave the way for the discovery of radium and radon, both of which had enormous commercial value. The creation of a radium market and the origins of radiation oncology during the inter-War period are detailed.

The sequential radium monopolies of Austria, Colorado and Belgium are examined as is their role in the procurement of uranium during World War II. The scientific work of the German scientists who identified the etiology of Miner's Plague while they overlooked the world's second largest uranium source is explained. The origins of the Cold War are traced to the Soviet obsession to procure enough uranium to achieve nuclear parity. This preoccupation alienated their G.D.R. and CZ satellites while helping to bankrupt the U.S.S.R. Documented are the combined efforts of the United States Public Health Service, Nuclear Regulatory Commission and the Environmental Protection Agency to finally gain control in 1990 over the toxicities of American uranium mining and radium products.

Contents

About the Author

Roger Frank Robison AB, MD FACP, FACR The author grew up in Indiana and Chicago. His family consisted of four generations of optometrists after his grandfather graduated from a 6 week correspondence school in 1907. Robison majored in chemistry at Indiana University in Bloomington. After graduation in 1958, he married Tiiu Raun, the daughter of a linguistics professor at the University and the Air Force language school. The Rauns had escaped Estonia in September of 1944 on a German troop ship, hours ahead of the Red Army. After graduation from Albany NY Medical College in 1963, the author served in the US Public Health Service. Dr. Robison trained in oncology at the Long Beach V.A. Hospital in CA and worked as a G.P. and an Internist at Kaiser Permanente before and after residency training.

Robison practiced oncology and taught at the Indiana University satellite medical school in Bloomington during 1970–1977. He trained in radiation oncology at M.D. Anderson Hospital in Houston during 1978–1980 under director Dr. G.H. Fletcher. Radium was then still the mainstay in the successful treatment of several cancers. Dr. Fletcher had graduated in 1941 from medical school in Belgium, where the school was amply supplied with Congo radium for brachytherapy and tele-gamma machines. In 1950 Fletcher and his physicist, L.G. Grimmett, published the first design for tele-cobalt (substitute radium) machines, which were later produced in Canada.

While practicing radiation oncology in Terre Haute and Vincennes IN, the author remained active with the MD Anderson-Fletcher Society and published articles on the radium business, tele-gamma machines, and early therapists H.A. Kelly, L. Wickham and M. Cleaves. Dr. Robison was invited to write a monograph on the history of megavoltage radiation therapy for Acta Oncology in 1995, the centennial for Roentgen's discovery. He served on the History Committee of the American Society of Therapeutic Radiology Oncology for ten years and in 2008 co-authored the fifty year history of the organization and the development of therapy as a separate specialty. During the research for that publication, the author came across so much intriguing material that he was encouraged by his mentor, historian Robert Ferrell, to put it into this book.

Chapter 1
Reich Thalers and Pechblende

1.1 Holy Roman Empire (Heiliges Römisches Reich)

By A.D. 400, the West Roman Empire was disintegrating and would not last the century. Various German tribes had occupied the Rhine–Danube border for 400 years and small groups had been slowly and peacefully immigrating into the empire. However, in 375, vast hordes of Asiatic nomads entered the German lands from the east. These forced whole German tribes, led by their kings, into a mass migration westward, looking for refuge within the Roman Empire. Within 100 years, 375–476, the Germans occupied essentially the entire West Roman Empire. After two sacks of Rome, in 410 and 455, the takeover of the residual empire by German mercenaries during 456–472 and the death of the last ersatz emperor in 476, the West Roman Empire was issued a death notice by the East Roman emperor, Zeno, in 480 [1–3].

Clovis, a German king, during 481–511, accepted Christianity, and for the next 14 generations (300 years), his tribal group, the Franks, was led by capable kings. This dynasty culminated in Charles the Great (Charlemagne: French) during 768–814. Charles (742–814) waged multiple military campaigns in an effort to unite all the German tribes into a single Christian state. This Christian objective resulted in him being crowned in 800 as "Roman Emperor" by the pope that he had returned to office, after rescuing him from a mob in Rome. On his borders, Charles (Karl: German) established a circle of fortified zones, known as Marches or Marks (e.g., Denmark), to protect against invasion or unrest in an area not completely under military control. He dealt with the chronic shortage of gold money in medieval feudal Europe by his promotion of the pound sterling silver monetary system, which was used in Britain for *1100 years*. The English pound currency sign, (£), is an abbreviation of the Latin "Libra" meaning scale or balance.

After Charlemagne, the German kings degenerated into rulers known as the Bald, the Fat, the Simple, the Stammerer, the Child, and the Pious. In *843,* this Christian German Empire was divided among the three contentious grandsons of Charlemagne (Karl the Great) by the Treaty of Verdun [1, 4] (Fig. 1.1). Louis (Ludwig: German) got everything east of the Rhine River—now Germany. Charles the Bald got everything west of the Rhone—now France. Brother Lothar got a

R. F. Robison, *Mining and Selling Radium and Uranium,*
DOI 10.1007/978-3-319-11830-7_1

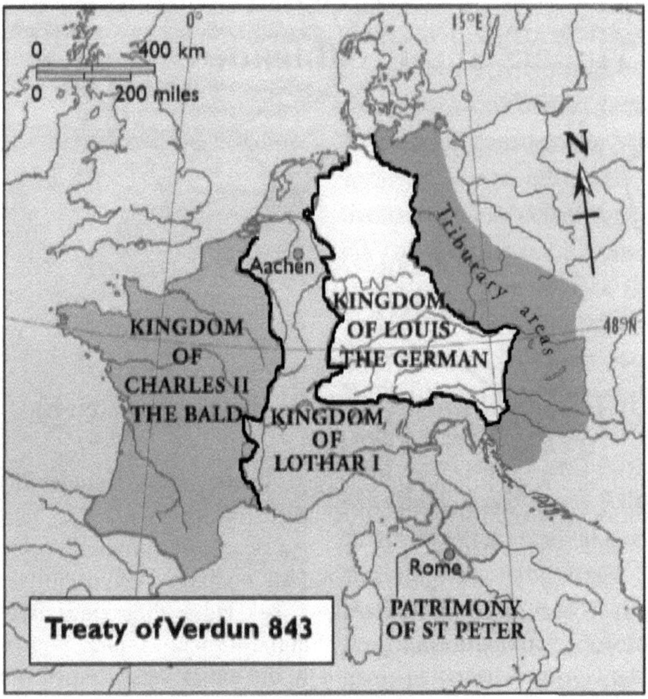

Fig. 1.1 The Treaty of Verdun: the division in A.D. 843 of Charlemagne's empire [1, 5]. (Public domain)

narrow corridor between them that extended from the North Sea to Rome. The previous year, Louis and his half brother Charles had been in armed revolt against their elder brother, Lothar. They had pledged to support each other against Lothar. They addressed each other's army, which required two languages; Romance and German [1, 5]. West Francia now spoke the Romance tongue, an early form of the French language. The Oaths of Strasbourg, which are preserved at the Biblioteque Nationale in Paris, were also written down in Latin as well as in Romance and German:

Latin translation: "And I shall never make any covenant/agreement with Lothair which/that would/shall (with my consent) harm/prejudice my brother Charles/Louis."
In Romance/Old French: "Et ab Ludher (and with Lothar) nul plaid nunquam prindrai (never make/plead any agreement) qui meon uol cist (which with my will/consent), meon fradre Karle in damno sit (places in harm my brother Charles)."
In High German: "indi mit Luthere in nohheiniu thing ne gegango (I will enter into no thing with Lothar) the minan willon (that by my will/consent) imo ce scadhen weerdhen (would cause harm to him)."

By 911, the succession of Carolingians, named for Charlemagne, had died out and the German electors opposed any king from West Francia (France). Instead, they first elected Conrad of Franconia and then his rival, Henry the Fowler of Saxony as king of the Germans. A Saxon dynasty followed. During 936–973, Henry's son, Otto I, was elected, and he agreed to investiture and coronation by the pope at Aachen. In 962, the pope crowned Otto I as Roman emperor.

Table 1.1 German kings and holy Roman emperors: 481–1806 [1–3]. (a) Merovingian Franks: 481–752. (b) Carolingian Franks: 752–911. (c) Conrad:911–919. (d) Henry: 919–936

SAXON FRANKS #1–9
1-Otto 936–973
2-Otto II 973–983
3-Otto III 983–1002
4-Henry IV 1002–1024
5-Conrad II 1024–1039
6-Henry III 1039–1056
7-Henry IV 1056–1106
8-Henry V 1106–1125
9-Lothair II 1125–1137
HOHENSTRAUFENS #10–16
10-Conrad III 1138–1152
11-Fred. Rothbard 1152–1190
12-Henry VI 1190–1197
Rivals for #13
13-Phillip[a] 1198–1208
14-Otto IV 1198–1215
15-Frederich II 1220–1250
16-Conrad IV 1250–1254
Interregnum
MOTLEY DYNASTIES #17–25
17-Rudolf Hapsburg 1273–1291
18-Adolph of Nassau 1292–1298
19-Albert Hapsburg[a] 1298–1308
20-Henry VII of Lux. 1308–1313
21-Ludwig IV of Bavaria 1314–1347
22-Charles IV of Lux. 1347–1378
23-Wenceslaus of Lux. 1378–1400
24-Rupert of Bavaria 1400–1410
25-Sigismund of Lux. 1410–1437
HAPSBURGS #26–38, 40–43
26-Albert II 1438–1439
27-Frederick III 1440–1493
Columbus 1492–1503
28-Maximillian I 1493–1519
29-Charles V[b] 1519–1556
30-Ferdinand I 1556–1564
31-Maximillian II 1564–1576
32-Rudolf II 1576–1612
33-Mathias 1612–1619
Thirty Years' war: 1618–1648
34-Ferdinand II 1619–1637
35-Ferdinand III 1637–1657
36-Leopold I 1658–1705
37-Joseph I 1705–1711
38-Charles VI 1711–1740
39-Charles VII 1742–1745
40-Francis I 1745–1765
41-Joseph II 1765–1790
42-Leopold II 1790–1792
43-Francis II 1792–1806

Lux. Luxembourg
[a] Murdered
[b] Abdicated

As the emperor was expected to protect the papacy, the pope and the emperor came to regard this as a *Heilige Romische Reich* or Holy Roman Empire. Later, Voltaire (1694–1778) famously announced that it was none of the above. However, the empire endured until the Napoleonic era (Table 1.1). Over time, the Holy Roman Empire was considered the First Reich. This would be followed by the enlightened Second Reich (1870–1919) and the infamous Third Reich (1933–1945). The Germans regarded their king as a kaiser, a corruption of Caesar. Kaiser Otto I was blessed with the discovery of an enormous ore body of silver as well as lead and copper in 938 in Lower Saxony. The silver ore was at Rammelsberg, near Goslar, in the Hartz Mountains and within his Saxon territory. Otto developed the mine and several mints to produce silver coins. Otto's son and grandson continued to mint coins during their tenures as emperor from 973–1002. The mine continued producing until 1988 when it became a United Nations Educational, Scientific, and Cultural Organization (UNESCO) World Heritage Site and tourist attraction [6].

Bohemia is somewhat diamond shaped, with its four sides surrounded by mountains which enclose an interior basin. The interior was once an inland sea which probably drained out by way of the Elbe River as it flows north to Hamburg. Bohemia measures about 20,000 mi^2 in area; the equivalent of New Hampshire and Vermont combined. Bohemia's natural mountain boundaries are formed by (a) the Erzgebirge on the northwest, (b) the Bohemian Forest range on the southwest, (c) the Giant Sudetes on the northeast, and (d) the Moravian Mountains on the southeast (Fig. 1.2) [7].

The western Sudetes Mountains of Central Europe, at latitude 50° north, are known as the Ore Mountains (German: Erzgebirge; Czech: Kusne Hory), and they have served to separate German Saxony from Czech Bohemia for over a 1000 years. They average a rather uniform 2500 ft. above sea level with the highest peak at 4182 ft. The Erzgebirge extend for a distance of about 93 mi. and have an average width of 25 mi. They run from northeast to southwest and thus face northwest. They are joined at their western tip by the Bohemian Forest Mountains, which run southeast to northwest. They also average about 2400 ft. altitude with the highest peak at 4500 ft. These two mountain ranges join to form an arrowhead configuration

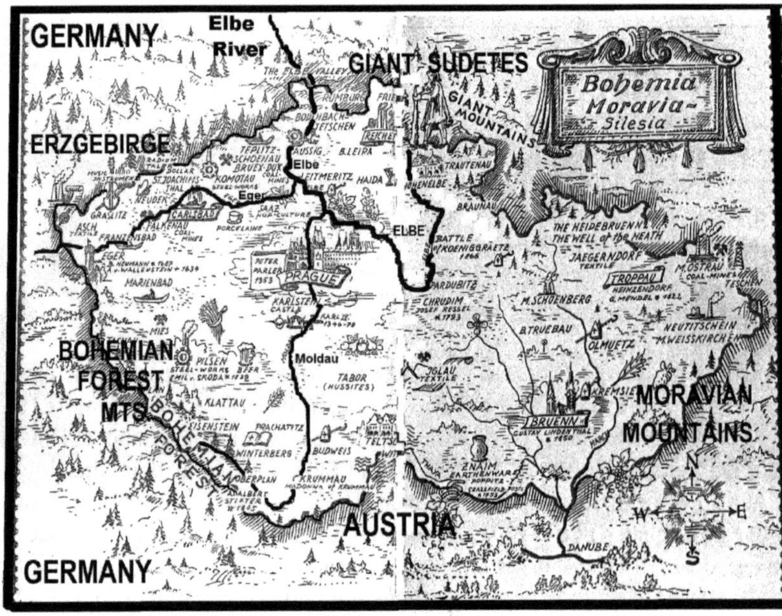

Fig. 1.2 Bohemia's mountain borders [7]. (Public domain)

pointing west. Adolph Hitler, in 1938, claimed it was a dagger pointed at the heart of Germany.

The Giant Mountains (*Riesengebirge:* German) form the eastern mountain border that extends from the Neisse River east to the Oder River. The *Riesengebirge* separate Poland's Silesia district on the north from the eastern Czech Republic on the south. As their name suggests, the Giants contain the highest peaks in the Sudetes range; 4890–5256 ft altitude. The Moravian Mountains are low and poorly defined leading to a natural connection with the state of Moravia [7–9].

The northern slopes of the Erzgebirge had been mined for centuries by Saxons for iron and lead. However, the mountains were covered with dense forest. The climate was regarded as too cold to support a growing season because of an average annual temperature of 37–41 °F (3–5 °C). Oberweisenthal, on the Saxon border, had only 140 days a year that were frost-free. As a consequence, the Erzgebirge were largely uninhabited, until the discovery of silver in *1168* in the Freiberg area of Saxony. On the Saxon side, the Erzgebirge slope gently and gradually. However, on the Czech side, they rise like a wall above the valley of the Eger River. The rocks are predominantly granite, followed by shale—phyllites and glimmer shale. The steep southern, Bohemian, slopes were also generally uninhabited [10].

The *1168* silver strike occurred at Christiandorf in the Mount Silberberg area of southern Saxony, near the northern slopes of the Erzgebirge. The town of Freiberg (free mountain) was eventually organized there in *1185*. Freiberg would become the dominant German mining center. The silver strike resulted in a rush of German prospectors to the Erzgebirge and the Sudetes Mountains. After 1200, there was a significant ethnic German migration across the Sudetes into Czech Bohemia. Over time, the Sudetes Mountain areas in the north and the Forest Mountains on the west and south, around and within the periphery of Czech Bohemia, became German settlements. Germans occupied both sides of the Erzgebirge. On the Bohemian side of the border, they settled mostly in the foothills, where they became a majority. These other peripheral areas became identified in the twentieth century as the "German Sudetenland," even though the Sudetes Mountains were only in the north [7–10].

Saxon miners became known throughout Europe for their expertise and their status as freemen (from free mountain) and not feudal serfs nor slave miners. Silver strikes in Europe continued during the thirteenth and fourteenth centuries. In 1410, mines opened in Schneeberg, Saxony, for copper and iron. Silver was later found there in 1470, and the mines were also worked for arsenic and cobalt. Shortly thereafter, there were 153 mines in the area [11, 12]. The late fifteenth century also saw silver strikes in areas of Aue (1479) and Annaberg (1496). Today, there is a 230-km scenic route in Saxony [13], called the "Silver Mines Route," which directs tourists to: Zwickau, Aue, Annaberg-Buchholz, Ehrenfriedersdorf, Geyer, Schlettau, Obernhau, Seiffen, Frauenstein, and Dresden (Fig. 1.3).

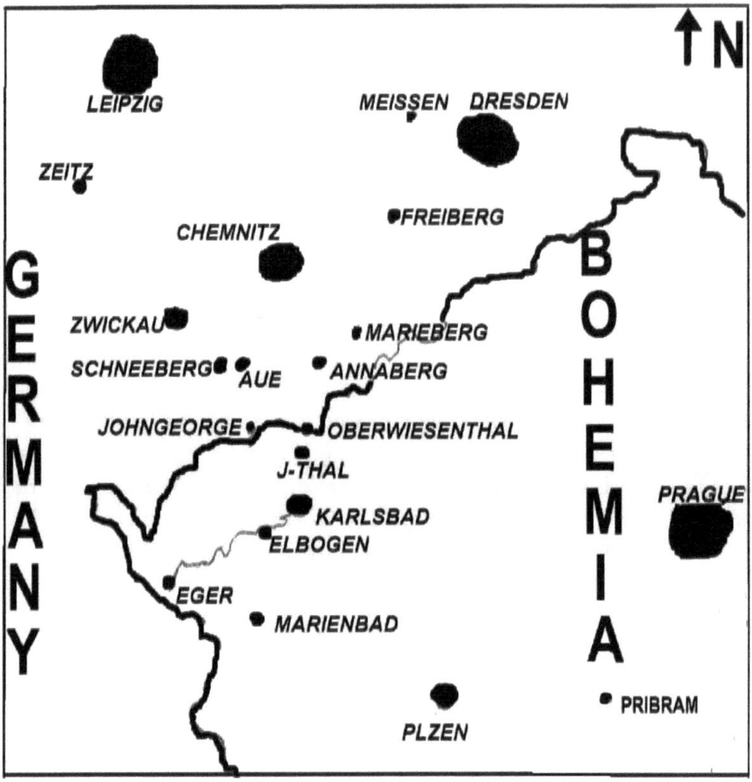

Fig. 1.3 Towns of the Erzgebirge. (Author)

1.2 Czech Bohemia Resists

After the Treaty of Verdun, a Great Moravian Slavic Kingdom was established during 830–896. It included Bohemia, Moravia, Slovakia, southern Poland, and western Hungary. This Christian kingdom was a buffer between the Roman Catholic German lands and the Greek Orthodox lands of the East Roman Empire (Byzantium) of Constantinople. In 863, Prince Rostislav of Moravia requested the Byzantine emperor Michael to send missionaries into the kingdom and help train the clergy to use the local language rather than German. The brothers Cyril and Methodus, Greek Orthodox monks or priors, arrived and created the Cyrillic alphabet (Slavonic script) for the Slavs of Eastern Europe. It is still used today by 250 million people in Russia, Eastern Europe (Ukraine, Bulgaria, Serbia), and Central Asia. It is the third official script of the European Union, after Latin and Greek [9, 14–17].

During *896–906*, the Moravian Kingdom was defeated by the Hungarian Magyars. For protection, the prince/duke of Bohemia was obliged to become a vassal of the East Frank (German) Kingdom. The remnant Czech Kingdom of Bohemia and Moravia was obliged by the pope to quit the Cyrillic alphabet and replace it with the Latin alphabet. The prince/duke, during 907–921, had to pay tribute to the German king, Henry the Fowler, and to fight off repeated Magyar attacks.

In 921, Wenceslaus (907–935) succeeded to the throne and developed a reputation for his acts of charity to the poor. He was murdered by his brother, probably under the influence of their mother (a pagan).

He became a Christian martyr and the patron saint of the Czech people. Emperor Otto I elevated him from duke to king posthumously. In 1853, the legend reached Victorian England where the venerable Christmas carol "Good King Wenceslaus" was written [18].

Bohemia's status improved in 955, when Otto I, the new H.R. emperor, awarded the March of Moravia to the Bohemian prince Boleslav I. Bohemia then became a Duchy or semiautonomous fief of the empire. A separate king of Bohemia, Otakar Premysl I (1198–1230), was acknowledged by the emperor in 1212. The second Bohemian king, during 1230–1253, encouraged extensive German immigration during the thirteenth century in order to help replace population losses after the brief Mongol invasion of Europe during 1240/1241. After the fall of the House of Hohenstaufen and the interregnum of 1254–1273, the authority of the emperor was weakened. Otakar Premysl II (1253–1278), became the third Slavic king of Bohemia, and he extended his domain from the Baltic to Croatia on the Adriatic. He seized Vienna and Austria as well.

In 1273, new Pope Gregory X and the German electors settled on a military man, Count Rudolph of Hapsburg, to be emperor #17 (1273–1291). Otokar II was chagrined that he had not been selected as emperor. Otakar II considered Rudolph a "petty count" from Alsace. Rudolph quickly laid siege to Vienna and Otakar had to withdraw to Prague. Otakar returned in 1278 to confront Vienna with a large army. In a surprise attack, Emperor Rudolph routed the Bohemians and personally killed Otakar. Emperor Rudolph then consigned the small duchy of Austria to his Hapsburg family in 1278. They subsequently concentrated on acquiring and dominating multiple small areas of the Germanic kingdoms [19].

Rudolph's son, Albert Hapsburg (1298–1308), was elected emperor #19. Bohemia was reduced to its natural mountain frontiers and Wenceslas II, king of Bohemia and Poland, agreed to be a fief of the empire. After he died of consumption in 1305, his heir, Wenceslas III, made peace with his uncle, Emperor Albert, by ceding the western Egerland to the empire. Wenceslas III was then promptly murdered in 1306. Albert then put his son Rudolph II on the Bohemian throne in 1307 but Rudolph died of dysentery the same year. Albert then decided to just go ahead and occupy Bohemia but he was murdered by a nephew in 1308. The electors had had enough of the Hapsburgs for now and Kaiser Albert (# 19) was followed by six emperors from either French Luxembourg or Bavaria [1–3, 19].

John (Jean) was the son of emperor (# 20) Henry VII of Luxembourg and John married a princess of the former Bohemian royalty in 1310. John joined his French relatives in battle against the English and died on the battlefield in 1346. His son Charles IV then became king of Bohemia and Luxembourg. He was selected to become emperor #22 during 1347–1378. Charles favored his mother's interests more than his father's, and he worked at improving commerce and culture in Czech Bohemia. He was born and died in Prague, which he made his imperial capital. He had been educated in France, and he rebuilt Prague along the lines of Paris and established a university. Charles University was posthumously dedicated in his name in 1383. The Czechs consider him the father of their country and his reign, as the Golden Age of Bohemia [9, 14–19].

Charles IV was widowed four times: 1348, 1353, 1362, and 1393. He had 13 children by the four wives who were French, Bavarian, German–Hungarian, and German–Polish. Charles was succeeded by his eldest son Wenceslaus IV, as emperor and the king of Bohemia in 1378. However, Emperor Wenceslas proved inadequate and unpopular. In 1400, the electors dethroned Wenceslas and elected Rupert of Bavaria (1352–1410) as emperor # 24. However, Wenceslas IV remained king of Bohemia until his death in 1419. Rupert (# 24) died of dysentery, and Sigismund, the second son of Charles IV, was selected emperor # 25, from 1410 to 1437.

Kings of Bohemia	Reign
Otokar Premysl I-Slavic	1212–1230
Wenceslas Premysl I-Slavic	1230–1253
INTER-REGNUM	1254–1273
Otokar Premysl II-Slavic	1253–1278= killed by emperor
Wenceslas II Premysl-Slavic	1278–1305= died of tuberculosis
Wenceslaus III Premysl-Slavic	1306–1307= murdered
Rudolph Hapsburg-son of emperor	1307–1307= died of dysentery
John of Luxembourg-son of emperor	1310–1346
Charles IV of Lux.-Emp. # 22	1346–1378
Wenceslaus IV-Emp. # 23	1378–1419
Sigismund-Emp. # 25	1410–1437
Emperors # 26, # 27	1437–1458
Jiri Podebrady-Slavic	1458–1471
Vladislav Jagelowski-Slavic	1471–1516
Louis/Ludwig Jagelowski-Slavic	1516–1526= died at Mohacs

Emperor Sigismund called the Council of Constance in 1414–1418 to resolve the Papal Schism or the Three Pope Problem. The three pretenders were all rejected and a new pope was elected. The council also found time to condemn to death the heretic Jan Hus, a reforming Bohemian cleric who preached ecclesiastical reform and Czech nationalism. Emperor Sigismund invited Hus to the council to explain his position and assured him safe conduct. Hus was then imprisoned and burned at the stake in July 1415, while Sigismund looked the other way. This led to the Hussite Wars (1419–1434) which pitted the Czechs against the Catholic Holy Roman Empire, the emperor and the pope for 17 years. During this time, the emperor was not welcome in Bohemia. The Hussites forced the emperor and the pope to recognize some degree of Czech nationalism and some religious freedom. This was confirmed at the *Council of Basel* in July 1436 and finally resulted in Sigismund being accepted as the Bohemian king, 1 year before his death [14–19].

Sigismund died without heirs in 1437, and the electors turned once again to the Hapsburgs, who would now control the empire until its demise in 1806; emperors #26–#43. Albert II, of Hapsburg, succeeded Sigismund for 1 year, 1438–1439, before dying of dysentery after eating too many melons. This was followed by the long reign, from 1440 to A.D. 1493, of his nephew, Kaiser Frederick III. His reign was described as 53 years of disaster and ignominy as he lost control of Bohemia, Hungary, Austria, and Vienna. Due to his imperial impotency and financial impoverishment, the Bohemians were encouraged to seek independence from the empire [19].

In 1458, the Czechs selected Jiri Podebrady, a pre-Protestant heretic and Hussite, to be the king of Bohemia. The angry pope, in 1462, declared the Basel agreement of 1436 to be null and void. He then excommunicated the king of Bohemia in 1465. However, before his death in 1471, King Podebrady and the Czechs gave the Crown to the Polish-Lithuanian prince Vladislav II Jagelowski (1471–1516). This successfully thwarted the Catholic Hapsburgs; temporarily. In 1490, Vlad became king of Hungary as well. His son Louis (Ludwig) II Jagelowski followed his father as king of Bohemia and Hungary in 1516, but he would be the last of the Slavic kings of Bohemia [9, 14–19].

1.3 Schlick Silver Mint

Towards the end of the thirteenth century, the German Schlick family became prominent in the northwestern Bohemian Sudetenland. They were based in Eger, now Cheb, in the Czech Republic, which was a favorite destination and travel stop for the emperors and kings of Bohemia. The Eger River continues east some 19 mi. (30 km) to Elbogen, where the river sweeps around the town in what was called an elbow. The Eger–Elbogen area had a German presence since the ninth century and had been ceded to the empire in 1305/1306 by the Bohemian king just before he was murdered. Elbogen is now Loket in the Czech Republic [10, 17]. Patriarch Heinrich Schlick was a businessman (Kaufmann) and councilman (Ratsherrn) in Eger, where his son Kaspar (1396–1449) was born. Kaspar attended the new University of Leipzig (founded 1409) in Saxony and in 1415 entered the diplomatic service of emperor #25, Sigismund (1410–1437), who was also the king of Hungary and the king of Bohemia. Schlick served on several diplomatic missions to the Ottoman Turks and to the Hussites during the Hussite Wars of 1420–1434. Kaiser Sigismund appointed Schlick as his representative to the Council of Basel in 1432/1433, when the Hussite claims were confronted by the papacy. Kaspar was appointed chancellor to the emperor in 1433. As such, he represented the emperor in negotiations with the pope in Rome. He also represented the emperor in negotiations with the Prussians, and the Poles and Lithuanians who contested for the Bohemian Crown after the Kaiser's death in 1437. Schlick continued as chancellor until his death in 1449; serving Kaisers Albert II (1438–1439) and Frederick III (1440–1493) as well [20–22].

Schlick rose through the ranks from commoner to count during the reign of Emperor Sigismund. In 1422, at age 26, he was made Baron of Weisskirken, now Holic, Slovakia. Schlick was also given an apparent honorary title to Passau in Italy: *Graf von* Passau (*richtig Bassano, bei Teviso* in *Ober-Italien*) The Italian property, while possible within the Holy Roman Empire, belonged to Venice. He was ennobled in 1437/1438 as Count (*Graf*) Schlick of Passau (*Bassano* in Italian) and Weisskirken. The properties in Italy were so far away as to be considered only honorary in nature.

Years	Rise of count (Graf) Kaspar Schlick: 1396–1449 [20–22]
1415–1417	*Attache* to emperor on trips to Spain, France, England
1416	Clerk, royal court. (Scribe)
1422	Made Baron *(Freiherrn)* of Weisskirken (Holic, Slovakia); age 26
1426	Secretary, royal court. (*Secketar*)
1427	Court official (*Protonotar)*
1429	Vice-Chancellor (*Vizekanzler*)
1433–1449	Chancellor to three kaisers: Sigismundd, Albrect II, and Frederick III
1434	Acquires Elbogen (Czech Loket) Domain; adjacent to Eger lands
1437	Acquires Schlackenwerth Domain in Erzgebirge
1437	Made *Reichsgrafen von Passaun (Bassaun,* Bassano) in Italy-honorary
1438	Made *Grafen* (Count) von *Passaun and Weiskirken.*

However, Count Slick did secure land that was not honorary. His service as chancellor to the emperor led to significant land acquisition in the Sudetenland of western Bohemia: (a) in Elbogen and (b) on the southern slopes of the Erzgebirge (1437). The Kaiser pledged the property to his chancellor. The original Schlick family estate in the Eger country was thus extensively enlarged by the acquisition of the Elbogen Domain, 19 mi. east of Eger, and the Schlackenwerth domain in the Northwest Erzgebirge. The Erzgebirge property was a relatively uninhabited forested mountain landscape that was obtained from Herrshaft Schlackenwerth and was 32 mi. (51.5 km) east from Eger. From 1437 and later, the large Schlick family was in possession of northwestern Bohemia from Eger to Elbogen to the Erzgebirge (Fig. 1.3).

The ennoblement allowed the Schlicks a coat of arms and it was a dandy; a shield divided into four quadrants with a feathered and crowned helmet atop the shield (Fig. 1.4) [23, 24]. The Schlick family estates were represented in the lower left and upper right quadrants, with three circles (annulets) on a trifurcated background (*tierced en chape*); basically a triangle with three rings. There was one red circle on a white background and two white circles on red backgrounds in the three parts of the quadrant. This logo was repeated superior to the gold crown at the top of the shield. The three rings might have represented the three properties of the Schlicks in Eger, Elbogen, and Schlackenwerth. The lower right and upper left quadrants featured a gold lion with *one* tail on a blue background. The lion is holding a red-roofed white church and is presumed to represent the property of WeissKirken, which was in Bohemia–Moravia but is now Holic in Slovakia. The escutcheon or small shield in the center contains the arms (charge) of Bassano, Italy; two red panthers holding a blue tower [23, 24].

Fig. 1.4 Schlick coat of arms [23, 24]. (Public domain)

Kaspar had married late in life, 1444, to Agnes, a daughter of the Osler Duke, Konrad V Kunstler (1381–1439). Agnes, who died at age 38 (1411–1448), and Kaspar, who died at age 53, were without known issues. Kaspar had been able to secure a Bishopric for his brother Heinrich II and his brother Franz took Holy Orders as well. However, the Schlick line continued through Kaspar's brother Mathaus (1413–1487), whose three sons inherited the Eger, Elbogen, and Erzgebirge properties [22]. By 1516, it was Count Stefan Schlick, the grandson of Mathaus, who inherited the Erzgebirge property as the *Herr auf Schlackenwerth* (Table 1.2) [20–22].

During *1512–1516,* Saxon miners were prospecting in the uninhabited forest valley that had previously been occupied by the settlement of Konradsgrun from 1300 to 1450. Due to the forests, harsh climate, and short growing season, the area had been found unsuitable for farming. The area was part of the Schlackenwerth property owned by the Schlick family and was only 3 mi. from the Saxon border at Oberweisenthal. The prospectors found some silver in an old abandoned drift in the deep valley on the southern slopes of the Erzgebirge. The silver was found in a black mineral ore, resembling hardened pitch, which was later known as *pechblende.* The mine had to be abandoned because of a lack of capital. After learning of the silver discovery, Count Stefan Schlick moved to form a mining group to rework the abandoned drift. After digging another 2 m, they discovered a major deposit in 1516 [20–22].

The strike brought a rush of miners and the new settlement was initially called "Tal" (Ger.: valley). Saxon miners were available because of declining mine work at

Table 1.2 Schlick properties [20–24]

Era	Schlicks	Heirs	Properties
????–1431	Heinrich I	Five sons: Kaspar, Heinrich II, Franz, Nikolas, Mathaus	Eger businessman and councilman
	Sons of Heinrich		
1396–1449	a-Kaspar	Childless	Eger, Elbogen, Erzgebirge
????–1448	b-Heinrich II	Took holy orders	Bishop of Freising 1443–1448
	c-Franz	Took holy orders	
	d-Nicholas	Heirs childless	
1413–1487	e-Mathaus	Inherited from Kaspar	Graf von Passau; Eger, Elbogen, and Erzgebirge
	Sons of Mathaus		Inherited from Mathaus in 1487
1437–1516	a-Nikolaus	Die Falkenauer	Inherited Eger
	b-Hieronymus	Die Elbogner	Inherited Elbogen
	c-Kaspar II	Kaspar II had eight sons	Graf von Passau and Herr auf Schlackenwert
	Sons of Kaspar II		
1480–1524	Heinrich III	Died at Mohacs	Graf and Herr auf Schlackenwert
1487–1526	Stefan	Died after Mohacs	1526: Graf and Herr auf Schlackenwert
????–1526	Lorenzo	Inherited via Stefan	Estates confiscated by Hapsburgs:1545–1547
1494–1550	Hieronlymus II		
1505–15??	Kaspar III, Mathaus		
	Burian, Wolfgang		

Schneeberg, Marienberg, Freiberg, and Annaberg as well as other areas in Bavaria and Bohemia. In 1517, Tal was renamed as *Tal des Heilige Joachim's* (valley of Saint Joachim) or simply *Sankt Joachimsthal.* The Saxon miners had named some of their mining towns in southern Saxony, on the slopes of the Sudetes, after the Virgin Mary and her mother Saint Anne—Marienberg and Annaberg. In Bohemia, the Saxon immigrants therefore chose "Sankt Joachimsthal" for the new mining town. *Sankt Joachim* (St. Jacob: English) being the husband of Saint Anne and the father of the Holy Virgin. Joachim and Anna were the parents of Mary, the mother of Jesus Christ, and thus they were the grandparents of Jesus [25].

The Schlick family purchased land in the new town and Stefan Schlick built a castle, *Schloss Freudenstein,* near the mine site as a security measure. Schlick recruited Saxon experts and issued new mining regulations modeled on those at Annaberg. The Schlicks had allegedly been granted minting privileges in 1489 from the king of Bohemia, Vladislay Jagelowski (1471–1516). Regardless, they started minting the first prototypes of their own large silver coins in the basement of "Schloss Freudenstein" in 1519. The coins were issued by the noble family of Schlick and not the Bohemian monarch. On 6 January, 1520, Ludwig (Louis) Jagelowski, the current king of Bohemia and Hungary (1516–1526), confirmed the mint privileges of the Schlicks. He also elevated the settlement to the status of a royal "free mining city," with all the rights and privileges of such, including a coat of arms [25–27].

Joachimsthal became one of the largest mining centers in Europe during the early sixteenth century. By 1520, the town had about 5000 people, and the silver strike resulted in the settlement of nearby areas as well. During 1520–1534, eight new mining towns were surveyed. The city population reached 13,000–15,000 during 1525/1526 and boasted 29 mines and a mint, while the entire valley had over 1200 houses, 12,000 miners, and 800 operating mines. The mining boom in Joachimsthal, at a zenith during 1516–1540, produced an estimated 3 million ounces of silver. By 1534, the town population was 18,500, second only to Prague (50,000) in all of Bohemia. During 1534–1540, Joachimsthal adopted Lutheranism, built the first ever Protestant-style church and developed into an important cultural center [26–29].

At first, the counts Schlick sold their silver to Augsburg and Nuremberg. They started to mint their own coins in 1519 and delivered the coins to Leipzig, a Saxon trading center. The Schlick silver coin was required to match the value and purity of the 1450 standard; a Saxon gold coin called the *Gulden* (German: gold) which contained 3.5 g of gold and was the equivalent of 30 g of silver. European silver discoveries had led to the minting of large silver coins weighing 30 g, which would be equal in value to the gold *Gulden.* In 1486, the Tyrol had issued the *first* of what was called a *Guldengroshen,* a gold-like silver coin. It weighed 31.9 g (1 Troy ounce = 31.1 g) and contained 29.93 g of pure silver [22, 30, 31].

The first German silver *Guldengroshen* were issued in 1500 in Annaberg, and they were meant for significant commercial circulation. They weighed slightly less than one Troy ounce. Their diameter of 43 mm (1.7 in.) gave the engravers an artistic opportunity to depict their monarchs in realistic portraitures. They also preserved for us the heraldic arms and standards of the various countries and principalities.

The Schlick coin was called a *Joachimsthaler Guldengroshen* after its place of origin. The name was corrupted, over time, to *Joachimsthaler,* and finally to *thaler. Thaler* became a general term for a large, 40+/− mm silver coin. They became known in different countries as thalers, talers, dalers, and dollars. The Schlick mint exported to Saxony, 2.2–3 million thalers, which supplied the international market from 1519 to 1538. The Schlick coins were the first large Bohemian silver coins and allegedly the first to bear the year of minting. Thalers from the Schlicks' mint had exceptional quality and a constant silver content compared to others [22–24, 30, 31] (Table 1.3).

The first Schlick coins featured a likeness of *Sankt Joachim* on the Obverse, with the Schlick coat of arms at his feet (Fig. 1.5) [23, 24]. In 1525, St. Joachim was identified by an S on his right side and an I (for J) on his left. And the mint date was also split with 15 on his right and 25 on his left. The text on the rim has abbreviated words, which start at 12 o'clock, just to the left of the figure's head; AR*DOMI:SLI:ST (the Schlick coat of arms) E:7:FRA*C(M):D:BA. This is believed to represent: ARgento (silver), DOminion of SLIick Stefanus Et (and) Seven FRAtre (Seven brothers); COMes (Counts) of Dominion BAssano (Passanau). The seven brothers were thought to be Burian, Hieronymus II, Mathaus, Wolfgang, Heinrich III, Kaspar, and Lorenz.

Table 1.3 History of the *Reich Thalers* [22–24, 30–32]

Year	Coin—issued by government of	Grams of silver	Coin weight in grams × % silver
1486	First Guldengroschen—Tyrol	29.93	31.9 × 0.938
1500	First Saxony Guldengroschen—43 mm	27.38	29.2 × 0.938
1519–1528	Schlick Guldengroschen—41 mm	27.21	29.2 × 0.931
1528–45	Royal Bohemian Joachimsthaler	27.21	29.2 × 0.931
1547	HRE Bohemian thaler	26.16	29.2 × 0.895
1566–1750	HRE Reichsthaler	25.96	29.2 × 0.889
1667	Saxony thaler	25.0	28.1 × 0.889
1690	New German thaler	23.0	25.9 × 0.889
1754	Convention-thaler: Austria and Bavaria	23.3	28.0 × 0.833
1857	Veriens-thaler: Austria and Germany	16.6	18.5 × 0.900
1872	Last Thaler minted in Saxony		
1908	Thaler withdrawn in Germany		

1794/1795: First USA silver dollar: 39–40 mm; 24.06 g silver (26.96 g wt. × 0.924 purity)
HRE Holy Roman Empire

Fig. 1.5 1525 Thaler obverse, St. Joachimsthal [23]. (Public domain)

Fig. 1.6 1525 Thaler reverse, Bohemian lion [23]. (Public domain)

Figure 1.6 is the 1525 reverse and features the magnificent Bohemia Lion with his double tail; *doppel-schwanziger*. The text starts at the Lion's head and reads as follows: (X)LVDOWIC*PRIM* DEI*GRA*REX*BOEMIE. Presumably, this represents Ludwig the First by Grace of God, King of Bohemia.

Ludwig (Louis) Jagelowski (1506–1526) was the heir of the Polish-Lithuanian kings, whom the Czechs had supported since 1471 to keep the empire and the Hapsburgs out of their country. His father was king of both Bohemia and Hungary due to the weak emperors Frederick III (1440–1493) and his son Maximillian (1493–1519). Louis was crowed as King of Hungary in 1508 at age two and as king of Bohemia at age three in an effort to thwart the Hapsburgs. Louis did manage to succeed his father at age ten. However, the wily Hapsburgs sensed an opportunity and forced the betrothal of Louis to the granddaughter of Maximillian, when both were aged nine! Emperor Maximillian then arranged their marriage when both parties were aged 16; in 1522 [14–19].

1.4 Battle of Mohacs

After the fall of Constantinople and the remnants of the East Roman (Greek) Empire in 1453, the Turkish Muslims had continued their military advance up the Balkan Peninsula. The brilliant general Suleiman (1494–1566) had started his reign in 1520 by capturing Belgrade (Serbia) on the Danube. This conquest opened the way to Budapest and Vienna and Western Europe. After the expulsion of the Christian Crusader states from the Levant in 1291, the Hospital Knights of St. John, one of only three Orders to survive the Crusades, had continued to oppose the Muslim advance into southern Europe from their naval base at Rhodes. However, on 1 January 1523, after a prolonged siege, Suleiman forced the Knights from the island of Rhodes, where they had been since 1306. Suleiman's forces were now masters of the western Mediterranean as well as the eastern. In 1526, Suleiman was ready to attack Europe again, and he returned to Belgrade with 200,000 troops and crossed the Danube into Hungary. They planned to take Buda and Pest and then the Austrian/Hapsburg capital of Vienna, the gateway to all of Western Europe. Meanwhile, in the Mediterranean, Emperor Charles V relocated the Knights of St. John to Malta, where they would inflict a rare defeat upon the aged Suleiman at the Siege of Malta in 1565 [1–3, 19, 32–34].

In 1526, Emperor Charles V, the defender of the faith against the Muslims, was having a hard time mobilizing a defense force against Suleiman. The king of France supported the Muslims against the empire and the newly emerging Protestants distrusted the Catholic emperor's call to arms. The Christian forces were to include soldiers from Hungary, Bohemia, Poland, the Papal States, the empire, Bavaria, and Croatia. They were outnumbered over 2:1 as half their forces did not arrive in time. Nonetheless, 19-year-old King Ludwig (Louis) of Hungary and Bohemia hastily assembled what forces he could muster and met Suleiman at Mohacs, Hungary, on the northern bank of the Danube on 29 August 1526. Accompanying him were the Counts Schlick: Stefan and Lorenzo.

The Europeans were soundly defeated, while suffering casualties of over 10:1 against the Muslims. Young king Ludwig was unhorsed and drowned in a swamp during the retreat. He was only 20 and childless with a young Habsburg wife, Mary. Count Stefan Schlick was one of the 15,000–20,000 Christian casualties on the battlefield. Count Lorenz Schlick survived the retreat but later died of his wounds. Stefan Schlick is still revered in Joachimsthal, now Jachymov, for his efforts and the city erected a memorial in 1924 acknowledging him as the great benefactor of the city. Under his sculptured portrait, the local Germans proclaimed him: *Dem Grossen Wohltäter Grafen Stefan Schlick. In Dankbarkeit. Gewidmet von der Stadt* (The Great Benefactor Count Stefan Schlick. In Gratitude. Dedicated by the city) (Fig. 1.7) [28, 29].

After the defeat at Mohacs and the death of King Ludwig, southern Hungary was lost to the Muslims for the next 160 years. Northern Hungary and Bohemia came under the domination of the Holy Roman emperor and the Hapsburgs. Emperor Charles V (1519–1556) had already made his brother Ferdinand, the Archduke of Hapsburg, Austria. Ferdinand was the brother-in-law of the late king Ludwig, whose sister, Princess Ann of Hungary, had been coerced into marriage with him. Now, the emperor made Ferdinand the king of Bohemia and king of what remained of Hungary; that which was not under Turkish rule. The Mohacs defeat ended the Jagelowski dynasty in Bohemia and brought 400 years of Austrian Hapsburg domination of Bohemia; 1526–1919 [19].

Fig. 1.7 1924 memorial to Count Stefan Schlick in Jachymov [28, 29]. (Courtesy J. L. and V. R. Marshall, "Rediscovery of the Elements," 2010, ISBN 978-0-615-30795)

In 1528, the new Bohemian king, Austrian, Archduke Ferdinand, in concert with the Council of Bohemian States cancelled the mint privileges of the surviving Counts Schlick. It was declared that mint privileges were reserved only for the king, meaning Ferdinand. However, the high quality of the Schlick Thaler and their international popularity begged otherwise. The Hapsburgs had to compromise. The Schlicks were appointed as administrators of the Royal Mint for the next 10 years; 1528–1538. However, the Schlick Arms and St. Joachim disappeared from the *Joachimsthaler*. Archduke Ferdinand replaced St. Joachim on the new Obverse for 1530 [32]. On the reverse, the Lion of Bohemia was reduced to sharing the field with the Hungarian standard. Coin production moved to a new Mint Building in *1536*.

In the latter half of the sixteenth century, the mining industry and the population of Joachimsthal decreased continually due to (a) the flood of Spanish-American silver, (b) diminishing silver deposits, and (c) the Protestant Reformation. Silver from the Americas began to appear in Spain after the conquests of Cortez in Mexico in 1520 and of Pizzaro in Peru in 1532. Spanish silver, obtained with slave labor, began to dominate the Bohemian competition. Then the silver began to run out even when the mines extended to a depth of 7–10 km. During 1631–1655, only 838 kg of silver was mined. The Joachimsthal Imperial Mint was moved to Prague in 1671 [25–29].

1.5 Pechblende

Finding increasing amounts of *pechblende* ore was considered bad luck for the Saxon silver miners as it meant that the silver deposits were running out. *Pechblende* is a black mineral ore, resembling hardened pitch, which is about 50 % uranium and 50 % copper, bismuth, barium, and lead along with small quantities of just about every other known element. The miners had discovered it while mining silver and cobalt. The miners had no use for this black pitch (*pech swartzen*) rock and discarded it. They described it as *pechartige sterile Blende* (English: a sterile mixture of a species of pitch). *Blende* is German for mix, blend, or mineral. *Pech* was German for black, sticky, dark, and/or bad luck; as "pech haben" is to be "unlucky" [25, 35]. *Pechblende* thus had a double meaning.

The first person to mention the mineral *pechblende* was Johannes Kentmann of Saxony in 1565 [36, 37]. He was born in Dresden in 1518 and died in Torgau, Germany, in 1574. The *Krankenhaus* (hospital) in Torgau is named in honor of Kentmann. Torgau is now remembered as the town on the Elbe River, where the invading armies of the USA and the Soviet Union met on 25 April of 1945. Kentmann studied at Leipzig, Wittenberg, and Nurnberg from 1540 to 1546, when he obtained his M.A. He then studied medicine at Padua and Bologna, where, in 1549, he obtained his MD. His interest in natural history developed during his stay in Padua, where he was involved with their Botanical Gardens. In 1550, on his trip home, Kentmann visited with the remarkable Conrad Gesner (1516–1565), who was a Zurich physician and naturalist in Basil. Gesner was also known for his work as a botanist, geologist, zoologist, linguist, bibliographer, and publisher, among other pursuits, including the invention of the pencil [38].

After returning to Saxony, Kentmann was appointed a state physician in Meissen. He relocated to Torgau in 1554 and held a similar medical position there until his death in 1574. His publications spanned the fields of medicine, biology, geology, and botany. On a commission from the Saxon Elector, Augustus, in 1563, he produced a book on herbs (*Krauterbuch*) that described some 600 plants. This was embellished by the illustrations of a renowned local artist, David Redtel. Shortly afterward, he published a medical treatise, with illustrations, on the 12 types of stones which originate within the human body: *Calculorum qui in corpore ac membris hominum innascuntur, genera XII.*

During his 20 years at Torgau he also began to collect and classify fossils, rocks, and minerals of that region of Saxony. Not content with internal stones, he also collected external ones. He assembled a large collection of these which he arranged and classified, more or less, according to the crude sixteenth-century system of Agricola. He kept the actual examples of his mineral collection in a *Mineral Cabinet,* which had 13 drawers, containing 26 different groups of minerals. In total, the cabinet held 1608 specimens from 135 locations; 1136 specimens were from Saxony and 472 from other countries. This Cabinet was allegedly the first geological museum in the Western world [39]. In 1565, Kentmann published his extensive scientific study of these fossils and "things dug up": *Catalogues Rerum Fossilium.* This *Catalogues* was initially published by his friend Conrad Gesner in 1565.

Fig. 1.8 Profile of J. Kentmann [40]. (Public domain)

Gesner also published in 1565 a *Compendium* on The Entire Subject of Mineral Classes (*De Omni Rerum Fossilium Genere*) which contained reprints of eight tracts by seven authors concerning fossils, metals, minerals, and gems. Both Kentmann's *Catalogues* and his booklet on *Calculorum* were included in this Compendium along with a woodcut (Fig. 1.8) [40].

In his *Catologues,* Kentmann relates the oldest known mention of *pechblende* [40]. His chapter XXII is concerned with *Molybdaena* and *Plumbago.* Mineralogists from the time of the Romans until a century after Agricola (1494–1555) were greatly confused about the multiple Latin terms used for lead; (a) *molydaena, (b) plumbago, (c) plumbum, (d) galena* and (e) *spuma argenti* [41]. Kentmann describes the various Latin terms including 26 varieties of *plumbago* and six types of *plumbago sterilis*. The Latin term *sterilis* was to signify unadulterated or unalloyed. Kentmann described a *Plumbago Sterilis pici similis*; a sterile lead that is similar to black pitch. He called this *Bechblende.* Between 1500 and 1600, there was a shift in German vowels and *Bechblende* became *Pechblende.* The French, among others, also use the German term *pechblende,* not the English corruption pitchblende.

1.6 Religious War

The citizens of Bohemia found themselves in the geographic center of the developing Protestant–Roman Catholic religious wars of the sixteenth and seventeenth centuries. The Latin Church was being attacked by Luther, Calvin, Zwingli, and Strum during 1517–1536 because of alleged abuses. The Lutheran Church was founded between 1530 and 1560. Luther followed the doctrine of John Hus, the Czech martyr. Lutheranism was embraced throughout northern Europe, north of the Pyrenees and the Alps. Saxon Germany and most of Bohemia adopted Protestantism. *Joachimsthal* had become an important Protestant cultural center and maintained a famous Latin School from the 1520s until 1625. The large Schlick family was divided but tended to favor Luther over Rome. *Joachimsthal* had erected their Protestant Church during 1534–1540 and the high alter was donated in 1545 by Count Schlick [25–29].

The Schmalkaldic League was formed in 1531 as an alliance of Lutheran princes within the Catholic Empire. They avoided direct confrontation with the emperor, while spreading Lutheranism throughout Germany and confiscating church land and expelling Catholic princes. The emperor was charged by Rome to take military action against them. The emperor was Charles V (1500–1556), also King of Spain and all her New World possessions. It was his lot to defend (a) the Roman Catholic faith against the Protestant Reformation of Luther (1483–1546); (b) Europe from the Muslim invasions of the Turk Suleiman (1494–1566) and (c) the empire from the King of France (1494–1547) [19, 33, 34].

After Charles achieved somewhat of a military stalemate against France and the Muslims, he attacked the League in what became known as the Schmalkaldic War of 1546/1547. He got a decisive victory at the Battle of Muhlberg and captured many of the leaders. During this war, Charles and his brother, Archduke Ferdinand I, invaded Protestant Bohemia and, this time, confiscated the Schlick mines at *Joachimsthal*. Many of the Schlick family had embraced Lutheranism. Ferdinand had demanded the submission of the recalcitrant in Bohemia and the confiscation of their properties. The *Joachimsthal* Royal Mint was confiscated. However, since the *Reich* needed the coinage, the Mint was quickly reopened as an Imperial or *Reich* Mint. After the Hapsburg takeover, the purity (% silver) dropped from the Schlick standard. In 1566, the empire established the *Reichsthaler* coin as the new standard with a purity of 0.889. These were minted at *Joachimsthal* until 1671 [22–24]; Table 1.3, History of the Thaler.

The Peace of Augsburg in 1555 achieved a 60-year truce in the Protestant–Catholic wars. It legalized Lutheranism within the empire and temporarily ended two decades of religious war. In principal, each of the scores of the German princes of the empire was to be allowed to choose between Luther and the pope for his Principality; *cuius region, eius religo,* or the reigning sovereign dictates the religion of subjects. Denmark, Norway, Sweden, and much of northern Europe became Lutheran. Europe north of the Pyrenees and the Alps was in danger of being lost to the Catholic faith. This situation led to the formation of an aggressive Catholic Counter Reformation, which was promoted by the pope, the emperor, and the Jesuits.

After the Peace of Augsburg, Kaiser Charles abdicated as emperor in favor of his brother. Aged 56, he retired to a monastery in Spain, so crippled by gout that he had to be transported on a litter. Another problem for Charles and the Hapsburgs was the law of genetics. His grandfather, Emperor Maximilian I, had inherited a new problem. Maximilian's paternal grandmother was the Polish princess Cymburga. She was the alleged source of the macrognathia and deformed lower lip that became known as the *Hapsburg Jaw* [19, 32–34]. It was minimal in Cymburga but visible in Maximilian and prominent in his grandson, Emperor Charles V (Fig. 1.9).

This resulted in a severely impaired mastication; so much so that Charles was forced to eat in private, and suffered chronic indigestion. His speech may also have been impaired [42, 43]. A full beard was used to disguise it some. He died aged 58. His brother Ferdinand, the Hapsburg Archduke, succeeded Charles as emperor #30, during 1556–1564. However, Charles left his Spanish and American possessions to his son Phillip II (1527–1598), who inherited the genetic deformity as well. The last Hapsburg King of Spain, Charles II, (1661–1700), had a grossly deformed face and was mentally retarded as well, dying, thankfully, without issue [19].

Fig. 1.9 The Hapsburg Jaw of Charles Quint [32]. (Public domain)

The Christian religious dispute eventually degenerated into nationalistic acquisitions during the disastrous Thirty Years' War of 1618–1648. The war had four phases: (a) the Bohemian War of 1618–1625; (b) the Danish War of 1625–1629; (c) the Swedish War of 1630–1635; and (d) the Swedish-French War of 1635–1648 [3].

(a) The Bohemian War The Hapsburgs had been in control of Bohemia since 1526 but the Protestant Bohemian nobility did not trust any Catholic emperor. They wished to regain the privilege of electing their own King. In 1618, they started the war by throwing two representatives of the emperor out of a window, several stories high; 10–12 ft. per story. The victims fell 70 ft., into a refuse pile. This became known as the "Defenestration of Prague." The emperor, Ferdinand II (1619–1637), was not amused and answered in 1620 by sending a Catholic army under Count Tilly to destroy the Bohemian Protestants. In November of 1620, the Czech army was defeated at White Mountain, just outside Prague. An Imperial Court sentenced 45 rebels to death. On 21 June 1621, 26 of these were executed in front of the Prague Town Hall. Included was a *Count Schlick,* who was handed over by John-George, the Lutheran Elector of Saxony, who was an ally of the Kaiser against the Calvinists [17].

(b) The Danish War A second Catholic army, under General von Wallenstein, drove into Hungary before turning north to join Count Tilly in attacking the Lutherans in northern Germany. This prompted the Lutheran King of Denmark to intervene but he was driven back home in 1629. Flushed with victory, Emperor Ferdinand then promptly enacted the Edict of Restitution, which restored to Catholic princes all lands lost to Protestantism since 1552. France, Austria, Poland, *Bohemia,* and Bavaria were returned to the Catholic faith.

(c) The Swedish War Lutheran King Gustavus Adolphus of Sweden allied with the northern Germans and entered the war with a well-organized force. Sweden was joined by Catholic France and Cardinal Richelieu, who feared a Hapsburg domination of all Europe. The Swedes took Leipzig and Count Tilly was later killed in action. The Swedish army of G. Adolphus stalemated the juggernaut forces of

Table 1.4 Devastation of Bohemia by Thirty Years' War: 1618–1648 [17]

	Prewar	Postwar
Bohemian population	2.5–3.0 million	700,000 –800,000
CZ Landed peasant families	150,000	30,000
Bohemian villages and manors	36,000	6000
Bohemian towns	782	230
Olomouc Moravia population	30,000	1675
Brno intact houses	1356	168
Kralove Hradec inhabited homes	990	495

CZ Czech Bohemians

General von Wallenstein but the Swedish King was killed in action. General Wallenstein, against the wishes of the emperor, sought an armistice but was assassinated for trying. The emperor Ferdinand had now lost his two best generals.

(d) The Swedish–French War Now everybody got involved. Spain joined the Holy Roman emperor. The French and Dutch joined Sweden and north Germany. When the later alliance drove into southern Germany, Emperor Ferdinand III (1637–1657) called it quits. The Treaty of Westphalia was long and complex. One of the many provisions was that the 300 or so German princes of the empire were free to make their own religious decisions. The pope's response to the treaty was to declare it: "null, void, invalid, iniquitous, unjust, damnable, reprobate, inane, and empty of meaning and effect for all time" [3].

Bohemia was devastated by the Thirty Years' War (Table 1.4). The Joachimsthal Imperial Mint was moved to Prague in 1671. The Protestant majority in Bohemia was forced to emigrate or convert to Catholicism. They crossed the Erzgebirge during 1648–1663 to Protestant Saxony [17]. In 1654, they founded a new town, 10 mi. (16 km) north of Joachimsthal, in the northern foothills of the Erzgebirge. This town site was approved in Annaberg by the resident elector of Saxony, Johann George. It was, therefore, named *Johanngeorgenstadt* (John-George). The new town was at an altitude of 2600–3000 ft., north latitude 50° 26′ 00″ and east longitude 12° 42′ 50″. With strong winds and a winter snow cover lasting half the year, John-George became known as the Saxon Siberia.

However, by 1680, there were 100 ore mines in the area. The George Wegsfort Mine had opened in 1670 and produced 265 kg of silver. Silver and tin mining reached a zenith in 1715 at John-George and then declined. After the Silver Mining Era in the Erzgebirge ended, the mining of nonprecious metals such as arsenic, cobalt, nickel and bismuth followed. The George Wegsfort Mine became famous in 1789 when uranium was discovered in *pechblende* ore from this mine and from one in *Joachimsthal* [44]. This discovery would bring new life to the old silver mines at John-George and *Joachimsthal* during 1830–1847. The Uranium Mining Era of the nineteenth and twentieth centuries started in 1830 and continued in Czech Bohemia until 1964 and in Saxony until 1990 [11, 12, 17].

Sankt Joachimsthal had been occupied by the Swedish forces of Count Mansfield during the Thirty Years' War. Most of the city including Castle Freudenstein was destroyed by the military. One remaining castle tower still stands today and is

Fig. 1.10 Sankt Joachimsthal, circa 1900 [29]. (Courtesy J. L. and V. R. Marshall, "Rediscovery of the Elements," 2010, ISBN 978-0-615-30795)

known as *Schweden-schanze* or Swedish fortification [10, 28, 29]. The *Svornost* mine at *Sankt Joachimsthal* (SJT) dated back to 1530. In 1864, the *Svornost* had become flooded when a strong current of spring water was encountered at a depth of 500 m. This led to the establishment of a mineral water health spa in *Joachimsthal* (see Chap. 4). A photographic tour of the current village of 3500 souls, now *Jachymov* in the Czech Republic, is available on a DVD (digital versatile disk) [29] (Fig. 1.10).

References

1. Thompson JW, Johnson EN (1937) An introduction to medieval Europe. Norton, New York
2. Hayes C, Clark JH, Frederick F (1966) Medieval and early modern times. Macmillan, New York
3. Burns EM, Ralph PL (1955) World civilizations: from ancient to contemporary, vols 1 & 2. Norton, New York
4. Thatcher OJ, McNeal EH (eds) (1905) The treaty of Verdun. A source book for mediaeval history, Section I, Chapters 17–18. Charles Scribner's Sons, New York, p 62. http://www.oll.libertyfund.org/titles/2277. Accessed 22 Aug 2014

5. Thatcher OJ, McNeal EH (eds) (1905) The Strasburg oaths. A source book for mediaeval history, Section I, Chapter 16. Charles Scribner's Sons, New York, p 60. http://www.oll.libertyfund.org/titles/2277. Accessed 22 Aug 2014
6. Cowen R (1999) Medieval silver and gold. http://www.mygeologypage.ucdavis.edu/cowen/~GEL115/115CH7.html. Accessed 5 July 2014
7. Turnwald WCH (1949) Sudeten German picture book. Mandruck Dietz, Munich
8. Encyclopedia Britannica (eds) (2014) Ore mountains. http://www.britannica.com/EB-checked/topic/192171/Ore-Mountains/. Accessed 5 July 2014
9. Monroe WS (1910) Bohemia and the Czechs. L.C. Page, Boston
10. Sommer JG (1847) Elbogen district. Kingdom of Bohemia, vol 15 F. Ehrlich, Prague, pp 106–120. (English trans: Geiser U (2001)). http://www.home.comcast.net/~mhobein/genealogy/bohemia/contents.html. Accessed 5 July 2014
11. Hueper WC (1942) Occupational tumors and allied diseases. Thomas, Springfield
12. Schuttmann W (1993) Schneeberg lung disease and uranium mining in the Saxon ore mountains. Am J Ind Med 23(2):355–68
13. Ziethen H (2005) Entdeckungsreise durch das Erzgebirge. Ziethen-Panorama-Verlag, Bad Münstereifel
14. Encyclopedia Britannica (2014) Bohemia. http://www.britannica.com/EBchecked/topic/71528/Bohemia. Accessed 20 Aug 2014
15. Agnew HL (2004) The Czechs and the lands of the Bohemian Crown. Hoover Press, Stanford
16. Urban T (2014) Kingdoms of Eastern Europe. The history files. http://www.historyfiles.co.uk/KingListsEurope/EasternBohemia.htm. Accessed 6 July 2014
17. Seton-Watson RW (1965) A history of the Czechs and Slovaks. Archon, Hamden
18. Mershman F (1912) St. Wenceslaus. The catholic encyclopedia. Appleton, New York. http://www.newadvent.org/cathen/15587b.htm. Accessed 20 Aug 2014
19. Crankshaw E (1971) The Hapsburgs, portrait of a dynasty. Viking, New York
20. Fuchs F (2007) Kasper Schlick. Neue Deutsche Biographie (auf Deutsch). http://www.deutsche-biographie.de/sfz78496.html. Accessed 6 July 2014
21. von Krones F (2011) Kaspar Schlick. Allgemeine Deutsche Biographie. http://www.de.wikisource.org/wiki/ADB:Schlick,_Kaspar. Accessed 6 July 2014
22. Siegl K (1912) Zur Geschichte des Thalergroschen. http://www.genealogienetz.de/reg/SUD/bez-joachimsthal/geschichte-thalergroschen.html. Accessed 6 July 2014
23. Coin Archives (2014) Schlick coins. http://www.coinarchives.com/w/results.php?search=Schlick&s=1&results…. Accessed 30 April 2014
24. Cuhaj G (2009) Schlick coins. Standard catalog of world gold coins, Krause, Iola, books.google.com/books?isbn=1440204241. https://www.google.com/search?tbo=p&tbm=bks&q=isbn:1440204241&gws_rd=ssl. Accessed 22 Aug 2014
25. Goldschmidt B (1990) The valley of St. Joachimsthal. In: Atomic rivals. Rutgers, New Brunswick, p 28
26. Zeman A, Benes P (1995) St. Joachimsthal mines and their importance. Radiochim Acta 70/71:23–29
27. Veselovsky F, Ondrus P, Kominek J (1997) History of the Jachymov ore district. J Czech Geol Soc 42(4):127–132
28. Marshall JL, Marshall VR (2008) Jachymov (Joachimsthal) Czech republic. Hexag Alpha Chi Sigismundda 99(4):68–75
29. Marshall JL, Marshall VL (2010) Maps of cities: Jachymov. Rediscovery of the elements DVD. Jenny Marshall Computer Services, Denton. http://www.jennymarshall.xom/products.htm Accessed 6 July 2014
30. Aleksin A (2012) Austria silver guldiner guldengroschen. http:www.//coinscurrencyinvesment.blogspot.com/2012/07/austria-silver-guldiner-guldengroschen.html. Accessed 6 July 2014
31. Williams J (ed) (1997) Money: a history. The British Museum Press, London
32. Ertel V (2003) Coins of Charles V and contemporaries. http://www.coingallery.de/KarlV/index_E.htm. Accessed 30 April 2014

33. Encyclopedia Britannica (eds) (1910) Biography of Charles V, Holy Roman Emperor. In: 11th Ed. Vol XV. Cambridge University Press, Campbridge, p141. Available at: http://www.luminarium.org/encyclopedia/charles5.htm. Accessed 22 Aug 2014
34. Spahn M (1908) Emperor Charles V. In: The Catholic encyclopedia (eds), vol 3. Robert Appleton Company, New York. http://www.newadvent.org/cathen/03625a.htm. Accessed 22 Aug 2014
35. Anthony JW et al (2001) Uraninite. Handbook of mineralogy. Mineralogical society of America. http://www.handbookofmineralogy.org/pdfs/uraninite.pdf. Accessed 6 July 2014
36. von Gümbel W (1882) Kentmann, Johannes. Allgemeine Deutsche Biographie (ADB), vol 15. Duncker & Humblot, Leipzig
37. Helm J (1971) Johannes Kentmann 1518–1574. F. Steiner Verlag, Wiesbaden
38. Pettitt GA (2014) Conrad Gesner. Encyclopedia Britannica. http://www.britannica.com/EBchecked/topic/232042/Conrad-Gesner. Accessed 22 Aug 2014
39. Kinch BA (2011) The functionality of early modern collections: Kentmann collection. http://digitalarchive.gsu.edu/cgi/viewcontent.cgi?article=1069&context=art_design_theses. Accessed 6 July 2014
40. Gesner K (ed) (1565) De omni rerum fossilium genere, gemmis, lapidibus metallis…. https://archive.org/details/deomnirerumfoss00presgoog. Accessed 6 July 2014
41. Hoover HC, Hoover LH (1950) De Re Metallica by Georgius Agricola (translated from the first Latin edition of 1556). Dover, New York
42. PubMed Health (2014) Prognathism. http://www.nlm.nih.gov/medlineplus/ency/article/003026.htm. Accessed 7 July 2014
43. Norman B (1999) Facial prognathism in the hominid and human species. http://www.academia.edu/4981691/Facial_prognathism_in_the_hominid_and_human_species. Accessed 22 Aug 2014
44. Marshall JL, Marshall VR (2008) Klaproth. Hexag Alpha Chi Sigma 99(2):20–24

Chapter 2
Ionizing X-rays

2.1 Luminescence

The emission of light by a hot or heated object is called incandescence. In the nineteenth century, city, factory, and household lighting were dominated by the gas light industry. Efforts to produce better incandescent lighting, powered by electricity, started after the development of large batteries, known as voltaic piles, in 1800. Further interest in electric lighting was sparked when the electromagnetic generator became available after 1831. A practical incandescent lightbulb and electrical supply system was created by Thomas Edison during 1878–1880 [1, 2]. Another incandescent light source was the Welsbach gas mantle made in Germany from 1891 to the present (see Chap. 4). Chemist Carl Freiherr Aurer von Welsbach produced a knitted cotton fabric (mantle) soaked in thorium oxide with trace amounts of a "rare earth" element (REE) like cerium. This Aurer Mantle was placed in a gas flame and could burn for 3000 h compared to 1000 h for an incandescent bulb. The Aurer Company claims that 75,000 of their gas streetlamps are still in use worldwide. A portable gas mantle lamp is available for sportsmen and for outdoor family recreational use [2, 3].

A substance is luminous or luminescent if it emits visible light for any reason other than a rise in temperature or by heating. The element phosphorus was discovered in 1669 by the physician and alchemist Hennig Brand of Hamburg. He extracted it from urine as a waxy white substance that glowed (was luminous) in the dark! The element was named "phosphorus" after the Greek word meaning "light bringing" which was used to indicate the morning star, usually Venus, which appeared at daybreak. During the late eighteenth century, phosphorescent came to mean the property of being luminous at temperatures below incandescence, such as (a) from the slow oxidation of phosphorus or (b) after exposure to light or other radiation [1, 2, 4–6].

Different tetrahedral configurations of phosphorus and oxygen give us white, red, and black phosphorus. White phosphorus is luminous in the dark, poisonous and flammable, igniting spontaneously in air unless stored under water. Red phosphorus is less flammable and less poisonous and black is insoluble and least

R. F. Robison, *Mining and Selling Radium and Uranium*,
DOI 10.1007/978-3-319-11830-7_2

flammable. "Self-igniting matches," which used white phosphorus, were developed during the early nineteenth century. They required confinement in a glass vial that was then broken to produce a flame. By 1830, "friction matches" known as "Lucifer's" were invented. The match head, impregnated with white phosphorous, was drawn between folds of sandpaper to ignite. Airtight metal boxes were used to prevent spontaneous or accidental ignition. In World War I, the doughboys were still using Lucifer's to "light their fags."

White phosphorus was toxic for those workers manufacturing the matches as its chronic absorption caused bone necrosis ("phossy jaw"), and it was poisonous to those accidently or purposely eating the match heads. These matches were gradually outlawed in most countries between 1872 and 1925. "Safety matches" were developed which used red phosphorus on a separate specially designed striking surface instead of inside the match head [1, 2]. In the 1920s, "phossy jaw" was often confused with the jaw (mandible) necrosis seen in radium poisoning (Chap. 6). Both phosphorus and radium are deposited in bones.

"Fluorspar" had been coined in 1530 by Agricola (George Bauer, 1494–1555) as a Latinized version of the German "Flussspat," a mineral which was used as a flux for smelting metals. Fluorspar lowered the melting point and decreased the viscosity of solutions making them more fluid. "Fluo" or "fluores" means "to flow" in Latin. Fluorspar was made into an acid and was used for etching glass in the eighteenth century. By 1771, Swedish apothecary Carl W. Scheele began to study what was now called fluoric acid. By 1810, fluorspar was suspected to harbor a new element. By 1824, other fluorite minerals had been discovered. This confirmed that a new element would be found that would be similar to chlorine, iodine, and bromine [4–6].

Humphrey Davy (1778–1829), who had correctly identified chlorine as an element, suggested the name fluorine. By 1869, fluorine was included in the periodic table [7]. Because of its toxicity, many chemists were injured or died, trying to isolate fluorine. Henri Moissan (1852–1907), a chemist at the Sorbonne, isolated fluorine in 1886. He earned the December 1906 Nobel for his work. After returning from Stockholm, Moissan died, 2 months later, of what was thought to be an acute case of fluorine poisoning [8]. The term fluorspar was replaced by fluorite as the new mineral name and "fluorspar" was identified as the mineral compound calcium fluoride (CaF_2), which was first studied for its fluorescence in 1824.

The phenomenon of fluorescence had been observed for centuries before it came under scientific study in the nineteenth century. In 1824, a German mineralogist, Frederick Mohs (1775–1859), observed that, when viewed under ultraviolet light, calcium fluorite (CaF_2) crystals appeared a completely different color than when viewed in daylight. Alexandre Edmond Becquerel (1820–1891) in 1842 reported light emission from calcium sulfate crystals after exposure to ultraviolet light. He found that the emitted light was of a longer wavelength (visible light) than the inciting ultraviolet light, which was invisible. Sir John Frederick Wm. Herschel (1792–1871) found the same luminescence after quinine sulfate crystals were exposed to ultraviolet light. He dubbed it "epipolic dispersion"; a term which was fortuitously rejected [5, 6].

At Cambridge, George Gabriel Stokes (1819–1903) published a 100-page monograph on the subject in 1852. Since fluorspar, a fluorite mineral, also exhibited the luminescent phenomenon, Stokes coined "fluorescence" to describe the emission of visible light by a substance *during* exposure to external radiation such as invisible ultraviolet light. Fluorescence occurred because of the absorption of incident radiation and it persisted only as long as the stimulating radiation was continued. The noted shift in wavelength, from invisible to visible, became known as the Stokes shift. By 1989, it was estimated that 566 of the 4200 known minerals exhibit fluorescence [5, 6, 9].

Color in minerals was found to be due to the presence of trace elements inside the crystalline structure; a trace element being one with a concentration less than 1000 parts per million (ppm) or 0.1 % of a ***rock***'s composition [6]. Trace elements exist in an electronically unstable state as a result of not fitting perfectly within the crystal structure they currently occupy. When exposed to any kind of light, the atoms of the trace elements absorb energy and move into a higher energy level or state of excitation. This lasts for only a fraction of a second before the previously absorbed energy is released and the atoms return to their prior energy level or state of excitation. The excess energy is released as visible light. The color of the visible light, which is released by fluorescent minerals, will differ depending on the energy difference of the inciting light: daylight or ultraviolet [9].

The term "fluorescence" came into use in order to differentiate it from "phosphorescence." Fluorescence was defined as the emission of visible light that occurs only during the exposure to inciting energy. If the luminescence does not persist after the exposure has terminated, it is called *fluorescence.* Any luminescence which persists for any time after removal of the exciting agent is called *phosphorescent.* A mineral from Connecticut was reported to phosphoresce for 36,000 h after ultraviolet exposure ended [5]. Technically, if light persists over 10 ns (10^{-8} s), it is phosphorescence. If the light persists for only 10 ns or less, it is *fluorescence.* The distinction has become rather arbitrary and the terms are often interchangeable for the average person [5, 6, 9].

2.2 Fluorescence

The discovery of both X-rays and radioactivity during 1895–1896 was due to the observation of fluorescence, caused by unknown invisible light rays. The discovery of X-rays came about during the investigations of the discharge of electricity through gases which were confined within sealed glass vacuum tubes. After 1800, electricity from voltaic piles was used to separate chemical compounds into their individual elements, a process which became known as electrolysis. During 1807–1808, English chemist Humphrey Davy (1778–1829) passed an electric current through various molten or liquid mineral compounds and was able to isolate six new elements: *sodium*, *potassium*, *calcium*, *magnesium*, *boron,* and *barium* (Na, K, Ca, Mg, B, and Ba) [1, 4].

Davy found that his greatest discovery was his student, Michael Faraday, in 1813. Faraday went on to establish the basis of electrolysis, electrochemistry, and electromagnetism. In 1838, Faraday (1791–1867) passed electrical currents through rarefied or low-density gases. The gases were contained within a sealed glass tube which had been only partially evacuated due to the poor vacuum pumps of the time. Faraday described the luminescent glow that developed in the glass tubes when the current flowed through the gases. "The rarefaction of the air wonderfully favors the glow phenomena." The luminescence that Faraday described in his discharge tubes was considered to be due to electromagnetic radiation acting upon the *gases* in the tube in the same way as ultraviolet light acted upon various mineral crystals. Faraday's work sparked the interest of nineteenth-century physicists in the passage of electricity through rarefied gases in a vacuum tube [1, 10–12]. Those involved in vacuum tube research in Germany and the UK included:

1838	Michael Faraday (1791–1867), UK
1855	Heinrich Geissler (1814–1879), Bonn
1858/1859	Ulius Plucker (1801–1868), Bonn
1869	Johann W. Hittorf (1824–1914), Munster
1876	Eugen Goldstein (1850–1930), Potsdam
1877/1879	Sir Wm. Crookes (1832–1919), UK
1892	Heinrich R. Hertz (1857–1894), Germany
1893/1894	Phillip Lenard (1862–1947), Breslau, Kiel

In Germany in 1855, a Bonn instrument maker named H. Geissler invented a mercury vacuum pump which achieved a greater vacuum than previously possible. He was able to seal the electrical anode and cathode at opposite ends of the glass tube. Bonn physics professor J. Plucker utilized the Geissler tube in 1858/1859 and reported a bright green *fluorescence* glow in the glass wall of the tube, near the cathode. He was able to change the position of the glow inside the tube by the use of a simple magnet, which suggested an electric charge. Edmond Becquerel (1820–1891) in 1859 applied coatings of luminescent materials on the inner surface of the tubes to produce fluorescence.

J. W. Hittorf, a student of Pucker's, then made a significant advance by designing an L-shaped tube in 1869. The anode was moved off at a right angle from the cathode so that they were not in the same plane (Fig. 2.1, Tube I). The glow now appeared only within the cathode section and objects *placed in* front of the cathode cast a shadow. This showed that mysterious electromagnetic rays were coming in a straight line from the cathode end of the tube. That was evidence enough for E. Goldstein, an astronomer at the Potsdam Observatory, to coin "Kathodenstrahlen" (cathode rays) in 1876.

In England, Sir William Crookes was already renowned for originating the *Chemical News,* a science magazine, in 1859 and for the discovery of the element thallium in 1861 and its atomic weight in 1873 [10–12]. In 1873–1879, he adapted the L-shaped Hittorf vacuum tube for his investigations on the passage of electricity through rarefied gases. His experiments confirmed that the kathodenstrahlen had mass and charge. By 1897, the cathode rays would be identified as electrons. The fluorescence observed within the discharge tubes was found to be the result of ionization by the cathode electrons of the trace amounts of gas remaining in the incompletely evacuated tubes.

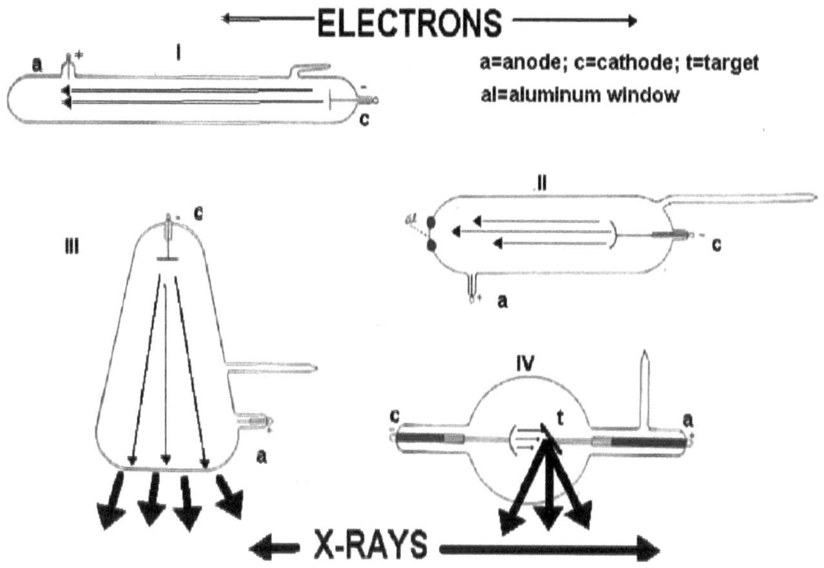

Fig. 2.1 Sealed glass vacuum tubes of Hittorf, Lenard, Crookes, and Röntgen. (Author)

The Crookes tubes used stronger batteries and had an improved vacuum, but they also produced a brilliant *fluorescence* within the tube. This occurred despite the better vacuums which left fewer gas molecules in the tube to interact with the cathode electrons. With less collisions to slow them down, the electrons accelerated more rapidly through the tube to strike the opposite glass wall, where they produced fluorescence by causing ionization when they struck the glass wall. And at the point of impact with the glass wall, the cathode electrons were also producing invisible X-rays, which were undetected until 1895. The Crookes and/or Hittorf tubes became ubiquitous in college laboratories the world over. After the discovery of X-rays, the Crookes/Hittorf gas tubes were used to produce X-rays until the invention of the Coolidge hot cathode tube in 1913 [11, 12]. The hot tungsten cathode generated the necessary electrons and made the gas tubes obsolete. Cathode ray tubes are used today to illuminate our television and computer screens.

The Geissler tube had a poorer vacuum than the Crookes tube and the cathode electrons could not travel far before colliding with a gas molecule. This led to a colorful fluorescent glow, which proved entertaining to nineteenth-century science students. Neon gas, discovered in 1898, was instilled within the Geissler tube to create brilliant red signs for commercial advertising after about 1910. Different gases, usually from the inert (noble) gases, were used to produce different colors: such as argon (green), xenon (blue), and krypton (white). Regardless of the color, the signs were all called Neon [1]. In 1923, Georges Claude (1870–1960) and his French company Claude Neon, introduced neon gas signs to the USA, when he sold two in Los Angeles. The legendary Packard dealership of E. C. Anthony bought two "Packard" signs for US$1250 each [13].

By 1901, a fluorescent lamp was invented that used the cathode electrons to excite atoms of mercury vapor. However, the emitted photons from the mercury atoms were mostly in the ultraviolet range and invisible. General Electric solved that problem by filling the tube with both mercury and argon and a "phosphor." Electrons strike the mercury vapor atoms and produce an invisible ultraviolet light. The ultraviolet light impinges upon a fluorescent and/or phosphorescent coating on the inside of the tube to produce visible light. The coatings are called phosphors, which emit photons that are visible to the human eye. The phosphor chemicals are of various metallic and REE salts, such as elements # 39 and # 57–71. The same phosphors were being studied by General Electric and Radio Corporation of America (RCA) for use in the screens for commercial television, which began in Germany in 1935, in the UK in 1936, and in the USA in 1939 [1, 14].

By 1934, General Electric was able to produce a prototype fluorescent lamp. General Electric started sales of fluorescent lights in 1938 and promoted them at the 1939 World's Fair, where they accounted for one third of all the exterior lighting. They gained acceptance in department stores, factories, and large office buildings during World War II. Eventually, they were marketed for kitchens, bathrooms and work areas, garages, and cellars. By 1951, more light was produced in the USA by fluorescent tube lamps than by the incandescent lightbulb. The fluorescent or "cold light" is reported to use one fourth the energy and emit one fourth the heat of an incandescent light of the same strength [1].

2.3 Discovery

W. C. Röntgen (1845–1923) was an only child, who was born in Germany but attended no schools there. His father was a textile merchant who had married his Dutch first cousin and then moved to Holland, where they all became Dutch citizens. After undergraduate education in Apeldoorn, Holland, Röntgen earned an engineering degree in Zurich in 1868, without much in the way of instruction in physics. However, he decided to pursue a career in experimental physics after he was befriended by the noted experimentalist Arnold Kundt. Röntgen earned his physics Ph.D. at Zurich in 1869 [15–20].

While in Zurich, he became enamored of (Anna) Bertha Ludwig (1839–1919), and it was not difficult to see why: Fig. 2.2. She was some 6 years his senior and had been a patient for several months in 1899 at a local Zurich sanatorium, presumably for tuberculosis. She was the daughter of a Zurich innkeeper, who was a German political activist in exile. They married in January of 1872 and remained childless. They later adopted a niece of Berthas, named Josephine, later, Mrs. B. Donges. Röntgen's uncle Ferdinand and his cousin Louise immigrated to America during 1872–1879. Louise married an Indianapolis, Indiana, minister J. G. Grauel. Another cousin was Johann Heinrich Röntgen, who in 1893 was a minister of the First Reformed Church of Cleveland and superintendent of the Bethesda German Hospital. Other relatives were A. E. Röntgen of Cleveland and Mrs. H. Grabhorn in Indianapolis [15, 16].

Fig. 2.2 Bertha Ludwig Röntgen (< 1923). (Public domain)

Röntgen became an assistant to Professor Kundt and accompanied him to Würzburg, Germany, in 1870 and then to Strasburg in 1872. Röntgen became an associate professor of physics there from 1876 to 1879. Then, after an essentially 10-year apprenticeship from 1869 to 1879, he was able to obtain a position as professor of physics. This was at Ludwigs University in Giessen from 1879 to 1888. The Institute of Physics at Julius Maximillians University in Würzburg called him to chair their Physics Department in 1888. By the end of 1894, he had published 50 papers in experimental physics and his excellent work was well known both inside and outside Germany. Twenty-nine of his papers were published in international journals and 21 in regional publications, such as in Munich and in Würzburg [17].

In 1892, German physicist H. R. Hertz reported that the cathode rays could penetrate thin metal, gold, and aluminum foil windows at the opposite end of the tube. His pupil, Paul Lenard, studied this penetration by the cathode rays. Lenard published three papers during 1893-94 concerning the behavior of cathode rays (electrons) penetrating outside the tube and their absorption in matter. He found irradiation that was absorbed only after traveling about 3–5 cm, 1 or 2 in., through matter [17]. It is possible that Lenard was indeed measuring cathode rays, but cathode rays (electrons) can be blocked by as little as 5 mm of aluminum foil. More likely is that Hertz and Lenard were unknowingly creating X-rays at the point of impact of the cathode rays upon the metal window (target) (Fig. 2.1, Tube II).

In the spring of 1894, Röntgen became interested in the experiments of Paul Lenard and the study of the cathode rays (electrons) which reportedly penetrated outside the tube. He wrote Lenard in Breslau for advice in May of 1894. Lenard obliged him with details of his tubes and furnished him with small thin metal windows for the tubes. After studying Lenard's publications, Röntgen sought to first duplicate Lenard's findings and then to explore his own ideas. Unlike many other significant discoveries, Röntgen kept no historical record of the actual discovery, and he had all his notes burned upon his death [17–20]. We do know that Röntgen was using a Crookes–Hittorf glass tube, pear shaped, with the cathode (negative) at the small end and the anode (positive) at the side (Fig. 2.1, Tube III).

The electricity was provided by the standard Ruhmkorff induction coil energized by batteries. The negative rays exited from the cathode and were attracted by the anode onto a path towards the opposite end of the tube. It had been noted by all investigators that as the cathode rays struck the opposite glass wall of the tube, it caused a greenish glow or fluorescence. In order to mask this tube fluorescence, Röntgen covered the tube with lightproof black cardboard [19]. This made it easier to detect any induced fluorescence outside the tube. To better observe any fluorescence *outside* the tube, observers had to become dark-adapted in a completely darkened room before turning on the tube. Lenard had been studying the range of the apparent penetration of the cathode rays by observing this fluorescence outside the tube. He detected it by creating a thin paper screen soaked with an organic chemical, penta-decylparatoly-keton, which was relatively insensitive to everything but cathode rays. Using these screens as detectors, Lenard had estimated the penetration of apparent cathode rays (electrons) outside the tube at 3–5 cm [19].

After repeating Lenard's work, Röntgen decided to use tubes with thick walls and without the thin metal windows used by Lenard. Presumably, he wished to see if the cathode rays could really penetrate a thick glass wall. And Röntgen used a cardboard screen coated with inorganic barium platinocyanide, which was highly sensitive to any light. Specifically, he was not using the relatively insensitive organic ketone chemical detector used by Lenard [19]. Lenard wanted to measure only cathode rays not any extraneous fluorescence.

Röntgen, after a childhood illness, was partially blind in one eye. And he was color blind. In spite of this, he enjoyed hunting and was allegedly a good shot [15, 16]. He probably had the deuteranomaly type of color blindness, which results in a mild deficiency in perception of the green spectrum. This is the most common type of color vision deficiency and affects about 5% of white males [21]. His choice of screen detector may have been influenced by his color blindness. Certainly, there was a large discrepancy in screen sensitivity between his barium platinocyanide and the penta-decylparatoly-keton of Lenard (Fig. 2.3).

With the tube on, Röntgen noted fluorescence of one of his screens which was estimated to be 100 cm or 39 in. away. This screen was too far away to be affected by the cathode rays and there was no light in the room. No matter how far he moved the screen away from the tube, it still fluoresced. He even observed the glow when he placed the detector screen behind a thick wooden door. He had embraced photography as a hobby and was pleasantly surprised to find that the invisible rays served to expose photographic glass plates and/or photographic film. He reported that this took 3–10 min/exposure. He had discovered a new invisible light presumably produced by the cathode rays as they impinged on the glass wall. Since their identity was unknown, he called them X-rays [15–20].

a Wilhelm Roentgen b Philipp Lenard

Fig. 2.3 W. C. Röntgen and Phillip Lenard (< 1923). (Public domain)

During November and December, Röntgen was able to characterize the new unknown rays as (a) causing fluorescence and exposing photographic plates in the absence of visible light; (b) traveling in a straight line, unaffected by a magnet (neutral charge); (c) affected by the geometric inverse square law: intensity inversely proportional to the square of the distance between tube and screen; and (d) penetration of most solid matter but attenuated by the density and thickness of metals. A thickness of 1.5 cm of lead served to stop the rays.

While working on this penetration of the rays and holding attenuating objects in front of the screen, he must have been startled to see the bones of his hand appear on the screen. He documented this effect by placing Frau Röntgen's hand on a photographic plate for a 15-min exposure to the X-rays. He used the rays to expose several other objects on photographic plates. He also found that the output of X-rays could be increased by directing the cathode rays onto a heavy metal *target* at the anode instead of onto the opposite glass wall (Fig. 2.1, Tube IV) [17].

2.4 Post-discovery Reports

He submitted his paper on 28 December to the local Würzburg science journal which was already in press for the end of year edition. He had not presented the paper at the December meeting, but it was added as the last article in the December Proceedings. He instructed the printer to place it at the end, before the 1895 annual report [17]. The Proceedings of the Physical-Medicine Society of Würzburg did not have the international reputation and prestige of the journals at the major academic centers in Vienna, Berlin, and Paris. However, Röntgen was undoubtedly in a hurry to submit his priority because physicists all over Europe had been working with the cathode ray tubes and had undoubtedly been producing X-rays for years without knowing it. By submitting his phenomenal paper locally, he was assured of immediate publication. However, it was at the risk of it being overlooked by prominent physicists.

Röntgen was a mature academic, age 50, with 50 prior publications [15, 16]. He knew what would be academically correct in reporting his phenomenal discovery. Since the local Proceedings would not be published and distributed until the first week of January 1896, or later, Röntgen arranged with the publisher, Stahel's Royal and University Book and Art Merchants, to immediately run off about 90 reprints of his article. Thus, the reprints were available before the Proceedings were published. Within the next 2 months, there would be five different printings of these reprints (*Auflage*), which Stahel would sell for 60 pfennigs each. Collectors are able to differentiate the five printings by changes on the yellow wrapper cover pages [17, 22]. The original reprints of Röntgen's first two papers were available in 2011 to collectors for US$19,500.00 [23].

On January 1, a Wednesday, Röntgen mailed out these 90 reprints [24]. Seventy were sent to friends and colleagues in Germany, Austria, and Switzerland and another 20 to other European countries. None crossed the Atlantic Ocean. Röntgen's paper consisted of ten pages from the Proceedings, pages 132–141, and contained no illustrations, nor photographs. However, he included a sample of some of the nine X-ray photographs, which he had taken, in the mailing to at least ten prominent European physicists [17, 20, 24]. The most astounding image, which was not even mentioned in the article, was of Bertha Röntgen's hand; the single-ring photograph. This startling picture of bony anatomy instantly demonstrated, to both scientific and laypersons, the practical medical value of the discovery (Fig. 2.4a). This image was among those sent out with the reprints to a select group:

F. S. Exner in Vienna
Voller in Hamburg
O. Warburg in Berlin
O. Lummer in Berlin
L. Zehnder in Freiberg
F. W. G. Kohrausch in Gottingen
H. H. Lorentz in Leiden
Lord Kelvin (Wm. Thompson) in Glasgow
Schuster (German born) in Manchester
H. Poincare in Paris

Physicist Franz Serafin Exner (1849–1926) in Vienna, the medical capital of the Western world at that time, received the reprint along with several photographs including that of the hand. Röntgen had worked with Exner in Professor Kundt's departments during the 1870s. At a Saturday night social engagement on January 4, he showed them to Dr. Ernst Lecher, the son of the publisher of the newspaper *Wiener Neue Freie Presse*. The story and the X-ray photo of the hand were then published

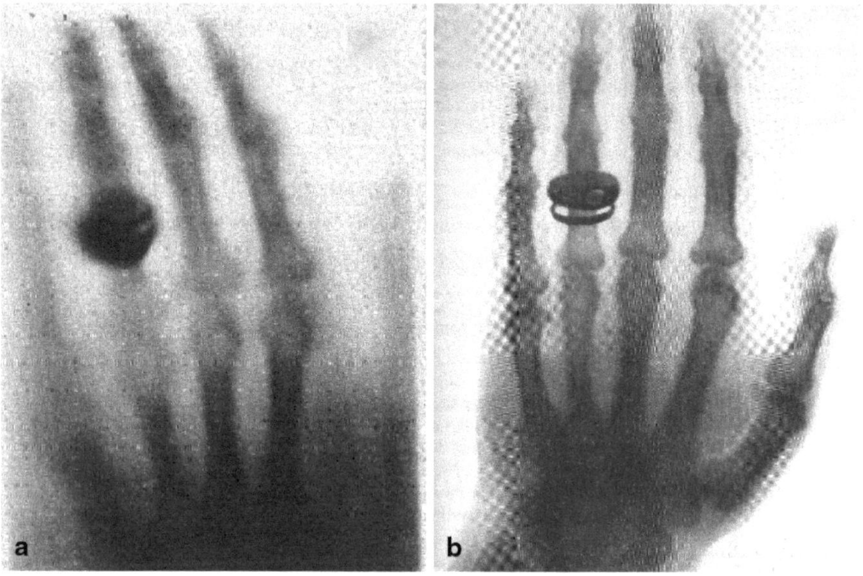

Fig. 2.4 **a** Single-ring X-ray of 1896; November/December (< 1923) (public domain). **b** Double-ring X-ray of 1897; January (< 1923). (Public domain)

Sunday morning on the front page of that paper. These were telegraphed to London and other cities on Monday the sixth and published in London and Frankfurt papers on Tuesday the seventh. The *London Standard* stated: "Among the practical uses of the new discovery...possible for surgeons to determine...the exact location of any bullet...or...to render visible any fractures...in bones...prior to operation" [12, 17, 20]. There was now a worldwide system of submarine telegraph cables connecting the world and especially the British Empire. London sent the story out on the world cable on the seventh and it was printed in New York City and elsewhere on Wednesday the eighth, 1 week after Röntgen slipped the reprints in the post. Scientific journals carried the story on the eighth through the tenth and medical journals followed on the 11th, 6 days after the first newspaper publication [12, 17, 20]. The significance of the discovery was easily appreciated (see first reports in Table 2.1)

Electrotherapists and neurophysiologists in 1896 already had most of the equipment needed to start producing X-rays [25–27]. Electric therapy had allegedly started in London in 1767–1777 when "static machines" were installed at Middlesex and at St. Bartholomew's Hospitals. Most electrotherapists enthusiastically incorporated Röntgen rays into their practices and some, such as Dr. M. K. Kassabian, became pioneer radiologists [26]. Neuropsychiatrists at Guy's Hospital in London had been using electrotherapy since 1840, and it was still popular 100 years later [25–27]. Also, most college physics laboratories had cathode ray tubes, and many professors now lamented their missed opportunity at discovery as their tubes had all been producing undetected X-rays. They all hastened to make their own X-ray pictures using apparatus similar to Röntgen's. However, Vienna and Berlin were the

Table 2.1 Reports of the discovery [12, 17, 20]

January	First reports of discovery	
01	90 reprints mailed from Würzburg	Wednesday
04	Vienna party	Saturday
	Warburg shows photos in Berlin	
05	Vienna (*Wiener Neue Freie Presse*) newspaper story	Sunday
06	Berlin demonstration of single-ring photo by Jastrowitz	Monday
07	*London Standard, Frankfurter Zeitung* reports	Tuesday
08	New York newspapers	Wednesday
	New York Electrical Engineer Journal	
10	*London Electrician* journal	Friday
	Vienna Medical Society: Exner shows single-ring photo	
11	*New York Medical Record*	Saturday
	British Medical Journal, Lancet	
13	Berlin demonstration by Jastrowitz	Monday
	Röntgen appears before Kaiser in Berlin	
17	Vienna Medical Society: Exner shows X-rays	Friday
20	Berlin demonstration of glass splinter X-ray by Jastrowitz	Monday
23	Röntgen at Würzburg Society takes double-ring image	Thursday
24	Vienna Medical Society: preoperative X-rays of bullet and extra toe	Friday
February		
7	John Cox (1851–1923) produces first clinical X-ray in Canada	
14	Dartmouth College: Collies fracture photo	
	Columbia University: hand photo with buckshot	
17	Chicago Medical Society: hand photo	
22	Liverpool: *Lancet* photo of preoperative bullet	

first to take clinical X-rays for medical purposes. They started in January, a month before the USA and the UK (Table 2.2) [12, 15, 16].

The Vienna Medical Society met every Friday, but did not resume in 1896 until 10 January. The neurophysiologist Siegmond Exner (1846–1926) was the brother of physicist Franz S. Exner. On 10 January, Siegmond addressed the Vienna Society by describing the discovery and showing the photo of Frau Röntgen's hand. On 17 January, the Exner brothers showed X-rays of their own, including the first angiogram, obtained on a cadaver. On 24 January, the surgeon Albert von Mosetig-Moorhof (1838–1907) demonstrated preoperative X-rays of (a) the location of a bullet in the hand of a forester and (b) the specific articulations of a supernumerary great toe in a 22-year-old female servant. These January sessions of the Vienna Society were recorded in both the *Wiener Klinische Wochenschrift* and *Münchner Medicinische Wochenschrift* [12, 15, 16].

In Berlin, the *Verein für Innere Medicin* met every Monday and neuropsychiatrist Moritz Jastrowitz (1839–1912) demonstrated the single-ring radiograph

Table 2.2 Vienna and Berlin demonstrations [12, 15, 16]

Vienna X-ray demonstrations	Author and publication date	Publication
10 January	S. Exner (16 January 1896)	"Demonstration of Röntgen pictures and their effects"
		Wien Klin Wschr 9:48
17 January	S. Exner (22 January 1896)	"Röntgen photographs"
		Munch Med Wschr 43:67
24 January	A. Mosetig-Moorhof (28 January 1896)	"Röntgen rays in surgery"
		Munch Med Wschr 43:90
24 January	A. Mosetig-Moorhof (30 January 1896)	"Röntgen rays to complete diagnosis"
		Wien Klin Wschr 9:83
Berlin X-ray demonstrations		
6 January	M. Jastrowitz (30 January 1896)	"Röntgen's experiments with cathode rays and their diagnostic value"
		Dtsch Med Wschr 22:65
13 January	M. Jastrowitz	"Use of Röntgen's method and of a photographic picture of a human hand with imbedded pieces of glass"
		Berlin Klin Wschr 4:89 (1896)
20 January	F. Koenig (20 February 1896)	"The importance of the Röntgen picture for the diagnosis of bone diseases"
		Dtsch Med Wschr 22:113
7 February	F. Koenig	"Demonstration of a preparation with a photograph taken with Röntgen rays"
		Berlin Klin Wschr 568 (1896)

of Frau Röntgen on 6 January. This was reported in the *Münchner Medicinische Wochenschrift* on 14 January and in the *Deutsche Medicinische Wochenschrift* on 30 January. The next Monday, 13 January, he exhibited Röntgenograms produced locally. On 20 January, he showed X-rays taken by the chief physicist at a Berlin industrial plant which demonstrated a glass splinter in the finger of a glassblower at the plant. On 7 February at the Berlin Medical Society, the surgeon Franz Koenig (1832–1910) became one of the first to correlate Röntgen images with the surgical and pathology findings. These demonstrations and photographic images were later reported in the *Berliner Klinische Wochenschrift* and the *Deutsche Medicinische Wochenschrift* [12].

In the USA, the Frost brothers at Dartmouth College reported in *Science* on 14 February the first X-ray taken in America, for the diagnosis of a Collie's fracture. The X-ray was taken in the physics laboratory by astronomer E. B. Frost, at the request of his brother G. D. Frost, the patient's physician. On the same date, a Dr. M. I. Pupin at Columbia University seems to have taken an X-ray image of a hand with buckshot. At the Chicago Medical Society on 17 February, a postoperative X-ray of a hand was shown. The preoperative X-ray had located small buckshot between metacarpals. The *Lancet* on 22 February published a "Röntgen radiograph" from Liverpool of a left wrist demonstrating a small bullet between carpal bones.

Thus, during the first 2 months of 1896, diagnostic radiology was established as a physician tool for the preoperative location of foreign bodies and the evaluation of fractures (and extra toes) [12].

Röntgen himself gave only two public demonstrations of the production of X-rays. On the 13 January, he demonstrated X-ray production to the Kaiser and Kaiserin in Berlin. He finally presented his discovery to the Würzburg Society on 23 January. At that lecture, Röntgen took an X-ray photograph of the hand of the biologist Rudolf Albert von Kolliker (1817–1905) (Fig. 2.4b). The hand photographs of Frau Röntgen and von Kolliker are often confused. The "double-ring" photo has had more modern reproductions, whereas the original "single-ring" negatives were destroyed during a World War II bombing. The demonstration at Würzburg was portrayed in a historically accurate painting by Robert A. Thom for the Parke Davis pharmaceutical company in 1962 [12, 28]. In January of 1896, Röntgen was interviewed by American journalist H. J. W. Dam, for *McClure's Magazine* (1893–1929), a popular monthly American illustrated periodical of 1893–1929 [29]. The complete interview from *McClure's* has been reprinted in the monograph by H. S. Klickstein [17] and is available online as well [30]. Röntgen declined all other invitations and interviews.

Röntgen's second paper [17] was submitted on 10 March 1896. The X-ray of von Kolliker's hand was not included but was sold separately by the publisher! Röntgen now recommended the use of a curved aluminum cathode and a platinum anode at a 45° angle (Fig. 2.1, Tube IV). Also Röntgen reported that the X-rays produced the conduction of ions that would discharge an electrometer: *ionizing radiation.* Since his first paper had appeared, this ionization had already been reported by J. J. Thompson, Ernest Rutherford, and others [24]. By February of 1896, several laboratories worldwide had observed that the X-rays would discharge a gold-leaf electroscope; they turned air into an electrolyte that could conduct electricity. They were ionizing. Röntgen stated that he had determined this in 1895 but he wanted to be sure before publishing this in his second paper in March of 1896.

It was finally conclusively demonstrated in 1897 by J. J. Thompson (1856–1940) that cathode rays were free electrons [10, 24]. They were produced when an electrical charge caused ionization of the gases remaining in the poorly evacuated vacuum tubes. The freed electrons were then accelerated to the opposite end of the tube. The X-rays were produced at the point of impact where the cathode ray particles (electrons) struck the wall of the glass tube at the opposite end. The electrons were slowed by interactions with the nucleus and orbital electrons of atoms *within the glass wall.* The deceleration of the electrically charged particles resulted in a transfer of their kinetic energy to the newly produced electromagnetic irradiation, which was in the form of X-rays (Fig. 2.1, Tube III).

Röntgen's third and last paper on X-rays was submitted a year later on 10 March 1897 [17]. In it, he addressed the superiority in radiotherapy of using a "hard tube." After continual use, a tube became "harder" due to the exhaustion of the gases within the tube. As noted earlier, this allowed the electrons to gain acceleration in reaching the point of impact against the opposite glass wall where the X-rays were produced. Röntgen wrote only the three papers on X-rays. This was the zenith of his career. He was awarded the very first Nobel Prize in Physics in 1901 when the first Nobel laureates were recognized [17, 31]:

Physics: Röntgen for X-rays by the Royal Swedish Academy of Sciences
Chemistry: Jacobus van't Hoff for chemical dynamics by the Royal Swedish Academy of Sciences
Physiology-Medicine: Emil von Behring for immunizations by the Karolinska Institute
Literature: Rene Prudhomme for poetry by the Swedish Academy
Peace Prize: (1) Henri Dunant (Red Cross), (2) Fred Passy (French) by the Norwegian Storting

However, that was just one of the many honors he received (see Supplemental Material 2.7 [15, 16]). For the next 28 years, Röntgen would be continually honored for his discovery. It would be difficult for anyone to be more honored during a lifetime than Röntgen was between 1896 and his death in 1923. He was awarded five honorary degrees: by Würzburg (an M.D.), Zurich, Odessa, Frankfurt am Main, and Munich. He was made a member of academic societies in Heidelberg, Amsterdam, Stockholm, and Munich. He was made an honorary member or corresponding member of 49 different academic societies all over Europe and in Mexico, Philadelphia, New York City, and Constantinople. At least five Röntgenstrassen were dedicated: in Cologne, Würzburg, Halle, Lennep, and Wilheim. He was awarded over 20 major prizes, medals, and orders of merit. The financial prizes included those from Vienna, Paris, and Hamburg, as well as the 1901 Nobel in Physics. The Nobel Prize money, US$54,000 in 1901, was donated to the University of Würzburg. However, Röntgen was able to maintain considerable investments in multiple countries until the economic depression of 1920. The medals included the Rumford (London), the Iron Cross, the Helmholtz Bronze and Gold, and the Prince Regent Silver. The Rumford medal was donated to the German government (for melting down) during World War I. The Orders of Merit were from the Prussian, Bavarian, and Italian Crowns, from the Franklin Society in Philadelphia, and from Columbia University (the Barnard).

Röntgen never gave the required Nobel lecture in Stockholm [31]. Biographer Otto Glasser [15, 16] thought that he arrived in Stockholm on 10 December, Alfred Nobel's birthday, with the other German laureates, E. von Behring from Marburg and J. H. van't Hoff from Berlin. However, in 1955, Ulf Rosenow claims that more recent evidence indicated that Röntgen never went to Stockholm [20]. He originally postponed his lecture until December of 1902. The Nobel Committee in October of 1902 notified him that the lecture was desirable but not unconditionally necessary. Röntgen concurred and received the gold medal by mail.

Phillip Lenard became increasingly bitter over the accolades given to Röntgen even though Lenard was awarded the 1905 Physics Medal. The deliberations of the Swedish Academy of Sciences in 1901 became public after 70 years and revealed that Lenard had been considered as a corecipient [32, 33]. Five *non-Swedish* physicists, whose opinion was solicited, had advised dividing the prize equally. Consequently, a special advisory committee of five leading *Swedish* physicists, known as the Nobel Committee for Physics, recommended an equal division of the money but with Röntgen listed first. That Committee included Ångström and Arrhenius. However, nominations were also solicited from 29 foreign experts. Röntgen received 12 votes, Lenard 1, and 11 votes were shared among nine other candidates, while five experts voted that both Lenard and Röntgen share the award [32, 33].

The 1901 Swedish Nobel Prize Board believed that Alfred Nobel wanted to honor only the one scientist who was most distinguished, and they voted accordingly. However, the Norwegians in charge of selecting the Peace Prize laureate selected two winners in both 1901 and 1902. The Swedes then voted for two Physics laureates from 1902 to 1904. Literature got double winners in 1904 and Medicine-Physiology in 1906. Currently, up to three candidates can share an award [33]. One can only speculate on what the effect would have been on both Röntgen and Lenard if they had shared the award (Fig. 2.3).

In 1900, at the request of the Bavarian government, Röntgen took over the new physical institute at the Luwig Maximilians University in Munich. He published only seven more papers, but he graduated 25 Ph.D. students at Munich before retiring in 1920. World War I had brought great hardship and sorrow as his wife became chronically ill, requiring multiple daily morphine injections all during the War. She died in 1919 at age 80. The German government and postwar inflation took all his funds in 1920. He died after a year of illness in 1923, age 78, of apparent colorectal cancer [15, 16].

2.5 Ionizing Radiation

Visible light rays can be separated by an optical prism into a spectrum of colors: red, orange, yellow, green, blue, and violet, which all have specific wavelengths, frequencies, and energies. Invisible light rays with both longer and shorter wavelengths than visible light were discovered during the nineteenth century. In 1800, Friedrich Wilhelm Herschel (1738–1822), the German-born British astronomer, discovered infrared light rays. One year later, Johann Wilhelm Ritter (1776–1810) found ultraviolet rays. The concept of an electromagnetic spectrum was developed by various physicists during the nineteenth century, starting with Faraday and culminating with Maxwell. English physicist Michael Faraday (1791–1867) in 1831 invented the electric generator, which used a moving magnet to produce electricity. Thereafter, scientific investigation of electricity was no longer limited by a dependence on batteries [2, 34].

During 1862–1864, James Clerk Maxwell (1831–1879) pronounced that visible light as well as the infrared and ultraviolet rays were all forms of electromagnetic radiation traveling at the speed of light: 671 million miles/h or 300 million m/s. The electromagnetic spectrum extended from the infrared, characterized by long wavelengths of low frequency and low energy, through to the ultraviolet waves of shorter wavelengths with a high frequency and high energy (Table 2.3). Maxwell's theory and his mathematical calculations suggested the existence of long length

Table 2.3 The electromagnetic spectrum [34–36]

Radiation type	Biological effects	Energy in electron volts (eV)	Frequency in Hz: cycles/second	Wavelength
Nonionizing radiation		>	>	<
Static: MRI, strong magnets	Vertigo, nausea	< 10^{-10}–10^{-10}	< 100 Hz	infinite
Radio: power lines	On nerve and muscle	10^{-10}	100 Hz	3000 m
Radio: TV, FM, AM, CB, shortwave, cell phone	Increase in body temperature	10^{-10}–10^{-4}	100–300 Hz	3 km–33 cm
Microwave: ovens, radar, Wi-Fi	Body heating	10^{-4}	300 GHz	33 cm–1.0 mm
Infrared: lasers, remote controls, telecommunications	Corneal, retinal, and skin burns; cataracts	0.0124–1.24	300 GHz (10^{9})–385 THz (10^{12})	1.0 mm–780 nm
Visible light: red to violet	Skin and retina burns	1.77–3.1	400–789 THz	750–380 nm
Ultraviolet A	Sun burn, skin cancer	3.10–3.94	750 THz	400–315 nm
Ultraviolet B	Sun burn, skin cancer, DNA damage	3.94–4.43	850 THz	315–280 nm
Ultraviolet C	DNA damage, skin cancer, germicidal	4.43–12.4	950 THz	280–100 nm
Ionizing radiation		>	>	<
Ultraviolet	Chemical bonds broken	10–124 eV	1014–1018 Hz	125–10 nm or 1250–100 Å
X-rays	DNA damage	124 eV–24 MeV	1018–1021 Hz	10–0.001 Å
Gamma rays	DNA damage	1.24 MeV	1018–1021 Hz	10–0.001 Å

Frequency: *Hz* Hertz = 1 cycle/s, *M* mega = 10^6 ; *G* giga = 10^9 , *T* tera = 10^{12}
Wavelength: *m* meter, *cm* centimeter = 10^{-2} m, *mm* millimeter = 10^{-3} m, *μm* micrometer = 10^{-6} m, *nm* nanometer = 10^{-9} m, *Å* angstrom = 10^{-10} m (1 Å = diagnostic X-ray)

radio waves, which were artificially produced in 1887 by Heinrich R. Hertz. During the twentieth century, the electromagnetic spectrum was then enlarged again by the creation of the long wavelength–low energy waves of radar, lasers, telecommunications, microwave ovens, television, cell phones, Wi-Fi, and MRI [34–36].

Invisible light rays of higher energy than ultraviolet light were unsuspected until 1896/1897 when X-rays were produced and also found to possess the energy to cause ionization. Ionizing radiation consists of electromagnetic waves or atomic particles with a high enough energy, 10 eV or more, with which to separate or detach orbital electrons from atoms or molecules. This detachment results in the formation of two ions: (a) the now free electron with a negative charge and (b) the remaining atom or molecule with a positive charge. These ions are very chemically active and create free radicals which damage DNA and prevent repair. Thus, all ionizing radiation is potentially damaging to biological systems and all X-rays are ionizing [34–36].

The very high energy end of the ultraviolet range is ionizing. However, all ultraviolet light can cause significant chemical molecular damage even without causing ionization. Fortunately, most of the damaging ultraviolet rays are filtered out by the earth's atmosphere and the ozone layer. Otherwise, life at the earth's surface would be untenable. Ultraviolet light was shown to be bactericidal in 1877. Ultraviolet rays from sunlight at high altitudes became a mainstay for the treatment of tuberculosis for the next 70 years. In 1896, Niels T. Finsen (1860–1904) began using ultraviolet light to treat lupus vulgaris, a tuberculosis infection of the skin. Finsen was an invalid Dane from Iceland who died of Pick's disease 1 year after being awarded the 1903 Nobel in Physiology and Medicine for his ultraviolet therapy system [25–27, 31]. At the Finsen Institute in a Copenhagen suburb, his apparatus used electrical carbon arc lighting to circumvent the Nordic lack of sunlight. This light was passed through quartz rock crystals and colored water to remove the red and yellow wavelengths. The concentrated ultraviolet light was then focused by lenses onto a small area of the skin, about the size of a dime. The machine treatments required an elaborate cooling system, a complex lens system, and an attendant nurse for each patient. The concentrated light radiated through four telescopes so that four patients could be treated at the same time [25–27].

Treatments lasting 60–70 min, were given daily for 20–25 days. Treatment was described as tedious, expensive, and requiring skilled personnel [27]; however, the cure rate was reported to be 50–60 % [37]. The Finsen apparatus that was used in Copenhagen was found impractical in the USA and a smaller American model was introduced (Fig. 2.5) [26]. Finsen therapy successes and his Nobel Prize led to many years of attempted phototherapy with sunlight and ultraviolet lamps in the twentieth century. And this led to a lot of sunburns and skin cancer. All ultraviolet light is now considered carcinogenic for skin cancer, but it continues to be used as a germicide and disinfectant.

2.6 Haarausfall and Therapy

Many people had access to Crookes tubes, electrical sources, and photographic plates. Experimentation by scientists and amateurs abounded along with the medical use of diagnostic X-rays.

By the end of 1896, it has been estimated that 995 scientific papers and 49 books or pamphlets about X-rays had been published [15, 16]. During 1896, the medical literature immediately began to report the untoward effects of "Röntgen-dermatitis,"

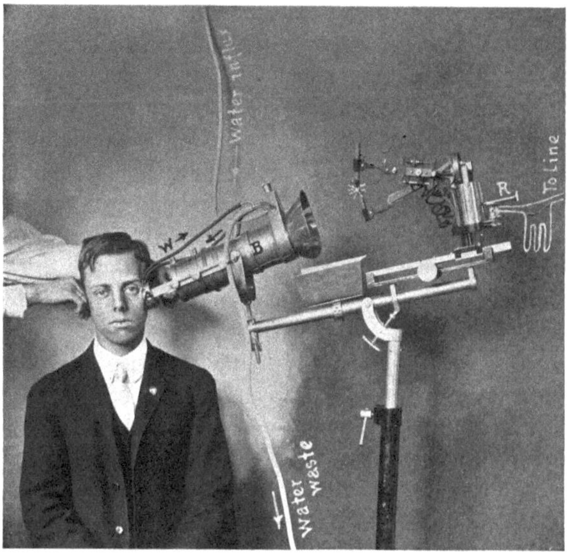

Fig. 2.5 Finsen Telescope; American modification [26]

which resembled severe sunburn and then got worse as the nail beds, hair follicles, and subcutaneous tissues were damaged. During 1896 alone, over 25 cases of Röntgen-dermatitis or skin damage were first reported [16, 38]. By 1902, a total of 147 cases of X-ray burns (dermatitis) had been documented: 55 cases in 1896, 39 in 1897, 10 in 1898, 15 in 1899, and 2 in 1901 [39].

In July of 1896, epilation (alopecia) was also reported (Table 2.4) [16, 37–39]. Usually, the hair loss (German: *Haarausfall*) was confined to the area of severe skin burn. A typical example was the report of H. D. Hawks, a student at Columbia University, who was giving daily 3-h demonstrations of X-rays for the edification of patrons at Bloomingdale Brothers' Department Store in New York City. After his facial skin was scalded, the hair on his temples fell out, then his eyebrows and eyelashes [16]. Another report from the State University of Minnesota described an attempt to X-ray the skull from the right side using 100,000 V. The victim's right

Table 2.4 Reports in 1896 of Röntgen epilation [16, 37–40]

Report	Author	Journal
18 July	A. Broca	Comp Re Fran Phys
22 July and 15 October	W. Marcuse: Berlin	Dtsch Med Wschr
	W. Marcuse: Berlin	Dtsch Med Wschr
16 September	H. D. Hawks	Elect Eng NY
10 October	W. Downie: Glasgow	Lancet
13 November	Jones: State University of Minnesota	Brit J Photog
November	E. E. King	Canadian Prac

ear swelled to twice normal size while blisters and then sores developed on the lips, mouth, and throat so that he was unable to eat for 3 weeks. However, only the right side of the head became bald [16].

In April of 1896, alopecia alone without any skin damage was reported [40]. A nonphysician investigator, J. Daniels, at Vanderbilt University in Nashville, Tennessee, was using an old Crookes tube and a Ruhmkorff coil of *weak energy* to take X-rays of foreign bodies in the hand and arm. He was asked to X-ray the head of a child to locate a bullet. As Daniels was not sure his apparatus was of sufficient energy to take a skull X-ray, he persuaded his dean, William L. Dudley, to be a test subject for a 1-h exposure at a source to skin distance of 0.5 in. No image was obtained. However, 21 days later, the dean developed a bald spot of 5 cm diameter, the exact size of the X-ray field. However, the scalp skin was perfectly normal due to the low energy of these X-rays. Here was epilation without skin injury! This report from the USA was commented upon in the 31 October 1896 issue of the *Wiener Medizinische Wochenschrift* [41].

In Vienna, at this time was 28-year-old physician Leopold Freund (1868–1943), who was training in the Dermatology Department of Dr. Eduard Schiff (1849–1913). Freund was familiar with the various reports of epilation after X-ray exposure. In early November, a 5-year-old girl was seen at the Dermatology Clinic with a 36×37 cm hairy nevus (birth mark) on her back, extending from the neck to the sacrum. Her parents considered it terribly disfiguring and wanted it removed. Freund was refused the use of the X-ray apparatus at the University of Vienna by physiology professor Siegmund Exner, the brother of physicist Franz S. Exner. However, Professor Eder at the Institute for Research in Photography and Reproduction Procedures allowed Freund to use their Röntgen equipment.

Freund started with the *superior* half of the nevus on 24 November and treated for 2 h daily for 10 days, 20-h total exposure. Epilation occurred from day 10 to 18 accompanied by minimal erythema and moist desquamation on day 18 which quickly healed completely. Then Freund decreased the exposure time for a second nevus on the girl's right arm but no effect was observed. He then returned to the *inferior* half of the large nevus and changed the technique again by prolonging the exposure for 21 days: 42-h total exposure. This time epilation progressed from day 8 through 32 and moist desquamation started on day 35. This progressed to a necrotic ulcer by day 51 which healed by day 82 with a deep scar. The ulcer would recur repeatedly and eventually required a skin graft [41, 42].

This patient was seen in follow-up *until age 75* in 1966. She had experienced normal menarche at age 13 and childbearing. The radiation appeared to have been superficial only; however, it resulted in chronic recurring sacral skin ulcerations. She developed a dorsal kyphosis spinal deformity that appeared to be due to osteoporosis and not X-ray damage to her vertebrae (Fig. 2.6). This detailed study of the effects of three different treatment techniques resulted in Freund being acknowledged as the first physician to practice radiation therapy. His treatment differed markedly from the many brief reports of one or two fractions of treatment given to incurable and/or dying patients without any prolonged follow-up observations of the radiation effects. He published on 6 March of 1897 [43] and reported the follow-up on 8 May

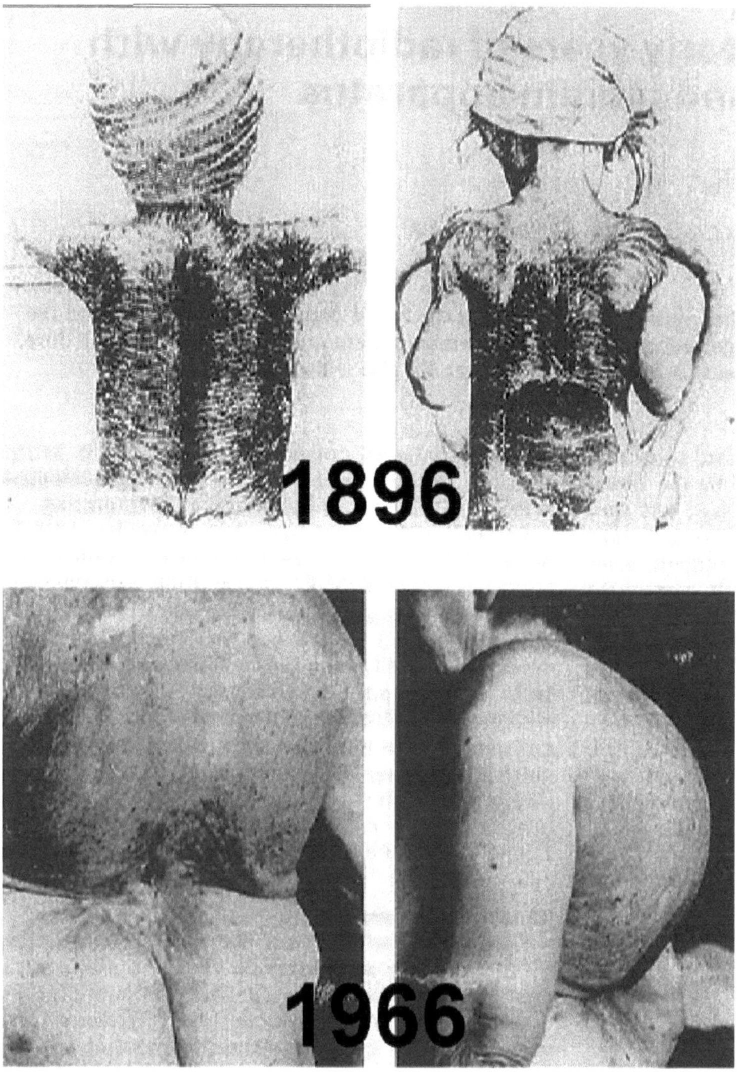

Fig. 2.6 Freund's patient in 1896 and 1966 [42]. (Courtesy of R. F. Mould, UK)

of 1897 [44]. These articles gave proof of the biological effectiveness or X-rays for the first time. The continued observation of this patient and the differences in the treatment results of the three areas allowed Freund to champion the use of fractionated (protracted) low-dose treatments over the rival technique of a single large dose. This conservative approach was considered prudent in this era before dosage could be accurately measured.

Additional reports of Röntgen therapy cases and multiple articles by Freund and his dermatology director E. Schiff interested European dermatologists during the waning years of the nineteenth century, 1897–1899. The early therapy reports,

Table 2.5 X-ray therapy first reports: 1897–1899 [26, 37–49]

Date	Diagnosis	First reports
1897	Hypertrichosis and hairy nevus	L. Freund (Vienna)
1897	Tinea fungus infection = ringworm = favus	L. Freund (Vienna)
1897	Lupus vulgaris = cutaneous tuberculosis	E. Schiff; L. Freund (Vienna)
1897	Acne vulgaris infection	Gautier (Paris)
1897	Sycosis barbae = infection of hair follicles	Schiff, Freund (Vienna)
1897/1898	Breast cancer	H. Gocht (Germany)
1898	Lupus vulgaris	H. Kummel (Hamburg)
1898	Lupus erythematosus (LE)	E. Schiff (Vienna)
1898	Nevus vascularis	J. Jutassy (Hungary)
1898	Eczema	R. Hahn (Hamburg)
1898	Systemic tuberculosis	Bergonie and Teissier (France)
1899	Epithelioma (skin cancer)	Sjogren and Stenbeck (Sweden)

1897–1899, were predominantly for fungus or bacterial infections (Table 2.5). Various skin infections were improved because the X-rays, while not truly germicidal, inhibited their growth enough for the normal body immunity to take over. Freund published the *first* comprehensive textbook devoted solely to radiation therapy in 1903, 416 pages. An English translation followed in 1904. Freund was eventually acknowledged as the "inaugurator of radiotherapy" in 1924 (Fig. 2.7). In 1937/1938, he was forced to leave Vienna along with Freud and other Jews. He died in Brussels under Nazi persecution in 1943 [41, 42].

Lupus vulgaris was a scourge of the nineteenth and early twentieth century in Europe. It was a localized tuberculosis of the skin that has since essentially disappeared. In the USA, lupus vulgaris was more prevalent in the poor and in recent immigrants [37, 45]. The *Journal of the American Medical Association* carried

Fig. 2.7 L. Freund (1920). (Public domain)

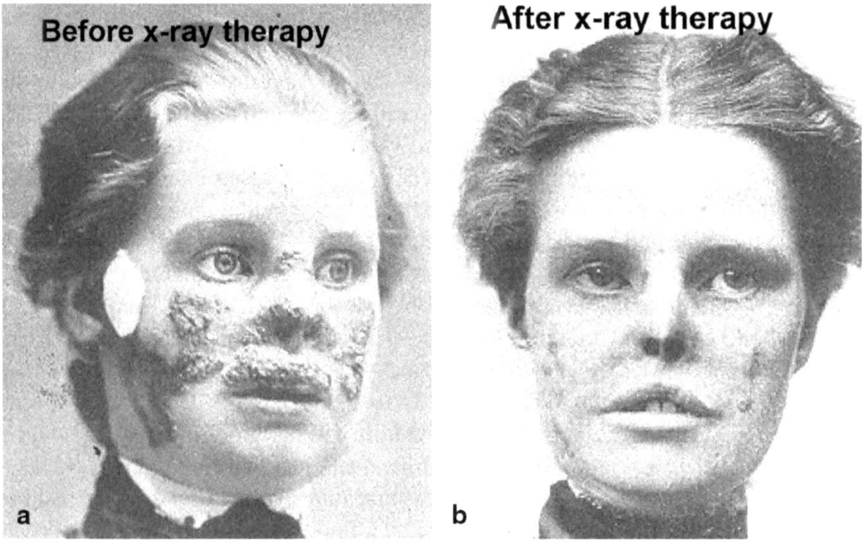

Fig. 2.8 Lupus vulgaris of the face; [37]. (Public domain)

no citations about X-ray therapy until 1898. For the next 4 years, lupus vulgaris was the single most common condition being treated. It accounted for 11 of the 32 therapy trials during 1898–1901 [46]. Lupus vulgaris usually began in childhood or adolescence and was more common in women. The clinical course was extraordinarily slow and could extend over decades. The patient's general health was preserved unless the tuberculosis disseminated systemically. It could occur anywhere on the body, but the face was the most common site, usually beginning on the nose, cheeks, or ears. The mucous membranes of the mouth and nose were frequently involved as well (Fig. 2.8) [37]. Röntgen therapy was found to achieve the same results as were obtained with the Finsen treatment. X-rays were also cheaper and easier to administer and to tolerate but were more damaging to the skin [37, 45].

In 1901 Dr. F. H. Williams (1852–1936) at the Boston City Hospital [47] published a general radiology text that included radiotherapy and had the subtitle "As an aid in diagnosis and as a therapeutic agent." It went through three editions: 1901, 1902, 1903. The 1902 edition of 683 pages had 48 pages on therapy. X-rays were tried for a variety of benign and malignant skin conditions for which physicians had no other answer. Some of these conditions were improved and X-ray therapy became a part of dermatology therapy for the next 100 years. The greatest benefit was seen in patients with skin cancer (epitheliomas), where cure rates would eventually approach 98 %. The first cure of skin cancer with X-rays came in Sweden in 1899–1901. At first, only superficial skin diseases were treated, but then (a) metastatic breast cancer involving the skin, (b) subcutaneous malignant lymph nodes from head and neck cancers, and (c) lymphosarcomas were included in the trials (Appendix II) [42, 47–49].

However, all of these early attempts at Röntgen therapy had to be considered primitive until both the quantity and quality of the X-rays could be measured. Attempts to measure the tube output (beam quality) and the dose delivered to the patient started in 1904 but were never standardized for international use. After 1912, the beam quality was measured by the half value layer (HVL) which indicated the thickness of copper or aluminum that would reduce the beam intensity by 50 %. It was not until 1928 that the Röntgen(R) unit, as measured by an ion chamber, was accepted to define and standardize the dose [42, 49].

One of the early American radiologists who suffered from unmeasured and grossly excessive exposure was George Henry Stover (1871–1915). He was an 1893 graduate of the Denver College of Medicine. He first practiced electrotherapy. Then he was the first professor of radiology at both the Denver College and the Gross College of Medicine in Denver from 1897 to 1915 [50, 51]. Both these schools were absorbed by the University of Colorado in 1911. Before dying at age 44, he underwent over 150 skin grafts on his left hand and lost two fingers. This was due to the early attempts of radiologists to assess tube output by putting their left hand in front of the tube while checking the image on a fluorescent screen held in the right hand. Dying of radiation toxicity and heart disease as well, Stover began visiting the Hawaiian Islands in 1912. With nothing to lose, he even did some Röntgen therapy at the Hawaiian leper colony on Molokai Island. He wrote the lyrics for several local songs by mixing the Hawaiian words into mostly English lyrics. Shortly before his death, he finished "On the Beach at Waikiki" and he requested that his ashes be scattered there [51]

Apparently, one of the first trials of X-ray therapy for metastatic breast cancer and lupus vulgaris never appeared in the literature. The treatment was allegedly delivered by a first-year student at the Hahnemann Homeopathic Medical School in Chicago in January of 1896. Emil H. Grubbe (1875–1960) was one of the minority of medical students who had a college degree (Fig. 2.9). Before the Flexner Report of 1910, only a high school degree or its equivalent was required by most schools [52]. He had a degree in pharmacy (1891–1895) from the Northern Indiana Normal School, now Valparaiso University. After graduation, he opened an office in Chicago doing mining assays. Then, with the help of a recent German immigrant, he started a factory for making Geissler and Crookes vacuum tubes and utilizing platinum from his assay business for wiring [53].

In the fall of 1895, he had enrolled in medical school. While experimenting with his tubes, he developed a severe dermatitis on the dorsum of the left hand in December of 1895. On 27 January 1896, after the reports of the discovery, he consulted with Drs. J. P. Cobb, R. Ludham, A. C. Halphide, and J. F. Gilman at Hahnemann for treatment. He was diagnosed as probable Röntgen-dermatitis; perhaps the first case to occur as he was being inadvertently exposed to X-rays before they were discovered. It was suggested to him that the destructive X-rays might be useful therapy for cancer or lupus.

On the 29th, Dr. R. Ludlum sent him Mrs. R. Lee, a patient with an inoperable ulcerated cancer of the left breast with systemic and nodal disease, stage IV. The next day a Mr. A. Carr, a patient from Dr. Halpide, arrived at the tube factory with

Fig. 2.9 E. H. Grubbe (1896). (Public domain)

an ulcerated lupus lesion involving the right cheek and extending into the neck. Grubbe started treatment of both patients with the Crookes tube almost touching the lesions and with exposures of 1 h/day. Mrs. Lee received 18 treatments and Mr. Carr, 21. Both died within a month, and no conclusions could be drawn except for a subjective decrease in pain. As he was a first-year medical student from a homeopathic school, treating moribund patients with a new unproven modality, it is not surprising that Grubbe did not attempt to publish the account.

However, Grubbe continued seeing patients, but not at the tube factory. He opened a radiology laboratory and office in April of 1896, while still a first-year medical student. He developed the first X-ray department in Chicago, at Hahnemann the same year. In 1902, he did report on 139 cases of skin cancer. Grubbe, described as a difficult person, was divorced in 1911 and suffered multiple surgeries and amputations before dying alone at age 85. In 1949, he wrote an interesting and illustrated book on the origin, birth, and early history of X-ray treatments. At the end of the book, he gives his list of the radiology pioneers who became martyrs, dying of X-ray-induced toxicity: 65 in the USA, 18 in the UK, and 25 in Europe [53].

Grubbe is still remembered at the Chicago Medical Society, where he left his fortune which funds an annual gold medal in his name. However, he waited too late to try and prove his priority [54]. He sent his two referral letters, handwritten with pen and ink in January of 1896 by Drs. Ludham and Halphide, to the Smithsonian for preservation. In 1961, their authenticity was questioned and the Federal Bureau of Investigation (FBI) was asked to examine them for fraud. The FBI found the paper and the ink used on the written referral messages were compatible with that time frame—1896. However, in the process of testing with chemicals, the FBI destroyed everything on the letters but the letterheads (Fig. 2.10).

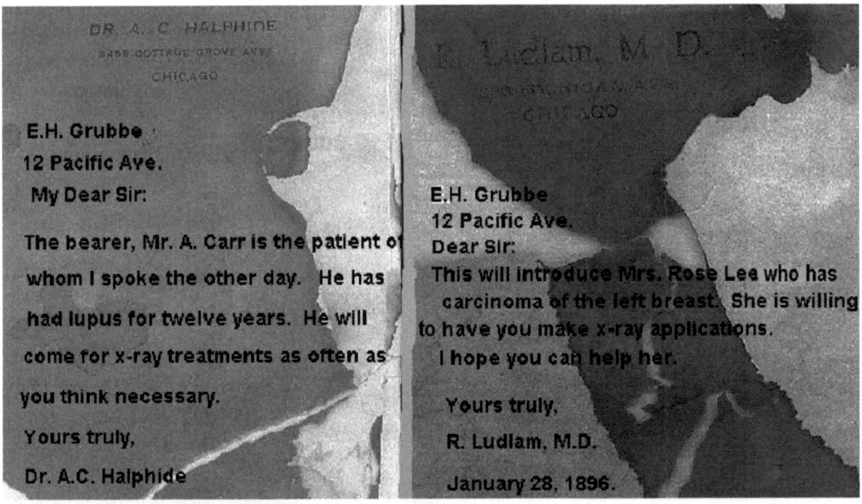
DR. A. C. HALPHIDE
CHICAGO

E.H. Grubbe
12 Pacific Ave.
My Dear Sir:

The bearer, Mr. A. Carr is the patient of whom I spoke the other day. He has had lupus for twelve years. He will come for x-ray treatments as often as you think necessary.

Yours truly,
Dr. A.C. Halphide

R. Ludlam, M.D.
CHICAGO

E.H. Grubbe
12 Pacific Ave.
Dear Sir:
This will introduce Mrs. Rose Lee who has carcinoma of the left breast. She is willing to have you make x-ray applications.
I hope you can help her.

Yours truly,
R. Ludlam, M.D.

January 28, 1896.

Fig. 2.10 Grubbe referral letters of 1896; original messages superimposed on FBI remnants. (Author)

2.7 Supplemental Material: Partial List of Röntgen Awards and Honors [15, 16]

Awards	Honorary memberships
1896	*1896*
Prussian Order of the Crown, II Class	Prussian Acad. Sci., Berlin
Honorary Dr. of Medicine, University of Würzburg	Bavarian Acad. Sci., Munich
Royal Order of Merit of the Bavarian Crown	Naturforscher Gesell., Freiburg
Rumford Gold Medal of the Royal Society, London	Soc. Sci. Antonio, Mexico
Baumgaertner Prize of the Vienna Academy	Physical Soc., Frankfurt a. M.
1897	Chester Soc. Nat. Sci.
Elliot-Cresson Medal: Philadelphia Franklin Institute	Soc. Nat. Sci. Math., Cherbourg
Mattencei Medal, Rome	Wissenschaftliche Gesell.
Prize Lacaze; Acad. Sci., Paris	1897
Member American Philosophical Society of Phil., PA	Swiss Naturforscher Gesell.
1898	Physical-Med. Soc., Erlangen
Prize Otto Wahlbruch Stiftung, Hamburg	Röntgen Soc., London
1899	Soc. Med. St. Petersburg
Title of Royal Geheimrat from the Bavarian Govt.	Soc. Med. Constantinople
Diploma University of Zurich	Alumnae Federal Polytech. Sch. Zurich
1900	Reale Acad. Geographici, Florence
University of Munich; Professor and Dir.	1898
Grosskomturkreuz Royal Order of Merit: Bavaria	New York Med. Soc.

Awards	Honorary memberships
Order of Merit: St. Michail, I Class	Reale Acad. Lincei, Rome
Silver Medal of Prince Regent Lutipold	Reale Inst. Sci., Venice
Member of the Academy of Sciences, Munich	Soc. Holland., Harlem
Barnard Medal, Columbia University, New York	*1899*-Rotterdam, Stockholm
Member Maximillian Order Sciences with Decoration	*1900*-Hamburg
1901-Nobel Prize Physics, first	Munich Soc. of Physicians
1907-Member Royal Acad. Sci. Amsterdam	Munich Acad. of Sci.
1908-Title of Excellency, Bavaria	*1901*-Stockholm
1912-Russian diploma, Odessa	*1902*-Inst. De Coimbra
1915–70th birthday: honors	*1903*-Cambridge, Turin, Berlin x-Ray
1919–50th anniversary of Ph.D. honors	*1904*-Vienna Soc. of Physicians
1920–75th birthday and retirement honors	*1905*-London, Edinburgh
1921–1930: continued honors	*1906–1921*: continued memberships

References

1. Brox J (2010) Brilliant: the evolution of artificial light Mariner. Houghton Mifflin Harcourt, Boston
2. Derry TK, Williams TI (1993) A short history of technology before 1900, 3rd edn. Dover, New York
3. Indo A (2013) Auer gas mantles light. http://www.auerlicht.com/aboutus/aboutus.htm. Accessed 9 April 2013
4. Marshall JL (2002) Discovery of the elements. Pearson Custom Printing, Boston
5. Wilkins A (1999) Ultraviolet light and its use with fluorescent minerals. http://www.uvsystems.com/articles/uvlightandfluorescentminerals.pdf. Accessed 17 March 2013
6. Bonewitzm RL (2005) Rock and gem. DK Publishing, New York
7. U.S. Dept. of Energy (2014) Fluorine. Los Alamos national laboratory. http://www.periodic.lanl.gov/9.shtml. Accessed 22 Aug 2014
8. Nobel Lectures (1966) Henri Moissan—biographical. Elsevier, Amsterdam. http://www.nobelprize.org/nobel_prizes/chemistry/laureates/1906/moissan-bio.html. Accessed 23 Aug 2014
9. Davidson MW et al (2009) Basic concepts in fluorescence. http://micro.magnet.fsu.edu/primer/techniques/fluorescence/fluorescenceintro.html. Accessed 22 Aug 2014
10. Segre E (1980) From x-rays to quarks. Freeman, San Francisco
11. Brucer M (1990) A chronology of nuclear medicine. Heritage, St. Louis
12. Grigg ERN (1965) The trail of the invisible light. Thomas, Springfield
13. Dixon L (2013) The real story of the first neon signs in America. Packard Concormant LX (4):2–9
14. Riehl N, Seitz F (1996) Stalin's captive: Nikolaus Riehl and the soviet race for the bomb. Chemical Heritage Foundation, Philadelphia
15. Glasser O (1945) Dr. W. C. Röntgen. Thomas, Springfield
16. Glasser O (1934) Wilhelm Conrad Röntgen and the early history of the Röntgen rays. English edition by C.C. Thomas, Springfield, 1934. Reprinted by Norman Publishing Co., San Francisco, 1993; (1993), San Francisco
17. Klickstein HS (1966) Wilhelm Conrad Röntgen. Mallinckrodt classics of radiology, vol 1. Mallinckrodt Chemical Works, USA

18. del Regato JA (1985) W.C. Röntgen. In: Radiological physicists. American Institute of Physics, New York, pp 1–10
19. Harder D (1987) Röntgen's discovery—how & why it happened. Int J Radiat Biol 51(5): 815–839
20. Rosenow UF (1995) Notes of the legacy of the Röntgen rays. Med Phys 22(11):1855–1867
21. Guyton AC (1991) The eye II ch 50 color vision. In: Textbook of medical physiology, 8th edn. Sanders, Philadelphia, p 552
22. Norman JN (ed) (1991) Morton's medical bibliography, 5th edn. University Press, Cambridge
23. Norman JN (2011) Catalogue 40-Jeremy Norman's HistoryofScience.com. https://www.yumpu.com/…/catalogue-40-jeremy-normans-historyofscien. Accessed 5 July 2014
24. Almond PR (1993) The x-ray centennial—Thompsons & Thomsons. Med Phys 20(2): 275–281
25. Aronowitz JN (2007) Ethereal fire: antecedents of radiology and radiotherapy. Am J Rad 188:904–912
26. Kassabian MK (1907) Electro-therapeutics and Röntgen rays. Lippincott, Philadelphia
27. Kovacs R (1938) Electrotherapy and light therapy. Lea & Febiger, Philadelphia
28. Bender GA (1966) Great moments in medicine. Northwood Institute, Detroit
29. McCullough H (1998) The rise of McClure's. In: The breakup of McClure's magazine, Chap. II. http://www.tarbell.allegheny.edu/mc2.html. Accessed 22 Aug 2014
30. Dam HJW (1896) A visit to prof. Röntgen in his laboratory in Wurzburg. McClure's Mag 6(5):403–415. McClure's Magazine archives for April 1896. http://www.onlinebooks.library.upenn.edu/webbin/serial?id=mcclures. Accessed 5 July 2014
31. Feldman B (2000) The Nobel prizes. Arcade, New York
32. Kraft E, Finbi N (1980) Röntgen's discovery of x-rays and Lenard's priority claim. NY State J Med 80(10):1623–1625
33. Knutsson F (1974) "Röntgen and the Nobel prize"; the discussion at the royal Swedish academy of sciences in Stockholm in 1901. Acta Radiol 15(5):465–473
34. Asimov I (1988) Understanding physics, vol I–III. Dorset Press, New York
35. Nave CR, Nave BC (1985) Physics for the health sciences, 3rd edn. Saunders, Philadelphia
36. Johns HE, Cunningham JR (1983) The physics of radiology, 4th edn. Thomas, Springfield
37. MacKee GM (1938) X-rays and radium in the treatment of skin diseases, 3rd edn. Lea & Febiger, Philadelphia
38. Gilchrist TC (1897) A case of dermatitis due to the x-rays. Bull Johns Hopkins Hosp 18: 17–23
39. Codman EA (1902) A study of the cases of accidental x-ray burns hitherto recorded. Phil Med J 9:438–442
40. Daniels J (1896) Letter to the editor: depilatory action of x-rays. Med Rec 49:595–596
41. Kogelnik HD (1997) Inauguration of radiotherapy as a new scientific specialty by Leopold Freund 100 years ago. Radiother Oncol (Eur) 42(3):203–211
42. Mould RF (1995) Invited review: the early years of radiotherapy with emphasis on x-ray and radium apparatus. Brit J Radiol 68:567–582
43. Freund L (1897) Ein mit Röntgen-strahlen behandelter fall von naevus pigmentosus piliferus. Wien Med Wochenschr 47:428–433
44. Freund L (1897) Nachtrag zu dem artikel Ein mit Röntgen-strahlen behandelter fall von naevus pigmentosus piliferus. Wien Med Wochenschr 47:856–860
45. Becker SW, Obermayer ME (1943) Modern dermatology and syphilology. Lippincott, Philadelphia
46. American Medical Association (1898–1901) Current medical literature. J Am Med Assoc Vol. XXX: 1–1538; Vol. XXXI: 1–1592; Vol. XXXII: 1–1466; Vol. XXXIII: 1–1662; Vol. XXXIV: 1–1686; Vol. XXXV: 1–1692; Vol. XXXVI: 1–1898; Vol. XXXVII 1–1734
47. del Regato JA (1993) FH Williams. In: Radiological Oncologists, Chap. 2. Radiological Centennial Inc., Reston, pp 13–24

48. Mould RF (1993) A century of x-rays and radioactivity in medicine. Institute of Physics (London), Bristol
49. Glasser O (ed) (1933) The science of radiology. Thomas, Springfield, pp 210–241, 276–304
50. Sherlock T (2013) History of Colorado's healthcare heritage, p 443. Google Books. http://www.books.google.com/books?isbn=1475980264. Accessed 20 April 2014
51. Landa ER (1987) Buried treasure to buried waste-the rise and fall of the radium industry. Colo Sch Mine Qly 82:1–77
52. Flexner A (1972) Medical education in the United States and Canada. Arno Press, New York
53. Grubbe EH (1949) X-ray treatment; its origin, birth, and early history. Bruce, St. Paul
54. Grubbe EH (1933) Priority in the therapeutic use of x-rays. Radiology 21:156–162

Chapter 3
Radioactive Rocks

3.1 Becquerel Rays

On 20 January 1896, Henri Poincare (1854–1912) presented Roentgen's discovery at the weekly meeting of the French Academy of Sciences. Poincare was a theoretical physicist, mathematician, and engineer who excelled in all the scientific disciplines. He ventured his opinion, shared by some others, that the Roentgen rays were somehow being produced as a result of the fluorescence, which was seen in the Crookes tubes at the opposite end from the cathode. At that time it was not appreciated that the X-rays were being produced simply by the cathode rays (electrons) striking the opposite glass wall (a target) of the tube. The observed fluorescence was only a byproduct of the ionization of the gases at the far end of the tube.

In the audience was the Chair of Physics at the Musee d'Histoire Naturelle, Henri Becquerel (1852–1908), who was an expert on fluorescence. Becquerel was the third in four generations of a family of physicists who studied fluorescence and phosphorescence. The Becquerels were particularly interested in the emission of visible light by crystalline substances after they had absorbed electromagnetic radiation of a shorter wavelength: fluorescence. When the light emission from the crystal persisted after the radiation exposure had ceased, it was called phosphorescence.

Henri Becquerel immediately took up the challenge of investigating the relationship, if any, between roentgen rays and fluorescence. As the Academy met weekly, they published a weekly journal. In six papers written between 24 February and 18 May, Becquerel reported his fantastic discovery of ionizing radiation being emitted from rocks containing the mineral uranium [1–3]. Becquerel's six papers of 1896 all appeared in volume 122 of the Academy journal, the Comptes Rendus de l'Academie des Sciences [4]:

R. F. Robison, *Mining and Selling Radium and Uranium*,
DOI 10.1007/978-3-319-11830-7_3

24 February	Sur les radiations emises par phosphorescence, pp. 420–421
02 March	Sur les radiations invisibles emises par les corps phosphoresce, pp. 501–503
09 March	Sur quelque proprieties nouvelle des radiations invisibles emises par divers
	Corps phosphoresce, pp. 559–564
23 March	Sur les radiations invisibles emises par les sels d'uranium, pp. 689–694
30 March	Sur les properties differentes de radiations invisibles par les sels d' uranium
	Et du rayonne ment de la saroi anticathodiques tube de Crookes, pp. 762–767
18 May	Emission de radiations nouvelle par l' urnai um metallique, pp. 1086–1088

Becquerel wrapped photographic plates in heavy black paper and then placed different fluorescent rocks on the plates while exposing the rocks to sunlight. His first paper announced that only *fluorescent* uranium emitted enough radiation to darken the plates. His second paper revealed that sunlight was *not* necessary to activate the uranium rays. On 9 March he reported that the uranium rays were *ionizing radiation* and were more penetrating than roentgen rays! He went on to characterize the rays as a specific property of all uranium compounds, whether they were *fluorescent* or not. It was enough to earn the 1903 Nobel in physics [1–3, 5–7].

Becquerel continued to study the uranium rays but he favored the photographic emulsion plate over the gold leaf electroscope for the detection and crude measurement of the ionizing radiation. Few others published on the Becquerel rays during the frenzy that followed the 1895 discovery of Roentgen rays. Even Becquerel's interest had waned until the Curie discoveries of 1898 [1–3]. *Becquerel's publications* markedly increased after that:

1896 = 7 papers	1900 = 8 papers	1904 = 1 paper, 1 book
1897 = 2 papers	1901 = 8 papers	1905 = 3 papers
1898 = 0	1902 = 3 papers	1906 = 1 papers
1898 = Po and Ra discovered	1903 = 6 papers	1907 = 2 papers
1899 = 4 papers	1903 = Nobel Prize	1908 = 3 papers

3.2 Klaproth

Becquerel's uranium came from the uranium factories in Bohemia, where it was being used as a coloring agent for porcelain and glass. Uranium oxide (U_3O_2) had been discovered in 1789 by Martin Klaproth at his Berlin laboratory. He readily recognized it as a new element. With an atomic mass of 238 it was the heaviest element known. Klaproth had found it in the Erzgebirge mineral known as pechblende and in its secondary mineral, tobernite; Cu $(UO_2)_2(PO4)_2$-11 H_2O.

Pechblende is a black mineral ore, resembling hardened pitch, which is about 50% U_3O_2 and 50% *lead*, copper, bismuth, and barium along with small quantities of just about every other known element [7–9].

Martin Heinrich Klaproth (1743–1817), was a Berlin apothecary (pharmacist) as well as a distinguished German mineralogist and chemist. He had discovered uranium in pechblende ore samples from John-George (German: Johanngeorgenstadt) in Saxony. He made the initial discovery examining tobernite, which was more soluble and easier to work with than pechblende. The later, however, furnished the bulk of the ore that he subsequently utilized. Klaproth's ore samples came from both John-George and from Sankt Joachimsthal in Bohemia. The two towns are on opposite sides of the Erzgebirge about 16 km (10 miles) apart. The source of the tobernite was the George Wegsfort Mine in Saxony, which had opened in 1670. The pechblende ore from Sankt Joachimsthal proved to be the richest source for uranium [10–13].

Klaproth was considered one of the best analytical chemists of his day. He was the first Professor of Chemistry at the University of Berlin, which opened in 1810. Legend has it that he first learned chemistry while working as a clerk in an apothecary [7]. He discovered both uranium-92 and zirconium-40 in 1789 and he helped other chemists substantiate their discoveries. He separated strontium-38 from calcium and barium in 1793 after Charles Hope had found $SrCO_3$ in Strontium, Scotland, in 1790. He confirmed in 1795 the 1791 discovery of element 22 by Reverend William Gregor, and his nomenclature suggestion of titanium was retained. Klaproth in 1798, at the request of Muller von Reichenstein, whom he credited, confirmed the previously obscure discovery of tellurium-52 (1782) and furnished the name [9, 10]. He had many discoveries and confirmations:

- 1789: uranium-92 discovered and named
- 1789: zirconium-40 discovered
- 1791: yttrium-39 confirmed
- 1793: strontium-38 separated from barium
- 1795: titanium-22 confirmed and named
- 1797: chromium-24 confirmed
- 1798: beryllium-9 confirmed and named
- 1798: tellurium-52 confirmed and named
- 1803: cerium-58 co-discovery with Swedes

In 1803, he independently discovered cerium-58 simultaneously with the Swedes Berzelium and Hesinger, but he graciously ceded the discovery to them. Cerium, the most abundant and second most reactive rare earth element (REE) was utilized by Carl von Welsbach (1858–1929) for the incandescent thorium gas mantle and the pyrophoric alloy used to make cigar/cigarette lighters, Chap. 4 [14]. Cerium was named after Ceres, the first (and largest) asteroid, which had just been discovered. In the twenty-first century, Ceres has come to be defined, along with Pluto, as a dwarf planet.

The planet Uranus had been discovered in 1781 by a German, Sir Friedrich Wilhelm Herschal (1738–1822) who just happened to prefer living in England. He was a talented musician who became a famous astronomer as well. Klaproth named uranium after the recent discovery, by his fellow German, of the invisible planet Uranus (Sky God). The Romans had named the seven visible planets after some of their favorite deities and these names were transferred to days of the week (Table 3.1) [15]. Astronomers continued the Roman tradition for naming planets and Klaproth started naming elements after planets. In 1940, physicists at Berkeley followed suit with elements # 93, Neptune (Sea God), and # 94, Pluto. By an eerie coincidence, plutonium-94 became associated with the Roman god of death [16], Appendix I: Periodic Table.

Little value was found for uranium until the early nineteenth century. By 1819 uranium ore was being processed in John-George, Saxony, for use as a yellow dye to color glass and porcelain. Pure uranium metal was finally isolated in 1841 by Eugene Peligot (1811–1890), who was professor of chemistry and glass-making at the Central School of Arts and Manufacturing in Paris. Bohemian metallurgical chemist, Adolph Patera (1819–1894) investigated the commercial possibilities for uranium. In 1852, he prepared a uranium yellow dye for the colorization of glass and china. The value of uranium for coloration of Bohemian ceramics and glass was thus demonstrated during 1852–1856. Glass could be colored yellow with fluorescent green reflections while porcelain could be glazed in yellow, green, orange, red, and black [7].

A dye factory opened in 1855 in Sankt Joachimsthal. By 1871 the uranium dyes were being produced at the K. K. Uranfarbenfabrik (royal uranium color factory), an old silver smelting building. In Bohemia, uranium colored glass became internationally famous and sought after. The factory produced six different yellow colors from light greenish yellow to deep orange and also black for use on glass and porcelain. The uranium was extracted from pechblende when the crushed ore was roasted with carbonate of soda and then washed with warm water followed by dilute sulfuric acid. The acid *solution* contained the uranium, therefore the insoluble *residue* was discarded as it was considered worthless [11–13].

Table 3.1 Nomenclature of the elements [15]

Roman deities	Visible planets and Latin days of the week	Known elements and alchemical meaning	German days of the week
Apollo	Solis	Gold-radiance	Sunnandaeg
Diana	Lunae	Silver-light	Monandaeg
War	Martis	Iron-war	Tuesdaeg
Messenger	Mercuri	Mercury-messenger	Wodendaeg
King	Jovis	Tin-supreme	Thorsdaeg
Love	Veneris	Copper-beauty	Frigadaeg
Time	Saturni	Lead-old age	Saeterdageg

3.3 Marie and Pierre

In November of 1897, Becquerel was faced with an unusual graduate student seeking a Ph.D. in physics. She was Marie Sklodowska Curie (1867–1934), age 30 on 7 November, a foreigner, the wife of a colleague and 2 months post-partum. The Sorbonne accepted women students but few if any had advanced degrees in any science field. Working mothers were equally rare but she had engaged her widowed father-in-law for child care. As a native of partitioned Poland she had been denied any advanced education at home and had come to the University of Paris in 1891. As a buffer zone between the Russian, German, and Austrian Empires, Poland had been forced to cede territory to all three belligerent neighbors in 1772 and 1792. Then in 1795 the entire country was partitioned and occupied until 1918 [20]. Russia occupied 73 % including Warsaw. Prussia occupied 15 % including Gdansk. Austria occupied 12 % including Cracow.

Sklodowska's education in Warsaw, the Russian Sector, was in Russian and had ended when she finished secondary school at age 15.5 in 1883. She began a program of self-study and home tutoring from her father, who was a high school physics teacher. Her formal education was as limited as were her finances when she arrived in Paris in November of 1891. Nevertheless, in 2 years she was able to obtain a college degree in physics, placing first in her class. This led to a scholarship from Poland for students studying abroad and she returned to the Sorbonne for 1893/1894 to study for a degree in mathematics [17–19, 21–23]. She obtained her mathematics degree in 1894 finishing second in her class.

During the 1893/1894 school year she was hired to do research on the comparison of the magnetic properties of French steel. The stipend was provided by the Society for the Encouragement of National Industry. The steel samples were provided by chemistry Professor Henry Louis Le Chatelier (1850–1936). The work was to be done in the Physics Department of the Sorbonne under its Director Professor Gabriel Lippmann (1845–1921). Space was severely limited in the Sorbonne Physics Laboratory and there was little room for the equipment Marie required for her industrial study. In addition, Professor Lippmann also expected Sklodowska to help him analyze and catalog his mineral collection. Marie suffered through the mineralogy duties during 1894/1895 and 1895/1896, never suspecting its future application to her career.

In March of 1894 she was complaining about her situation at the Sorbonne Physics Lab to some Polish friends. They introduced her to physicist Pierre Curie (1859–1906) who was known for his work in magnetism and for his instruments to measure ionization. Curie did not have an immediate answer for her problem but he did start providing social engagements. Although the son and grandson of practicing physicians, Pierre had been considered a dreamy sensitive child, immune to organization and discipline. He could not adjust to public schooling and had been home tutored through to his secondary school degree in 1875 at age 16 [17–19, 21–23]. He then entered the Sorbonne; where he earned a college degree in physics 2 years later.

Pierre then spent 6 years as a lab assistant at the Sorbonne, where his brother Jacques (1856–1941) was working in the laboratory of chemist *Charles Friedel*, who had become Professor of Mineralogy at the Sorbonne in 1876. Friedel became Professor of Organic Chemistry in 1884. He was to identify and name a newly discovered uranium mineral from the USA in 1898; see Chap. 5. The Curie brothers collaborated on the study of crystals. They invented the crystal electrometer for the *quantitative* measurement of ionizing radiation; the Curie piezoelectrometer. When Jacques left to become physics Chief at Montpellier in 1883, Pierre took a job as Laboratory Chief at the less prestigious *Ecole de Physique et de Chimie Industrielles de la Ville de Paris* (EPCI); Industrial School of Physics and Chemistry. He was to remain there for 21 years, from 1883 to 1904, with only four publications prior to 1898 [17–19].

Marie Sklodowska spent 1894/1895 in Paris (a) working on her industrial project at Lippmann's lab, (b) studying for the qualifying exam to teach in a French secondary or professional school, (c) being romanced by Pierre Curie, and (d) convincing him to complete his Ph.D. which had been pending since 1877. His thesis was published in 1895 on "the magnetic properties of materials at various temperatures." After he did this, they got married in July of 1895. The Ph.D. may have been more of a prenuptial requirement than a coincidence. Also, Marie was now able to switch her industrial job to her husband's lab at the EPCI. It would be here at the EPCI, and not at the Sorbonne, that radioactivity and radium would materialize [17–19, 21–23].

After their marriage, Marie S. Curie worked during 1895/1896 and 1896/1897 in her husband's lab at the EPCI, to complete her study on the magnetism of steel. She earned a stipend for the 3 years of work and published her findings in 1897; her first publication. As Marie had no status at the *Academie des Sciences*, her paper was presented there by Professor A. Potier before its publication in the *Comptes Rendus*. In September of 1897, Marie delivered her first child, Irene. At that time, Marie was diagnosed with probable pulmonary tuberculosis, a disease which had killed her mother and which would later plague Irene. Marie refused a sanatorium.

In November she started her Ph.D. program in physics by studying the "Becquerel rays." Marie was advised to examine all known mineral elements for the presence of ionizing radiation (Becquerel rays); an idea that had evidently occurred to only one other investigator; G. S. Schmidt in Berlin. Marie chose to do the work at the EPCI. With her background in mineralogy and magnetism, she began by testing the large number of minerals and rocks at the EPCI collection; Appendix III, Uranium Minerals. She also secured samples from the prominent chemists and mineralogists in Paris, who are further identified in the Supplement 3.6 [24].

Up until 1897/1898, the presence of uranium rays was readily detected, but inadequately measured, by examining photographic plates for blackening or by the gold leaf electrometer. Marie began by measuring the *quantity* of ionization by using her husband's piezoelectric electrometer, a primitive ion chamber. Fortuitously, she had decided to also include various mineral compounds as well. She compared these minerals to a standard of pure uranium metal. Her results were astonishing. Pierre Curie abandoned his projects on 18 March to join her in this promising research [24].

In April of 1898, Marie published her first paper on Becquerel rays; with the title "*Rayons emis par les composes de l'uranium et du thorium*" (On rays emitted from compounds of uranium and thorium). The ionizing radiation activity was found only in uranium and thorium minerals. She listed the ionizing activities of the uranium and thorium minerals in pico-amperes (pA = 10^{-12}) (Table 3.2). In her 1903 Ph.D. thesis she reported the ionization intensities in 10^{-11} A instead of in 10^{-12} A [24]. This paper was presented at the *Acadamie* by Professor Lippman.

Her study showed that pure uranium metal had an activity of only 24 pA. Moreover, various uranium minerals and oxides of uranium were also measured at only 11–27 pA. However, pechblende from both Saxony and Bohemia gave much higher readings: 83–67 pA. Pechblende from Saxony was 3.5 times more active than uranium; 83 versus 24! Another uranium mineral ore, now called tobernite (previously chalcolite), also gave a reading twice the value of metallic uranium; 52 versus 24 pA. After publication, the Curies found that the activity of thorium had already been discovered by Gerhardt S. Schmidt (1865–1949). He had published the activity of thorium on 4 February 1898, 2 months before [24].

This April 1898 study suggested that *one or more* elements, with greater activity than either uranium or thorium, were hidden within pechblende and tobernite. The Curies began the search on 14 April with a 100 g sample of pechblende that was 2.5 times more radioactive than uranium metal. This sample is assumed to have

Table 3.2 Ionizing activity of radioactive minerals [24]

Minerals tested	Report of April 1898	1903 thesis
Uranium ores	10^{-12} A	10^{-11} A
Pechblende: John-George	83	8.3
Pechblende: Bohemia	67	6.5 (J)—7.0 (P)
Tobernite (chalcolite)	52	5.2
Autunite	27	2.7
U_2O_5 black	27	
Uranium metal	24	2.3
U_3O_8 green	18	
Pechblende: Cornwall	16	1.6
Samarskite	11	1.1
Carnotite: USA	–	6.2
Clevite	–	1.4
Thorium ores		
Thorium oxide; 6.0 mm thick	53	–
Thorium oxide; 0.25 mm thick	22	–
Orangite	20	2.0
Various thorites	02–14	0.10–1.4
Thorium sulfate	08	–
Monazite	–	0.5
J = Joachimsthal, P = Prizban		

been pechblende from Prizban, Bohemia, which measured 65 pA, which was 2.7 times more radioactive than uranium. Pechblende was known to contain uranium oxide plus numerous other metals such as; copper (Cu), arsenic (As), antimony (Sb), bismuth (Bi), barium (Ba), and lead (Pb). The Curies used the standard chemical separation techniques that had been used for separating the various metals from pechblende. However, they additionally used the Curie electrometer to measure the ionizing radiation as an aid in the separation process [25, 26]. It was basically a six-stage separation process to obtain the radioactive unknowns, labeled X1, X2, and X3:

1) 100 g of crushed/milled pechblende is put into solution by HCL acid.
2) K_2CO_3 and NaOH added for fusion with the various elements.
3) H_2S acid added to above solution to precipitate (ppt.) out Pb, Cu, As, Sb, barium (Ba), and bismuth (Bi) as sulfides. This precipitate (ppt.) was more radioactive than the oxides of uranium and thorium in solution along with a possible radioactive unknown, *X3*. This solution was disregarded for now.
4) Ammonium sulfide was added to the ppt. to remove the As and Sb which ended up in solution. The remaining ppt. contained Pb, Cu, *Bi, and Ba*.
5) The insoluble sulfide ppt. was next dissolved in nitrous acid, HNO_3. This resulted in a solution containing radioactive Bi, Pb, and Cu. Radioactive *Ba* (*and X2*) remained in the ppt. for later evaluation.

6a) The above solution was then treated with sulfuric acid to ppt out the Pb and then ammonium hydroxide used to ppt. out the radioactive *Bi* (and X).

6b) The ppt. that was not dissolved by nitrous acid (step 5) contained radioactive barium [and X2], which was to be addressed later.
(X1 = polonium, X2 = radium, X3 = actinium).

Since pure bismuth and barium were both known to have *no* radioactivity it was obvious that the associated activity in both was due to new unknown substances, X1 and X2, that were radioactive. There was also a suspicion that the solution of uranium and thorium in step 3 had more radioactivity than expected and might contain another new element, X3. The Curies decided to concentrate on X1 first. However, the separation of the radioactive substance, X1 (polonium), which was associated with bismuth, turned out to be exceedingly difficult. Indeed, it would take 12 years to prove its existence. They were working with trace amounts of these new elements that were invisible and detectable only by their radioactivity [25, 26]. To prove the existence of a new chemical element they would need to (a) extract enough mass to determine an atomic weight and (b) demonstrate a unique flame emission spectrum with a characteristic band of lines.

Nevertheless, the Curies published their first joint publication on 18 July 1898; *Sur une substance nouvelle radio-active contennue dans la pechblende* (a new radioactive substance contained within pechblende), even though they could not prove it! The results of their radiochemical separation of X1 from bismuth had identified X1 as having a radioactivity 400 times as great as uranium. As neither had trained in chemistry, the Curies sought help from Gustave Bemont (1867–1932), who was in charge of practical chemistry training for the students at EPCI; the Chef de Travaux. Even with his help, they could not separate enough material to get an atomic weight [25, 26]. Bemont and the Curies are pictured in Fig. 3.1.

They also asked Eugene Demarcay (1852–1904), a specialist in thermal or flame emission spectroscopy for chemical identification, to search for the unique spectral lines (electromagnetic radiation) of a new element. Demarcay could *not* distinguish any new characteristic lines. Meanwhile, in a politically provocative move, the Curies stated: *Si l'existence de ce nouveau metal se confirme, nous proposons de l'appeler polonium, du nom du pays d'origin de l'un de nous*. Even though the Polish state no longer existed, they were going to name the new element for Poland, "the name of the place of origin of one of us." And so X1 became polonium (Po).

The Curies had probably separated about 6 ng of polonium, which is below the limit of detection by spectral lines. It would be 1910 before Marie Curie and Ander Debierne could separate enough polonium; 0.1 mg or 0.45 Ci; to get a characteristic spectral identity. A ton of uranium ore contains only about 100 μg of polonium, which has a half-life of only 138.4 days. Its radioactivity is dominated by alpha ray production; a milligram of Po emitting as many alphas as 5 g of radium [25, 26].

This paper was presented to the Acadamie by Becquerel himself. The Curies thanked Bemont in a footnote and Demarcay in the text. This paper also marked the first use of the new word "radioactive." "Radioactive" was used only once in this paper, in the title. In their next paper it was used sixteen times. This terminology quickly became accepted universally by the new radiochemists. By 1900 the Curies had eliminated the hyphen but it continued in English translations for a while. In 1923 Marie noted that she had coined the term to define this new property of matter; *the emission of ionizing* radiation [24].

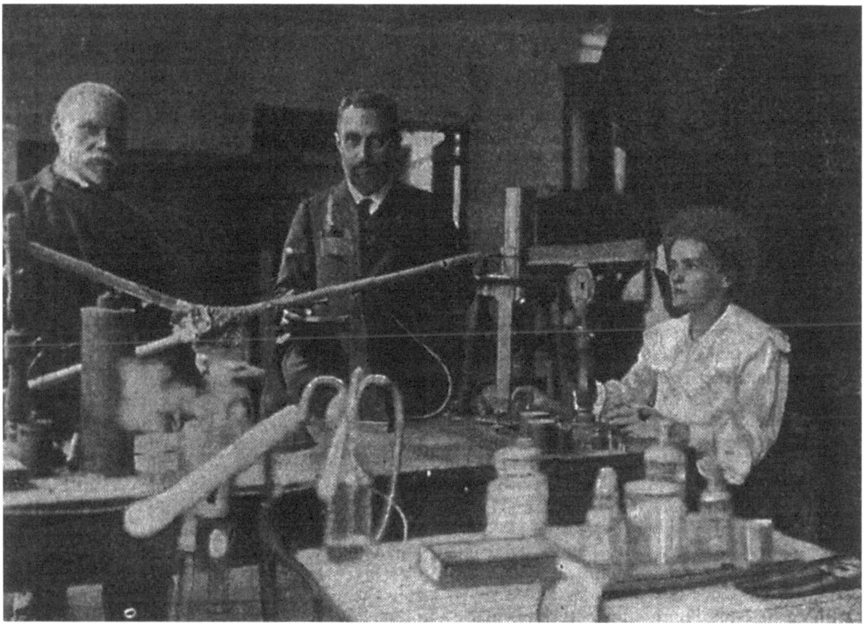

Fig. 3.1 Gustave Bemont, Pierre Curie, Marie Curie [45]. (Public domain)

Their second joint paper announced the discovery of radium (X2) on *26 December 1898*. The new element was described in the title only as *very radioactive*; *Sur une nouvelle substance fortement radioactive, contenue dans la pechblende.* Indeed it was measured as 900 times that of uranium and they named it radium. The chemical solution from step 6b with radioactive barium connected to the new element, X2, was much easier to work with and there was more of it [24–26]. The concentration of radium in pechblende is about 5000 times that of polonium [9] (Table 3.3). Radium is chemically similar to barium but radioactive radium could be separated from nonradioactive barium.

It is estimated that they separated 30 μg of radium and that was enough for Demarcay to identify a spectrum wavelength of 3814.8 Å, which was not associated with any known element. Bemont was honored as the third author on this 1898 paper, which was again presented by Becquerel. Demarcay published separately a one page note about his work [24–26]. The four papers of Marie Curie during 1897–1898 [17, 18, 25] were:

- 1897: "Proprietes magnetiques des aciers temps.", Compt. rend., 125, pp. 1165–1169. - Note de *Mme Sklowdowska Curie*, presentee par M. A. Potier.
- April 1898: "Rayon semis par les composes de l'uranium et du *thorium*." Compt. rend., 126, pp. 1101–1103. - Note de *Mme Sklodowska Curie*, presentee par M. Lippman.
- July 1898: "Sur une substance nouvelle *radio-active* contennue dans la pechblende." - Compt. rend.,127, pp. 175–178. Note de *M. P. Curie et de Mme S. Curie*, persentee par M. Becquerel.
- December 1898: "Sur une nouvelle substance fortement radio-active, continue dans la pechblende." - Compt. rend.,127, pp. 1215–1217. Note de M. P. Curie de Mme P. Curie et de M. G. Bemont, presentee par M. Becquerel.

During 1898 the Curies had exhausted their original 100 g pechblende sample. Between July and December they realized their need for a vastly increased supply of

Table 3.3 Relative quantities of radioactive elements in Pechblende [9]

Z #	Discovery	Atomic weight = A #	Relative quantities	Grams
92	1789	Uranium-238	10,000 mg = 10 g	10
92		Uranium-235	70 mg = 0.070 g	7×10^{-2}
92		Uranium-234	0.5 mg = 0.005 g	5×10^{-4}
88	1898	Radium-226	3.0 μg = 0.003 mg	3×10^{-6}
89	1899	Actinium-227	2.0 ng	3×10^{-9}
84	1898	Polonium-210	1.0 ng	1×10^{-9}
86	1900	Radon-222	Gas	2×10^{-11}
87	1939	Francium-223	Less than one oz. on earth	1×10^{-20}

high grade pechblende ore to complete their studies. The Curies had found radioactivity in every known mineral that contained either uranium or thorium. However, the greatest activity was noted in the pechblende that came from both sides of the Ore Mountains, which separate German Saxony from Czech Bohemia.

The Curie pechblende samples from John-George (Johanngeorgenstadt) in Saxony measured 83 pA of radioactivity and those from Sankt Joachimsthal in Bohemia measured 70 pA. These samples contained 50 % or more uranium, which was commercially valuable and which made the ore expensive. The uranium was used for coloring glass and porcelain. The Curies surmised that the residue left after the uranium had been removed ought to be "dirt cheap" [11–13].

The Curies found that the largest supply of pechblende residue was located at the uranium factory in Sankt Joachimsthal, Bohemia. Indeed, it was considered worthless by the factory, which had extracted the uranium from the ore and then dumped the residue in the forest. This factory was the K. K. Urangelbfabrik (Royal Uranium Yellow Factory) (Fig. 3.2). They were inundated with this residue of pechblende, which remained after the uranium had been extracted. It was this worthless residue, now mixed with various forest debris that the Curies needed. In 1898 Bohemia was still part of the Austria-Hungarian Empire and negotiations would have to go through the Austrian Academy in Vienna. The Curies, in order to get this residue, became involved with the imperious and Imperial Austrian government, one of the three oppressors occupying Poland [20, 27, 28].

Fig. 3.2 K. K. Urangelbfabrik at Sankt Joachimsthal circa 1900 [46]. (Courtesy J. L. and V. R. Marshall, "Rediscovery of the Elements," 2010, ISBN 978-0-615-30795)

In November of 1898 the Curies opened correspondence with the Austrian Academy of Science (*Kaiserliche Akademie der Wissenchaften*) in Vienna and its director Eduard Suess. They requested and received *100 kg* (100,000 g or 220.0 lbs) of uranium-free residue from Sankt Joachimsthal, Bohemia, at no charge. This residue had been considered worthless anyway. In appreciation, the Curies sent a radium sample to Vienna in December 1898. During 1899, the Curies got another *1000 kg* for nominal shipping charges. The major separation work was done at a Paris extraction factory and the final work at the Curie laboratory, Chap. 4. The Curies used about 1000 kg (1.1 t or 2200 lbs) of residue to isolate 90–120 mg of radium by 1902. It was necessary to get enough mass for an accurate atomic weight for radium (X2) in order to convince skeptical chemists that radium existed and to get her Ph.D. in physics.

The Austrians continued to supply the Curies through to *1903*, when Marie presented her Ph.D. thesis. After 45 months of work, she calculated an atomic weight of 225.9. The current accepted value is 226.0. Marie Curie presented her Ph.D. thesis on this work in June of 1903. It is doubtful that any other Ph.D. in history received more attention than hers [29]. Between 1903 and 1905, her thesis would appear in 17 issues of five different editions and in five different languages: French, German, Russian, Polish, and English [24]. Nor has any scientific illustration been reproduced more often than the famous Fig. 3.3 about radium rays [29, 30].

During 1898–1900 four new elements, which were all radioactive, were discovered. In 1899, the Curies had Andre Debierne follow up on the radioactivity of X3, which had been noted in the uranium and thorium solution obtained in step 3 of their separation process. Attached to the thorium was actinium, which is 150 times

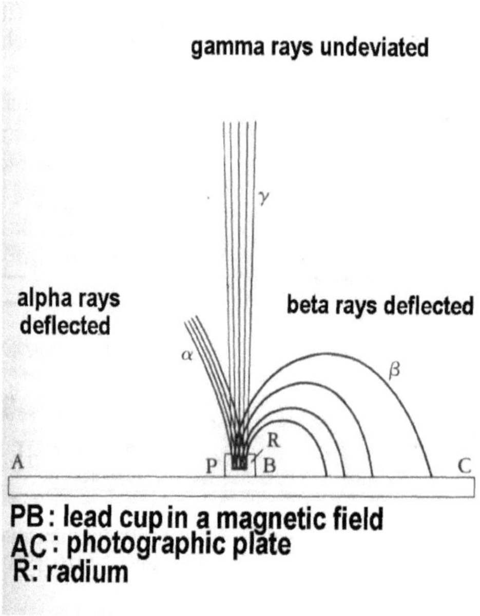

Fig. 3.3 Deviation of the positive, negative, and uncharged rays of radium by a magnet [29, 30]. (Public domain, labeled by author)

more radioactive than radium. The discovery was independently confirmed in 1902 by Frederick Otto Giesel (1852–1927) [9, 31]. By 1900 the first and only radioactive gas, radon ($^{222}Rn_{86}$), was discovered by several investigators including (a) the Curies, (b) Ernest Rutherford (1871–1937), and (c) Frederick Ernest Dorn (1848–1916) [32–34]. Thus, polonium-84, radon-86, radium-88, and actinium-89 were to be added to the periodic table, Appendix I. It would be discovered that all the elements beyond #83 were unstable and emitting ionizing radiation (Table 3.4).

Meanwhile, in November of 1903, Becquerel and the Curies were announced as the 1903 Nobel laureates in physics. The prize money was divided two ways: Becquerel got 50 % and the Curies shared 50 %. Since the Roentgen-Lenard controversy (Chap. 2) there was no longer a problem in sharing the award. There may have been some discussion about whether to give the award to Pierre alone or to both Curies. The Curies shared one diploma or certificate [18] and probably one medal as well. Becquerel alone went to Stockholm in December of 1903 to give the obligatory lecture [35]. The Curies did not go to Stockholm in 1903. Marie had suffered a miscarriage in August and both Curies were described as appearing ill; pale and weak. They were undoubtedly suffering from radium toxicity. And as introverted, introspective, and socially ill at ease scientists, they were also uncomfortable with the accolades. Marie Curie wrote that: "the overturn of our voluntary isolation was a cause of real suffering for us and had all the effect of disaster" [21].

The Curie gold medal was received from the King of Sweden by the French minister in Stockholm. He and Becquerel observed the proper protocol for the occasion. The Curie share of the prize money was 70,000 francs or US$ 20,000 in American dollars, which was estimated to be *ten times* their combined salary over the next 10 years. Usually the prize money was not to be awarded until a lecture was given, but the Curies were able to secure the money as had Roentgen. Pierre agreed to come at a later date for the lecture, which he did 18 months later in June of 1905. Marie Curie attended but did not speak and might not have been asked? Pierre spoke in his usual reticent style and Henri Poincare lamented that: "he rose to glory with

Table 3.4 Unstable radioactive elements # 84–92 [43]

Discovery	Element-Z#	Isotope A#	Relative $T^1/_2$
1898	Polonium-84	209	103 years
1940	Astatine-85	210	8.3 h
1900–1903	Radon-86	222	3.8 days
1939	Francium-87	223	22 min
1898	Radium-88	226	1620 years
1899–1902	Actinium-89	227	21.2 years
1828	Thorium-90	232	1.4×10^{10} years
1918	Proto-actinium-91	231	32,480 years
1789	Uranium-92	238	4.5×10^9 years

Z# = proton # = electron #. A# = mass # = protons and neutrons the spirit of a whipped dog" [21]. However, both Curies were suffering from the effects of chronic radium exposure and neither they nor the audience were aware of it. Indeed, Pierre personally carried radium samples to Stockholm to demonstrate at his lecture.

3.4 Widowed at 38

Ten months after his Nobel lecture, on 19 April 1906, Pierre was killed in a traffic accident. He stepped in front of a team of horses pulling a wagon of military uniforms. He had left a luncheon meeting around 2:30 P.M., walking in a heavy rain with a large umbrella. He was headed for a library on a crowded street, the rue Dauphine. The wagon, pulled by four horses, was moving in the right hand lane. A horse cab, in the left lane, was going in the opposite direction and passed the wagon. Directly behind the horse cab, Pierre Curie walked into the path of the horse drawn wagon on the opposite side of the street. Curie, at last aware of the danger, stopped suddenly but slipped on the wet pavement. Falling down, he grabbed at the harness of the lead horse on the left, which startled the horse. The horse reared up and Curie lost his grip, falling underneath the four horses. He escaped the horse's hoofs and the front two iron-rimed wheels of the wagon. However, the left rear wheel shattered his skull. His face was left untouched but he died instantly [22].

Pierre was known to be absent-minded and careless while crossing streets. At the news of his death, his father, Dr. Eugene Curie remarked "What was he dreaming of this time?" However, radium poisoning may have been more of a contributing factor. His colleague, Paul Langevin, had noted that the 47-year-old Pierre was prematurely grey, was losing weight, and had such pain in his extremities as to make walking difficult. Pierre's physical condition suggests he was suffering from the effects of chronic radiation exposure. The Curies were notorious for ignoring the systemic effects of radiation, having only recognized the local skin burns. Marie Curie's notebooks and cook books remain radioactive (in museums) to this day [17–19, 23]. Pierre's academic production had disappeared after 1904 and chronic illness is suspected.

Publications	Pierre	Marie
<1898	4	1
1898–1904	21	22
>1904	0	18

At the death of her husband, Marie Curie was appointed a Professor at the Sorbonne, the first woman so honored. In November of 1910 she was persuaded to also apply for admission to the French Academy that had rejected Pierre in 1902 before finally accepting him in 1905. The application process was demeaning enough but Marie also found herself attacked by the fourth estate. The daily press was beginning to feel its power. With the invention of the linotype in 1881 it was easier and less expensive to furnish the masses with sensational entertainment for only pennies. For some reason the Press decided that it was unseemly for a foreign-born woman to aspire to the lofty French Academy. They decided she must be Jewish as well as a liberal university free-thinker. Allegedly, her father had converted to Christianity from Judaism. Her rival, a good Catholic family man, beat her out by a vote or two in the election of January 1911. Marie Curie would send no more of her papers to the Academy journal. It would take another 50 years before they admitted a woman; M. C. Perey in 1962 [21].

Since 1899, Pierre had been concentrating on the study of uranium rays while Marie continued her chemical separation studies on uranium ores [18, 19]. After her Ph.D. in 1903, she continued to isolate radium for a more accurate atomic weight of radium, which she arrived at in 1907: 225.0 +/− 1. However, with the death of Pierre in 1906 and Becquerel in 1908, the Curie lab was less involved with the study of radioactivity and more oriented to radium isolation and medical uses. Marie and Andre Debierne worked on the isolation of pure radium metal. They prepared a gram of radium chloride and then isolated the pure metal by electrolysis of the chloride salt in 1910 and arrived at an atomic weight of 225.35; compared to the modern value of 226.0254 [9, 33].

In September of 1910 an International Congress on Radiology and Electricity was held in Brussels. It was financed by the Belgian industrial chemist, *Ernest Solvay*. The international community, especially physicians and radium vendors, recognized the need for a single authoritative reference standard for radium. There was a growing commerce in radium sales and the radium preparations were still being labeled by how active they were compared to an equal weight of uranium: one gram of uranium contained 3.42×10^{-7} g of radium. An International Radium Standards Committee of ten was convened for the purpose of preparing a radium standard (Table 3.5) [33, 34, 36, 37].

First the Committee agreed to define a unit of radioactivity as equal to the amount of radon and its daughters (radium decay products) in equilibrium with one gram of radium. It was to be measured in disintegrations per second (dps), which was determined to be 3.67×10^{10}. This was later altered to 3.7×10^{10} dps from any radioactive source not just radium. In 1960, 1 dps was defined as a "becquerel." The Committee wished to honor Pierre Curie by naming the radioactivity unit the "curie."

As most laboratories were working with only milligrams (10^{-3}) or micrograms (10^{-6}) of radium, the Committee suggested the "curie" be defined as the amount of radon in equilibrium with a milligram of radium. Marie Curie argued that the

Table 3.5 International standards committee [33–37]

Marie Curie: Paris Andre Debierne: Paris	Ernest Rutherford: Manchester	S. Meyer: Vienna Radium Institute O. Honigschmidt: Vienna Radium Institute
B. B. Boltwood: Yale	F. Soddy-Glasgow	Egon von Schweindler: German physicist
Arthur S. Eve: McGill	Otto Hahn: Berlin	Hans Geitel: Buchler Co.

"curie" needed to be a larger unit. She convinced them to define the "curie" to be in equilibrium with a gram of radium. She was asked to prepare a standard at her laboratory and the Vienna Radium Institute, which had more than 4 g of radium, was asked to prepare three standards. Vienna had the world's largest supply due to their mines at St. Joachimsthal.

The Committee met again in March of 1912 to compare the samples. Marie Curie had prepared a standard of 21.99 mg. of *pure* radium chloride. Otto Honigschmidt from Vienna had prepared three different amounts of *radium chloride*; 10.11, 31.70, and 40.43 mg [37]. The four were compared and found to be within the statistical range of error [33]. This was dubbed the International Radium Standard and the standards were preserved in Paris (Sevres) and Vienna. Duplicate standards were then prepared, using St. Joachimsthal pechblende, for Austria, France, Germany, North America, and the UK. The USA got theirs in December of 1913 and it is preserved in Washington D.C. The certification was signed by M. Curie, E. Rutherford, and Stefan Meyer [36]. Meyer (1872–1949) had taken his Ph.D. in physics with F. S. Exner in Vienna and was the Director of the Vienna Institute of Radium Research from 1910 to 1938. Although forced to resign after the Anschluss, he survived the Holocaust and returned as Director.

After Pierre's death when she was 38 years old, Marie Curie wore black for the next 5 years. However, in the spring of 1911, she was beginning to wear white dresses with a rose at her waist. She had become involved in a romantic affair with Paul Langevin (1872–1946), a former student of Pierre's, who was 5 years her junior and a world class physicist. He had left his wife and four children and moved into a bachelor flat across the street from the EPCI where he worked. It was just 10 min from the Curie Lab. His estranged wife had been nagging him to quit the university and get a better paying job in industry to support his family. Rather than do that, he moved out, but he forgot his love letters from Marie. His wife broke open his desk and obtained the letters. On 26 October, Mrs. Langevin started leaking the love letters to the Press in order to pressure Langevin into a financial settlement [21].

The affair came into the open in early November of 1911 when both Langevin and Marie attended the first Solvay Conference, also known as the International Physical Institute, in Brussels [34]. Ernest Solvay paid for 20 or more of the world's best physicist to attend an annual meeting in Brussels. He did this after finding the 1910 Brussels meeting such a success. The Solvay was by invitation only and in 1911 included the 23 most accomplished physicists of the time. Marie was the only woman invited and, like Roentgen's address book, there were no North American physicists on the invitation list:

W. Nernst	J. Perrin	R. Goldschmidt
M. Brillouin	W. Wien	M. Planck
H. Lorentz	M. S. Curie	H. Rubens
E. Warburg	H. Pointcaire	A. Sommerfeld
H. Kamerlingh	A. Einstein	P. Langevin

G. Hostelet	F. Lindemann	F. Hasenohrl
E. Herzen	M. de Broglie	E. Rutherford
J. H. Jeans	M. Knudsen	

While Marie Curie and Langevin were in Brussels the Paris newspapers went public with the *Affaire* on 4 November. She was portrayed as a foreign (and probably Jewish) concubine and home-wrecker who had seduced Langevin away from his family. On 23 November, the weekly magazine L' Oeuvre, published extracts from a dozen pages of the love letters. L'Oeuvre was known as a salacious scandal sheet which was xenophobic and virulently anti-Jewish and anti-foreigner. The extracts revealed little but a lonely widow giving advice on a bad marriage. The letters that were quoted were never dated, not particularly incriminating and suggested more of a companionship than carnal knowledge [21]. In the best French tradition, three duels ensued. On the 23rd, Henri Chervet, pro-Curie, took on Leon Daudet, anti-Curie, using epees. Daudet suffered a 6 cm cut at the elbow. On the 24th, editor Pierre Mortier of the pro-Curie *Gil Blas* paper was challenged by Gustave Tery, the *L'Oeuvre* editor who published the letters. Mortier was slightly wounded in the forearm. Then on the 25th Paul Langevin took on Tery with pistols at 25 m. Langevin was going to shoot but Tery refused to fire his weapon.

Alfred Nobel (1833–1896) was born on 21 October. He died on 10 December, which is when the Nobel rituals are observed and the Prizes awarded. The announcements are currently made in October but early on they were made in November [35]. On 8 November, at the beginning of the Langevin *Affaire*, Marie Curie received a telegram informing her that she was the 1911 Nobel laureate in chemistry for her work with radium and polonium separations. This time she was glad to go to Stockholm and she gave the required lecture. On the 9th of December, the Langevins settled out of court and were separated [21].

After World War I, 1914–1919, MSC would concentrate on radium chemistry and the medical applications of radium at the Curie Institute. While the deaths of Pierre and Becquerel had brought an end to experimental and investigative radioactivity research, the workers at the Curie Lab continued their radio-chemistry work:

Pierre Curie (1859–1906)	Nobel in physics 1904. First Director of the Curie Lab
Marie S. Curie (1867–1934)	Nobel in physics 1904. Second Director of the Lab
	Nobel in chemistry 1911
Andre Debierne (1874–1949)	Discovered Ac-89 in 1899; Asst. to MSC; then #3 Director of the Curie Lab
	Isolated Ra metal with MSC; Director Curie Institute after MSC
Gustave Belmont (1857–1932)	Chemistry lab chief at ESPC; retired in 1916
Eugene Demarcay(1852–1904)	Spectroscopist, who identified lines for radium
	Discovered Eu-63 in 1901
Jacqus Danne (1882–1919)	Course instruction 1904–09. Entered radium industry and edited house organ

Jan Danysz 1852–1914)	Polish assistant with radium biology research, killed WW I
Albert Laborde	Physicist collaborator supervised measurement and purification service
	Husband of Simone Laborde MD, radium therapist
Fred and Irene Joliot-Curie	Nobel in physics in 1935 for discovery of artificial radioactivity
Mlle. Margarate Perey	Discovered Fr-87 in 1939

A return to past glories occurred in the early 1930s. Irene Joliot-Curie and her husband Jean-Frederic (Fred) Joliot narrowly missed several important discoveries in nuclear/particle experimental physics: the neutron and the positron [5, 38]. They did win the 1935 physics Nobel for their discovery of artificial (laboratory created) radioactivity. They were about to achieve the first chain reaction from a uranium pile in June of 1940 when Paris was occupied by the German Army [5–7, 38].

Marie died at age 67 of a bone marrow disorder consistent with pre-leukemia. Pierre and Marie were buried in the Curie family cemetery in Sceaux. Their daughter Irene Joliot-Curie (1897–1956) died of acute leukemia. Her husband, Jean-Frederic (1900–1958), died 2 years later of apparent chronic hepatitis. Irene and Fred were also buried in Sceaux [38]. The cemetery at Sceaux was undoubtedly radioactive and Pierre, Marie, and Irene were likely all radium victims. In 1995 the bodies of Pierre and Marie were exhumed and re-buried in Paris at the Pantheon.

3.5 Rutherford

Experimental and theoretical physics had received a new stimulus from the Curie discoveries. These discoveries introduced radioactivity, radiochemistry, and nuclear physics. The periodic table was then empty from # 83 (Bi) to # 92 (U). Where did the new elements, if they were new elements, fit in the table? What was their true atomic weight? There seemed to be more elements being discovered than there were empty places in the periodic table. A frenzy of research activity and competition in the new disciplines of radiochemistry and radiophysics ensued. German chemists read the Curie publications and began preparing their own radioactive elements. Physicists in the UK began studying and measuring the amounts and sources of the radioactivity [5–7].

Ernest Rutherford was born and raised on a farm in New Zealand. After college graduation in 1895, he won an 1851 research scholarship to study for his Ph.D. in the UK. The scholarships were financed by the proceeds from the 1851 London Exhibition. Approximately eight young scientists or engineers of exceptional promise from the UK, Ireland, or a Commonwealth country are selected annually for a 3-year research scholarship. Rutherford actually finished second in the area competition but the winner decided to stay home. Rutherford elected to pursue his Ph.D. at the Cavendish Laboratory in Experimental Physics at Cambridge [5]. Rutherford ended up by creating his own school of physics, which utilized the team approach rather than individual research. The Rutherford School produced an amazing series of discoveries, which were to define the radioactivity first observed in Paris, as a natural process of spontaneous atomic disintegration and transmutation of unstable isotopes of the chemical elements [5–6, 33, 34]. He worked at three different institutions between 1898 and 1937:

- The Cavendish with J. J. Thompson: 1895–1898
- McGill with R. B. Owens, Fred Soddy, Otto Hahn, BB Boltwood: 1898–1907
- Manchester with Hans Geiger, Thomas Royds, Ernest Marsden, H. G. J. Moseley, C. G Darwin, Nils Bohr, G. Hevesey, F. A. Paneth: 1907–1919
- The Cavendish with J. Chadwick, J. D. Cockroft, E. T. S. Walton: 1920–1937

At the Cavendish in 1895 Rutherford and J. J. Thompson began studying the behavior of ionized gases. They began by measuring the ionization produced by X-rays and then, the ionizing radiation from uranium. By 1898, Rutherford had discovered, described, and named the alpha and beta rays being emitted by uranium. Thompson would identify the beta rays as electrons but the alpha rays remained a mystery. Gamma rays were identified as similar to X-rays in 1900 by P. V. Villard in France [5–7]. In 1898, Rutherford was made professor of physics at McGill in Montreal. Working with R. B. Owens, an electrical engineer, he investigated the so-called induced radioactivity from thorium, uranium, and actinium. They found this "emanation" to be a decay product of all three of these large unstable elements. They identified the one and only radioactive gas, now known as radon [32–34].

During *1900–1903,* Rutherford worked with the Oxford graduate and new McGill chemist Fred Soddy (1877–1956) on the spontaneous disintegration of the radioactive (unstable) elements and their transformation into new elements; the theory of "radioactive transmutation." They found this occurred by the emission of *alpha,* beta, and gamma rays. Rutherford suggested that the *alpha* rays were charged particles of ionized helium ($^{4}He_2$). The rate of transmutation (decay) occurred at an exponential rate that was specific for each radioactive element. The time taken for 50% of the element to decay was expressed as its "half-value", later renamed the "half-life", and expressed by the symbol $T^1/_2$ [5, 35, 39, 40]. Transmutation may have been the most important discovery of the twentieth century and it earned the 1908 Nobel in chemistry for Rutherford (Fig. 3.4). Eventually, the complete decay series for thorium, uranium, and actinium were finalized [5, 33, 34, 39, 40], Appendix IV: Radioactive Decay Tables.

Otto Hahn (1879–1968) had earned his Ph.D. in organic chemistry at Marburg in 1901. To learn English, he went to London to spend 1904/1905 with Sir William Ramsay. Working with thorium ($^{232}Th_{90}$) he discovered what he thought was a new radioactive element but it was an isotope of thorium ($^{228}Th_{90}$). Hahn went to Montreal in *1905/1906* to work with Rutherford and found another new radioactive element. Instead, this also turned out to be another thorium isotope ($^{227}Th_{90}$). Of course neither of these thorium isotopes could be separated chemically from thorium-232, the most stable thorium. Hahn and Rutherford both failed to comprehend isotopes [5–6, 34, 39, 40]. Hahn returned to Germany to work at the University of Berlin and in 1907 he found another isotope, mesothorium ($^{228}Ra_{88}$), which was radium-228. It

Fig. 3.4 Ernest Rutherford, circa 1908. (Public domain)

appeared that he might spend his academic career as the King of Isotopes. However, in 1918 he discovered element # 91 ($^{231}Pa_{91}$) and in 1937 discovered nuclear fission. [5, 6, 34].

Rutherford worked with Yale chemist Bertram Borden Boltwood (1870–1927) to find the correct activity ratios of uranium to radium. Between 1905 and 1908, they determined that the uranium to radium activity ratio was $1:2.94 \times 10^6$; the activity of 1 g of radium was 2.94×10^6 that for an equal amount of uranium. The radium to uranium ratio is $1:3.4 \times 10^{-7}$; 1 g of uranium contains only 3.4×10^{-7} g of radium. These ratios were used to evaluate the commercial sales of ore and radium [33]. Yale tried to hire Rutherford but he turned them down after a visit, complaining that: "they act like the University was made for the students!" [5].

In *1907* Rutherford transferred to the University of Manchester, where he inherited the German lab assistant, Hans Geiger (1882–1945). Manchester had less than 20 mg of radium but Stefan Meyer and the Vienna Academy loaned them 350 mg of radium [5, 34]. After obtaining the radium, Rutherford began working with physics graduate student Thomas Royds (1884–1955) on the spectroscopy of the alpha particle. During 1907–1909 they confirmed that the alpha was indeed an ionized helium atom ($^4He_2$) with a charge twice that of hydrogen and a mass of four. Rutherford and Geiger modified an invention of W. Crookes which he had called the spinthariscope (spinthar: spark). This device used radium alpha rays to illuminate a ZnS screen. This had been suggested originally by F. Giesel, who had begun using radium and ZnS to market luminous paint in 1906. In 1908, Geiger and Rutherford developed a scintillation counter to visualize and count alpha particle production. With this technique, the decay transformations of uranium, thorium, and actinium were identified [6, 39, 40], Appendix IV.

During *1909–1911,* Rutherford and Ernest Marsden (1889–1970) and Hans Geiger saw alpha particles being deflected as they passed through seemingly empty space. Some were deflected 180°. Rutherford decided that all the mass and positive charges of the atom must be concentrated in a central *nucleus* and that this nucleus was responsible for the deflections. He conceived that the positive electric charges of the nucleus would determine its chemical behavior. They were able to count the number of positive charges (named protons) and found that number to be to be roughly proportional to the atomic weight. They had discovered the nucleus and calculated its radius as $10–^{15}$ cm. as compared to 10^{-8} for the radius of the atom [5, 6, 34].

Graduate student Henry G. J. Moseley (1887–1915) came from Oxford to work with Rutherford during *1910–1913*. At the University of London, C. Barkla (Nobel 1917) had found that when X-rays struck different metal targets, electromagnetic wave lengths characteristic of the metal were emitted. W. H. Bragg and W. L. Bragg at Leeds also found X-rays diffracted or diffused into a spectrum when passed through crystals. Rutherford sent Moseley to Leeds to learn that technique so as to investigate the nuclear mass by its X-ray spectra. Moseley then developed a method to obtain and photograph the X-ray spectra of the nucleus of an element. Using the nucleus as a metal target, electrons struck the target and emitted characteristic X-rays which were diffracted through a crystal to give a characteristic spectrum, which could be photographed. He found that heavier elements produced a shorter (more penetrating) wavelength. The decrease in wavelength occurred in regular steps as the nuclear mass increased. Thus, the wavelength was inversely proportional to the nuclear mass and its positive charges and therefore characteristic of the nuclear positive charges.

The Manchester group established that the number of positive charges (protons) in the nucleus were also equal to the number of orbiting electrons, which controlled the chemical activity. This number would constitute the atomic or proton number, Z #, which correlated with the chemical nature of the element. Each element could now be assigned an exact proton number: the Z # [5, 34, 44]. The old periodic table based on atomic weights, the A #, had to be *somewhat* revised. After publishing this in 1913/1914, Moseley returned to Oxford. He apparently considered Rutherford rather provincial and Manchester, dirty and depressing. He refused noncombatant status in World War I and was killed at Gallipoli in August of 1915. His demise led to British scientists being removed from combat service [44].

The new elements that were chemically identical but had different radioactive characteristics were now explained. During 1910–1913, Kasimir Fajans at Karlsruhe and Fred Soddy, now at Glasgow, identified radioactive isotopes. For this and his work with Rutherford, Soddy was awarded the 1921 Nobel in chemistry (Fig. 3.5). The Greek word "isotopes", meaning the same place, was coined by Soddy for chemical elements that had the same proton number (Z #) but had a different atomic mass or weight (A #). The different atomic weight numbers were eventually found to be due to varying numbers of neutron particles in the nucleus, the *N* number [41, 42]. Therefore $A = Z + N$; the atomic weight is the sum of the protons and neutrons.

Fig. 3.5 Fred Soddy, circa 1921. (Public domain)

The large elements from # 84 to # 91 are all unstable and very radioactive (Table 3.4). However, # 90 and # 92 have such *long* half-lives, from 10^8 to 10^{10} years, that they are considered essentially stable. Since none of these larger elements are stable, their atomic weight, A #, is based on the isotope with the longest half-life. The disintegration of isotopes $^{238}U_{92}$, $^{232}Th_{90}$, and $^{235}U_{92}$ leads to the production of the other seven unstable elements from # 84 to # 91. Elements # 85 and # 87, however, have such *short* half-lives that they currently exist only in the laboratory, as do all the elements beyond #92 [9, 43].

During World War I (August 1914–November 1918), the Manchester Department was decimated. Before Marsden left, he and Rutherford used alpha particles as projectiles to bombard various elements in an effort to alter the nucleus. They artificially transmuted a nitrogen atom into one of oxygen with the release of a hydrogen nucleus; $^4He_2$ (alpha) + $^{14}N_7$ = $^{17}O_8$ + $^1H_1$ (proton). In 1919, Rutherford identified this as "transformation and disintegration of nuclei with the liberation of a hydrogen nucleus." For this nuclear particle he coined the name "proton." This brought an end to his active experimental work. He succeeded J. J. Thompson at the Cavendish in 1920 at which time he predicted the discovery of the neutron and of deuterium. Both were found in 1932 [5, 34, 39, 40]. Rutherford, now approaching 50, served as adviser and mentor to four more Nobel winners at the Cavendish Laboratory (Table 3.6) [5, 33, 34, 39–44].

Outside the Rutherford circle, other physicists at the Cavendish continued the discoveries in nuclear physics. Following up on the isotope studies, J. J. Thompson had brought in Francis W. Aston (1877–1945) from Birmingham in 1910. J. J.

Table 3.6 Rutherford School of Physics and the Cavandish Lab [5, 33, 34, 39–44]

Institutions	Nobels	Institutions	Nobels
MC Gill: 1898–1907		Cavandish: 1920–1937	
E. Rutherford-transmutation		J. Chadwick—neutron	1935 physics
F. Soddy—isotopes	1921 chemistry	P. M. S. Blackett—cloud chamber	1948 physics
O. Hahn—fission	1944 chemistry	J. D. Cockcroft—linac	1951 physics
Manchester: 1907–1919		E. T. S. Walton—linac	1951 physics
Rutherford—nuclear physics	1908 chemistry	Oliphant-Ellis-Goldhaber-Williams	
H. Geiger—detector		*Cavandish Non-Rutherford*	
E. Marsden—nucleus		J. J. Thompson—electron	1906 physics
H. G. J. Mosely—atomic Z #		F. Aston—mass spectroscope	1922 chemistry
N. Bohr—quantum theory	1922 physics	C. T. R. Wilson—cloud chamber	1927 physics
G. de Hevesy—tracers	1943 chemistry	P. Kapitza—superconduction	1978 physics

Thompson had found in 1914 that isotopes were not limited to just the radioactive elements, those heavier than bismuth-83, but were present with the stable elements as well. After the War, Aston developed the mass spectrograph in 1919 and this allowed him to identify over 200 of the naturally occurring isotopes. Thompson and Aston found that the number of unstable isotopes outnumbered the stable ones!

Almost all the lighter elements, from Z # 1 to # 83, have at least one stable isotope or one with such a long half-life ($T^1/_2$) that it is considered essentially stable. These 81 "stable" elements have about 200 unstable isotopes but they emit very little radioactivity as they convert to a stable state [9, 43]. The two exceptions to stability among the lighter 83 elements are Z # 43 and # 61, which no longer exist naturally on earth because of a $T^1/_2$ shorter than the age of the planet. They only exist when produced in the laboratory [43].

3.6 Supplemental Material: Parisian Chemists at the Time of the Curies [24]

Ferdinand Frederick Henri Moissan (1852–1907) was a French chemist who won the 1906 Nobel Prize in Chemistry for his work in isolating fluorine from its compounds. He was Professor of Chemistry at the *Ecole de Pharmacie* and at the Sorbonne. He was the first to isolate elemental fluorine in 1886; see Chap. 1.

Alfred Lacroix (1863–1948) was a chemist, mineralogist, and volcanologist. He was a graduate of the Superior School of Chemistry of Paris in 1887. His passion for mineralogy led him to study earth science and the eruption of volcanos on

Martinique in 1902. His knowledge in volcanology and mineralogy made him an expert who has been solicited to observe various volcanos throughout the world. Among his principal writings are five volumes of mineralogy and two volumes of his observations and conclusions about volcanos. He was a member of the Legion d'Honneur and professor at the Museum of Natural History since 1893. He was perpetual secretary of the Academy of Sciences for 34 years.

Eugene Demarcay [1852–1904] and Georges Urbain [1872–1938] were two of the gaggle of chemists involved in trying to separate out the various "rare earth" elements, which turned out to be not rare but quite widely dispersed. Eugene-Anatole Demarcay (1852–1904) was a French chemist who discovered europium. He also aided the Curies by proving the discovery of radium by spectroscopy. Demarcay would discover Eu-63 in 1901. In 1907, Urbain would share the discovery of Lu-71 with Aurer von Welsbach (Austria) and Charles James (USA).

George Urbain (1872–1938) was a French chemist who discovered and named lutetia, while sharing the discovery with Welsbach and James. He served as president of the Societe de Chimie. He specialized in the study of the complex REEs until the X-ray spectroscopy work of Mosley in 1914 simplified their analysis and identification.

References

1. Dutrex J, Dutrex A (1995) Henri Becquerel (1852–1908). Med Phys 22(11):1869–1875
2. Genet M (1995) The discovery of uranic rays. Radiochim Acta 70/71:3–12
3. Mould RF (2008) Antoine-Henri Becquerel 1852–1908. Nowotw J Oncol 58(6):258e–270e (Poland)
4. Ruth Lilly Rare Book Library. Indiana University Bloomington, Indiana
5. Segre E (1980) From x-rays to quarks. Freeman, San Francisco
6. de Hevesy G (1961) Marie Curie and her contemporaries. J Nucl Med 25(1):118–131, 169–182
7. Goldschmidt B (1990) Atomic rivals. Rutgers, New Brunswick
8. Anthony JW et al (2001) Uraninite. Handbook of mineralogy. Mineralogical Society of America. http://www.handbookofmineralogy.org/pdfs/uraninite.pdf. Accessed 6 July 2014
9. Marshall JL (2002) Discoveries of the elements. Pearson Custom Printing, Boston
10. Marshall JL, Marshal VR (2008) Klaproth. Hexag Alph Chi Sigma 99(2):20–24
11. Marshall JL (2011) Uranium mining & radium refining in St. Joachimsthal/Jáchymov. Nowotw J Oncol 61(1):11e–15e (Warsaw)
12. Zeman A, Benes P (1995) St. Joachimsthal mines and their importance in the early history of radioactivity. Radiochim Acta 70/71:23–29
13. Veselovsky F, Ondrus P et al (1997) History of the Jachymov ore district. J Czech Geol Soc 42(4):127–132
14. Marshall JL, Marshall VR (2002) Althofen Austria and Auer von Welsbach. Hexag Alph Chi Sigma 93(1):8–11
15. Duncan DE (1998) Calendar. Avon, New York
16. Rhodes R (1986) The making of the atomic bomb. Simon & Schuster, New York
17. Mould RF (2007) Radium history mosaic. Nowotw J Oncol 57(Suppl 4):99–127 (Warsaw)
18. Mould RF (1998) The discovery of radium in 1898. Br J Radiol 71:1229–1254
19. Mould RF (2006) Pierre Curie 1859–1906. Nowotw J Oncol 56(2):147–155 (Poland)
20. Pogonowski IC (1987) Poland-a historical atlas. Dorset Press, New York

21. Reid R (1974) Marie Curie. Collins, London
22. Brian D (2005) The Curies. Wiley, Hoboken
23. del Regato JA (1985) Marie Sklodowska Curie. Radiological physicists. American Institute of Physics, New York, pp 11–22
24. Klickstein H (1966) Marie Sklowdoska Curie, recherches sur les substances radioactives. Mallinckrodt classics of radiology, vol 2. Mallinckrodt Chemical Works, St. Louis
25. Adloff JP, MacCormick HJ (1995) The dawn of radiochemistry. Radiochim Acta 70/71:13–22. http://www.nucleonica.com/wiki/articles/Article03/Article3.htm. Accessed 4 July 2014
26. Adloff JP, Kauffman GB (2007) The troubled story of polonium. Chem Educ 12:94–101
27. Roque X (2001) Displacing radioactivity. In: Joerges B, Shinn T (eds) Instrumentation between science, state, and the industry. Springer, Heidelberg, pp 51–68
28. Rentetzi M (2008) Trafficking materials & gendered experimental practices: radium research in early 20th century Vienna. Columbia University Press, New York
29. Curie MS (1904) Radio-active substances; thesis. Chemical News Office, London (Reprinted Van Norstrand, New York)
30. Baskerville C (1905) Radium and radio-active substances. Williams, Brown & Earle, Philadelphia
31. Adloff JP (2000) The centenary of a controversial discovery: actinium. Radiochim Acta 88:123–127
32. Marshall JL, Marshall VR (2003) Ernest Rutherford. Br Hist Chem 28(2):76–83
33. Badash L (1979) Radioactivity in America. Johns Hopkins, Baltimore
34. Brucer M (1990) A chronology of nuclear medicine. Heritage, St. Louis
35. Feldman B (2000) The Nobel prize. Arcade Publishing, New York
36. Coursey B (1999) NIST: Marie Curie and the NBS radium standards. http://www.nist.gov/pml/general/curie/. Accessed 4 July 2014
37. Grigg ERN (1965) The trail of the invisible light. Thomas, Springfield
38. del Regato JA (1985) Jean Frederic Joliot. In: del Regato JA (ed) Radiological physicists, chapter 8. American institute of physics, New York, pp 97–116
39. del Regato JA (1985) Ernest Rutherford. In: del Regato JA (ed) Radiological physics, chapter 4. American Institute of Physics, New York, pp 37–50
40. Wilson D (1983) Rutherford. MIT, Cambridge
41. Romer A (1970) Radiochemistry & the discovery of isotopes. Dover, New York
42. Marshall JL, Marshall VR (2010) Soddy and isotopes. Hexag Alpha Chi Sigma 101(4):68–74
43. Asimov I (1988) Understanding physics, vol 3. Dorset, New York
44. Marshall JL, Marshall VR (2010) Moseley and atomic numbers. Hexag Alpha Chi Sigma 101(3):42–47
45. Danne J (1904) Le Radium, 1st edn, vol 1. de Lisle Radium Company, Pairs
46. Marshall JL, Marshall VR (2010) Photographs of cities: Jachymov CR. Rediscovery of the elements DVD. Jenny Marshall Computer Services, Denton, TX. http://www.jennymarshall.xom/products.htm. Accessed 4 July 2014

Chapter 4
European Radium Sales

4.1 Radium Factories

The Curies published openly their processes and techniques for the separation of radium and polonium. They also decided against seeking any patents in spite of their lack of financial support.

Chemists in Germany, which was more industrialized than the third French Republic of 1870–1940, were stimulated by the reports of radioactivity and new elements. Industrialized Germany had large chemical companies that could separate radium in larger quantities than the small laboratory of the Curies. German companies were the first to offer radium for sale, 1899–1901 [1–3]. The Curies worked to achieve industrial scale production by collaborating with various Paris companies [4–8]. After 1900, the demand for radium resulted in the emergence of several European radium factories (Table 4.1).

4.1.1 The Eugene de Haën Chemical Company

The Eugene de Haën Chemical Company (Chemische Fabrik of E. de Haën) extracted uranium from Bohemian pechblende for use as a coloring agent in the glass industry. It was founded by E. de Haën (1835–1911) in 1861 in Linden [14]. It was later moved to List, near Hanover, which is 82 mi. from the port of Hamburg. E. de Haën was processing tons of both uranium and thorium during 1892–1901. After reading the Curie publications, he was able to separate radium from his pechblende ores [3, 9].

On 14 July of 1899, de Haën published a short note, six paragraphs, about this [1]. Utilizing his factory resources he was able to obtain more radium than the Curies, who were doing the separation process manually, 20 kg of ore at a time. Moreover, he was the first to observe that a quantity of radium could produce self-luminosity (Selbst-leuchten). He was also the first to offer radium for sale, when at the end of his paper he stated: "Die Substanzen sind aus der Chemischen Fabrik E. de Haën, List vor Hannover, zu beziehen" or "These substances are obtainable from

R. F. Robison, *Mining and Selling Radium and Uranium*,
DOI 10.1007/978-3-319-11830-7_4

Table 4.1 European radium factories [1–13]

Sales	Factories for radium-226	Ore sources
1899–1902	E. de Haen Chemical Co.; List, Germany	Bohemia
1901–1958	Chininfabrik Buchler Co.; Braunschweig, Germany	Bohemia, Cornwall, USA
1902–1923	Societe Centrale de Produits Chimiques; Paris	Bohemia, USA, Portugal
1904–1926	Armet de Lisle Quinine Co.; Paris, E	USA, Portugal, Indo-China, Madagascar
1906–1958?	Auer Co. at Atzgersdorf; Vienna, Austria	Bohemia
1909–1947	K. K. Urangelbfabrik; Joachimsthal, Austria	Bohemia
1910–1926	Societe Anonyme des Traitments Chimiques; Paris N	Portugal, USA, Madagascar
1910–1918	British Radium Corp.; London, Limehouse	Cornwall; St. Ives
1911–1926	John S. Mac Arthur Co.; Liverpool; Scotland	Utah carnotite
1912–1930	Societe Francaise du Radium; Paris, SW	Cornwall; South Terras, Madagascar
	Factories for radium-228; mesothorium	
1907–	Dr. Oskar Knoefler & Co., Plotzensee (Berlin)	Thorium factory
1907–	Auer Company, Berlin	Welsbach factory waste
1909/1911–	Societe Miniere Industielle Franco-Bresilienne	Brazil monazite

the E. de Haën Chemical Company in List, near Hanover" [1]. In August 1902, he moved the company to Seelze, east of Hanover, and their product line concentrated on inorganic chemicals. However, by 1910 they were selling mesothorium, $^{228}Ra_{88}$, also known as German radium, which was produced in Berlin (Table 4.1). In 1926, de Haën was absorbed into a larger chemical firm [14, 15].

4.1.2 The Buchler Company

The Buchler Company, formerly a quinine factory, is still in business as Buchler GmbH and Amersham Buchler GmbH & Co. in Braunschweig, Lower Saxony, about 109 mi. south of the port of Hamburg. This venerable purveyor of quinine and radioactive substances was established in 1858 by Hermann Buchler (1815–1900) from Trieste, Hermann settled in London, where he established a sugar refinery and married a woman from Braunschweig. When the sugar business soured, he moved to Braunschweig and built the quinine factory or *Chinninfabrik* (1858–1958). The Company erected a monument on the site to honor Herman and also their chemist F. O. Giesel, who initiated the side venture of radium production and supplied Rutherford, Soddy, and Ramsey [16].

About 8 mi. south of Braunschweig, in Wolfenbüttel, lived two high school science teachers named Johann Elster (1854–1920) and Hans Geitel (1855–1923). They lived together for 40 years, 1881–1920, nurturing flowers and butterflies, while teaching science at the local Herzoglich Gymnasium. During that time they published 150–200 papers on the electrical conductivity of gases, atmospheric electricity, the photoelectric effect, and radioactivity (40 publications). They had both trained at Heidelberg in 1875 and at Berlin in 1877 before becoming teachers at the Gymnasium [2, 3, 9]. After the announcement of polonium in July of 1898, they worked with pechblende residue and obtained Pb-214, the decay product of Po-218, Appendix IV. They reported this to F. O. Giesel, the chief chemist at the Buchler Quinine Factory.

Friedrich Oskar Giesel (1852–1927), had studied industrial chemistry at the *Königliche Gewerbeakademie* (Royal Industrial Academy) in Berlin before earning his Ph.D. in organic chemistry at Göttingen [2]. Intrigued by the Curie discoveries, Giesel obtained 10 kg of pechblende residues (pechblende minus uranium) from the de Haën factory at List. He followed the Curie procedure for the separation of the bismuth-polonium precipitate but instead ended up with a radioactive substance bound to barium. He needed another 200 kg to identify this as radium. By 2 March 1899, less than 3 months after the Curies, he was able to demonstrate radium to the Chemical Society of Braunschweig where it was reported: "Herr Dr. Giesel, in connection with the recent lecture by Herr Professor Geitel, brought a large amount of radium-containing salts from uranium residues and showed the brilliant light which was produced on a fluorescent screen" [17].

Giesel published papers on both radium and polonium on 9 August and on 24 August of 1899 [18, 19]. He confirmed the self-luminosity (*Selbst-leuchten*) reported by de Haën. Giesel reported the differences in penetration by the rays of radium (gamma) and polonium (alpha) and other original observations. His first paper, "about the behavior of radioactive barium and polonium," was printed in the Leipzig Annals of Physics and Chemistry on pages 91–94. It was preceded on pages 83–90 by an article written by Elster and Geitel on "further experiments on Becquerel rays" and "attempts to find the energy source of the Becquerel rays." In his own paper, Giesel acknowledged their work and the generosity of the de Haën Chemical Factory [18]. Later, he developed radioactive luminous paint using crystalline zinc sulfide, ZnS, also known as Sidot's Blend. From Elster and Geitel he had learned that radium supplied the alpha particles to activate the phosphorescent ZnS in the absence of light [2].

Using the industrial facilities at Buchler, Giesel was able to improve on the separation technique of the Curies [2, 3, 10, 20]. Quinine production required some of the same techniques used for radium separation. Giesel used hydrobromic acid instead of hydrochloric acid in the final stages of separating radium from barium. The bromides were easier to separate by fractional crystallization and reduced the time of production from 90 to 30 days [20]. Giesel was able to produce 260 mg (four grains) of radium bromide per ton of Joachimsthal pechblende, whereas the Curies were getting only 32.5 mg of radium chloride for every 2 t [11].

In mid-1899, Giesel acquired 1000 kg of residues from de Haën [3, 9]. By 1901, he had radium for sale and the Buchler Quinine Co. became a recognized supplier of radium salts and medical sources. Giesel began providing radium to scientists E. Rutherford, F. Soddy, W. Ramsay, O. Hahn, B. Boltwood, the Braggs, Stefan Meyer in Vienna, and Dimitri Mendeleev in Russia at 10% of the regular price of 400 francs/mg [2, 4–6]. Soddy noted that he paid only 8 schillings for a milligram of radium bromide that was 50% pure (actually 53.6%) [17]: pure being 3.0 million times the activity of uranium. Buchler obtained their uranium ore from Bohemia, until the 1904 embargo by Austria [21]. Then they were limited to the low grade German ore from Baden-Baden, Bad Durkheim, and Kreuznich. Later, Buchler obtained ore, which came through the port of Hamburg, from South Terras in Cornwall and from America. They were still making and selling radium through 1945, Chap. 10.

4.1.3 The Society Centrale de Produits Chimiques

The Society Centrale de Produits Chimiques (SCPC) had marketed scientific instruments, such as the piezoelectric quartz crystals and electrometers, for the Curie brothers. In late 1898, Pierre and Marie Curie reached an agreement with Vienna to continue to procure ore from Sankt Joachimsthal, Bohemia. The Austrians would supply pechblende to the Curies for only nominal charges until 1903/1904. The Curies, however, needed help from French industry to improve their amateur academic "science lab" separation techniques. In July of 1899, the SCPC was enlisted to undertake the separation of radium from 1000 kg of pechblende residues. In return they would be able to profit from the sale of a portion of the production [4–8, 12].

The SCPC worked in collaboration with the Curies and under the direction of Andre Debierne, who modified the Curie Laboratory techniques for an industrial scale of production. The major separation work was done at the SCPC factory and the final work at the Curie Lab. The Curies would get 3500 francs, as a "titre de droits d'auteur" (a percentage of the rights of the creator), each time the company sold the radium produced from 1 t of pitchblende [4, 9]. The SCPC factory produced the initial 90–120 mg of radium used by Marie Curie in 1902 for her first attempt at determining the atomic weight of radium. As the Curies were reportedly buying radium from Giesel in 1902 [6], it appears that the SCPC did not have significant radium for sale until late 1902 or 1903.

The Austro-Hungarian Empire had placed the Joachimsthal mines under the control of the Vienna Academy. In late 1903, the Academy placed an embargo on the ore apparently to preserve it for scientific use rather than commercial (medical) use. The Austrians declared that "It is only scientific institutes and savants of repute who are eligible as purchasers" [22]. Uranium was embargoed and exports were halted in late 1903 [21]. The Austrian embargo of 1904 had essentially created a monopoly for Austria on radium production and sales.

The embargo became effective just as the first reports on radium therapy began to appear in the medical literature. The Paris and Braunschweig radium factories that had been supplied by Bohemia now had to search elsewhere for uranium ore. The minerals studied by M. S. Curie in 1897 served as a guideline for future exploration. The radium manufacturers began to import ore from Colorado and Utah in the USA, Portugal, Cornwall, and French Madagascar. *See Appendix III* for a detailed list of the European Factory Ores.

The Curies were not affected by the embargo as they had reached a new agreement with the Vienna Academy. The Joachimsthal mines produced only about 6000–7000 kg (6.0–7.0 t) of pechblende residue a year. Therefore, Austria would alternate shipments of 1000 kg (1 t) between Paris and Vienna. This limitation influenced the Curies to become involved in securing uranium ore from other countries and in aiding other commercial extraction plants in Paris [4–8, 11, 12].

4.1.4 The Armet de Lisle Radium Company

The Armet de Lisle Radium Company was formed in Paris in 1904, after the embargo became effective. In 1904, the Curies and their associates began collaborating with industrialist Emile Armet de Lisle (1853–1928), who was entering the radium business. De Lisle was the son of a quinine manufacturer and chemical industrialist and he had studied chemistry at the Faculty of Sciences in Paris [4, 5]. He inherited a quinine factory just east of Paris at Nogent-sur-Marne, in 1878. Using the quinine factory staff and facilities, he expanded into radium production in 1904 [4–8, 11, 12]. The factory used pechblende from Hungary, Sweden, Canada, and East Colorado. They obtained autunite from Portugal and France, chalcolite from Bohemia, carnotite from Utah and Portugal, thorianite from Ceylon, betafite from French Madagascar, and tobernite from French Indo-China [11]. De Lisle stayed in business until 1926 [4, 12].

Like the SPCP, de Lisle planned to sell radium to physicians for a profit [7]. De Lisle operated in close collaboration with the Curies, using their methodology and having the radium tested and certified by them in return for a percentage of the product. Ore from de Lisle allowed Marie Curie to determine her second atomic weight for radium in 1907. The Curie laboratory furnished assistants to serve as technicians at the company: Jacques Danne, Frederic Haudepin (from the SCPC), Georges Gabriel Robillard, and Paul Razet. Danne became the co-editor of the journal Le Radium (1904–1919) which was started by de Lisle as a house organ to promote his company [23]. Paul Razet was a chemist from the *Ecole de Physique et de Chimie Industrielles* (EPCI) and he began to work on the concentration of polonium. In 1907, he produced about 200 g of bismuth-polonium that was 3500 times the activity of uranium. Marie Curie and A. Debierne then continued the concentration and obtained about 0.1 mg of polonium. The emission spectrum was obtained in 1910 and this finally verified the existence of element # 84, $^{210}Po_{84}$ [9].

4.1.5 The Auer Company

The Auer Company at Atzgersdorf (Vienna) began radium extraction in 1904 [24]. After their embargo of their Bohemian ore, the Austrians planned on opening their own radium factories at both Vienna and St. Joachimsthal. The Austrian Academy of the University of Vienna and their physicist Stefan Meyer (1872–1949) lacked adequate facilities to process the gigantic amounts of pechblende necessary for radium extraction. They turned to industrialist Karl Auer for factory space and guidance on the extraction and refinement of radium from the ore at Joachimsthal.

In March of 1904, Auer responded to the Academy by offered them the resources of his factory at Atzgersdorf, now a Vienna suburb. His Auer Gas Mantle Factory started work on radium production as soon as the Academy obtained 10,000 kg of the 23,000 kg of residue from the Joachimsthal uranium mines. The factory manager was the chemist Haitinger, who in the 1880s had collaborated with Auer on the rare earth elements (REE) and uranium research. He was assisted by physicist Karl Ulrich in devising a radium extraction process for the residues at Joachimsthal. The Bohemian pechblende, with 53.4 % uranium, had been treated with sodium sulphate to produce sodium uranate, which was dissolved in dilute sulphuric acid. The radium was left in the residue.

Haitinger and Ulrich published their extraction process in 1908. They were able to extract a yield of 3 g of radium chloride and 0.236 g of radium bromide from 10,000 kg (11 t) of pechblende [24]. During 1907–1909, the Vienna Academy was able to furnish radium to the Royal Society of London as well as a number of Institutes in France and Germany, including Berlin, at nominal cost. The Academy, in addition to its aid to the Curies, would also loan radium for scientific research to William Ramsay in London and E. Rutherford in Manchester [4–7, 10, 11, 21].

Karl Auer (*1858–1929*) was both a noted scientist and an extremely successful industrialist [25–29] (Fig. 4.1). He was born in Vienna to a prominent family. His father Alois Auer (1813–1869) was ennobled in 1860 as Ritter von Welsbach. Karl studied the sciences at the University of Vienna from 1878 to 1880 and then went to the University of Heidelberg (1880–1882) for his Ph.D. in chemistry. His Ph.D. advisor was Robert Bunsen, from whom he learned the use of flame spectral analysis for the identification of chemical elements. Auer returned to the University of Vienna to investigate the chemical separation techniques for the rare earth elements (REE), the Lanthanides of atomic numbers 57–71 [30]; Appendix I, Periodic Table.

In 1885, using fractional crystallization, he was able to separate and identify REE #59 and #60. Twenty years later, after establishing an industrial career, he returned to the study of the REE. In 1905, using spectral analysis, he isolated #70 and #71 but he did not publish the spectra and atomic weights until 1907. Therefore, he lost the priority dispute in 1909 to French chemist Georges Urbain [25]. The REE are much in demand today for use in various industries and as components of electronics, and their number has been expanded to 17 with the addition of elements 21 and 39.

Fig. 4.1 Karl Freiherr Aurer von Welsbach, 1923. (Public domain)

The REE have some of the most confusing nomenclature in the Periodic Table, due to an erstwhile Swedish monopoly. They also proved to be extremely difficult to separate from one another as they have *almost identical chemical properties*. Yttrium-39 was discovered first in 1787 and named after Ytterby, Sweden, where it was found. That led to three more REE named for Ytterby: terbium-65, erbium-68, and ytterbium-70. Then there is scandium-21, named for Sweden; thulium-69 for Nordic Thule; and holmium-67, Latin for Stockholm. In Greek, lanthanum-57 means hidden and dysprosium-66 means hard to find. A suspected new REE was called didymia or "twin" as it was so difficult to separate. Didymia turned out to be a mixture of five different REEs. Auer had separated neodymium-60, "new twin" and praseodymium-59, "green twin" [25, 26, 30].

After 1885, Auer turned to industry for *three* significant inventions: (a) the gas mantle light, 1885–1890; (b) the metal filament lamp, 1898–1902; and (c) a small portable lighter for cigars and cigarettes, 1903. Electric arc lighting had been available for decades and a practical incandescent light bulb and electrical supply system had been created by Thomas Edison during 1878–1880. However, few homes were wired for electricity and instead burned candles, oil or gas lamps. Only after the introduction of radio and home appliances in the 1920s did a majority of homes in the USA and Western Europe have electricity installed. Auer invented a type of incandescent light source that required no electricity [25–29].

In 1885, Auer patented an incandescent gas mantle, which utilized a knitted fabric/gauze sleeve (*Glustrumpt* or glow stocking) that was impregnated with 60 % salts of magnesium and the REE lanthanum-57 and yttrium-39, comprising 20 % each. The mantle was hung above a gas flame to provide a bright white illumination. The REE oxides gave off a strong while light, glowing brightly while emitting little infrared irradiation. The Glustrumpt was patented as the Auerlicht but it failed commercially during 1887–1889 and the factory in Atzgersdorf closed.

Auer switched to a mixture of 99 % *thorium-90* and only 1 % cerium-58 for the Auerlicht in 1890. He used fractional crystallization to separate thorium oxide and the REE cerium from inexpensive monazite sand. Monazite is a phosphate mineral which occurs with various combinations of a REE but with one predominant, such as $CePO_4$, $LaPO_4$, $NdPO_4$, or $PrPO_4$ [31]. Monazite sand deposits are usually $CePO_4$, which contains 45 % cerium and lesser amounts of other REE. It also contains *5–7 % thorium, 0.3 % uranium*, and trace amounts of silica. The sand came from Brazil, where it was being used by German merchant ships as free ballast for the trip home [2]. This Auerlicht proved successful as a competitor to electricity for street lighting and home illumination. Production resumed at Atzgersdorf and in 1892 the company was organized as the Deutsche Gasgluhlicht AG, also known as Degea, DGA or Auergesellschaft (Auer Company) [26–28, 32].

Between 1891 and 1901, more than 100,000 Auerlicht mantles were used to illuminate the streets of Berlin. In 1901, the first DGA satellites were opened: English Welsbach, American Welsbach, and German Auer. In 1901, Auer was ennobled by Emperor Franz Joseph as a Baron (Freiherr), Karl Freiherr Auer von Welsbach. By 1915, the United States Bureau of Mines estimated the worldwide consumption of gas mantles at 300 million each year and the American consumption at 80 million in spite of the development of the electric light [31]. In 1930, half the street lights in London were still gas mantles. The Indo-Auerlicht Company in Surat, India, now makes the mantles and claims that 75,000 of their gas street lamps are still in use worldwide, including 64,500 in Berlin, Dusseldorf, and Frankfurt. The Auerlicht is reported to burn for 3000 h compared to 1000 h for an incandescent bulb and the light is softer and less glaring [32]. A portable gas mantle lamp is also available at all hardware stores for sportsmen and for outdoor family recreational use (Fig. 4.2).

Between 1898 and 1902, Auer developed an osmium metal filament for the electric light bulb. Edison was using a carbon filament. In 1902, this was marketed as the Auer-Oslicht. Auer and Koppel, the majority owner, formed the OSRAM Company at Berlin in 1906 and collaborated with General Electric on tungsten filaments. The name was concocted as a combination of Osmium and Wolfram (tungsten) as research was continuing with platinum, tantalum, and tungsten filaments. After World War I, the German light bulb manufactures all consolidated with the OSRAM brand [26].

During 1902–1903, Auer invented the cigarette lighter, which used a ferro-cerium alloy for a flint. By 1907 he had a working model for sale and by 1909 he was able to use the monazite residue from the gas mantle production to furnish the REE cerium. The lighters were made at the Treibacher *Chemische Werke* in Treibach, Austria, where he eventually retired to his castle [27]. Since 1998, there is a Welsbach Museum in the *Treibach-Althofen* area, where 700 of the lighters can be seen. And for only 15 € one can have a 91 page catalog (auf Deutsch) with 85 B/W illustrations mailed to their home, S & H included [33]. And if smoking tobacco was discoloring your teeth, the Auer Company in the 1920s decided to add some of their abundant thorium oxide to toothpaste to brighten your smile—Doramad Zahncreme (Fig. 4.3) [26, 34].

Fig. 4.2 Portable Welsbach incandescent outdoor kerosene mantle. (Public domain)

Fig. 4.3 Doramad radiothorium toothpaste, circa 1920s. (Public domain)

The Baron died in 1929 and his Jewish business partner, Leo Koppel (1843–1933), died in 1933, as the German Nazi government came to power. The Koppel family bank was the primary stockholder in Auergesellschaft and OSRAM Companies. Koppel had been a prominent philanthropist since 1905 when he established a Koppel Foundation (Koppel-Stiftung) for German scientific research and development. In 1911, he contributed 1 million Marks for the founding of the Kaiser

Wilhelm Institute for Physical Chemistry, which later became the Max Planck Institute. He stipulated that Fritz Haber (1868–1934), a Jewish Nobel laureate (1918 chemistry), was to be the Director. Koppel also donated toward the salary of Albert Einstein from 1913 to 1926 to keep him in Berlin during and after World War I [35–37]. The Nazi government forced the sale of the Koppel firms to non-Jews or Aryans in 1934. The Nazis called this Arisierung (Eng: Aryanization) of German industry and this was quickly enforced after they took power. The Aryan buyer was DEGUSSA (*Deutsche Gold Und Silbur Scheide Anstalt*), the German Gold and Silver Separation Institute. At the end of World War II, Auergesellschaft was still extracting radium, thorium, REE, and mesothorium while manufacturing the incandescent thorium gas mantles [38–40].

4.1.6 The St. Joachimsthal Radium Factory

The St. Joachimsthal Radium Factory was located in the *K. K. Urangelbfabrik or Uranfarbenfabrik*, which was known in English as the Imperial and Royal Uranium Yellow Factory or Uranium Dye Factory (see Chap. 3). By 1830 uranium had been found to produce bright colors when used as a dye for glass and ceramics. Adolph Patera (1819–1894), from Vienna, developed an improved method of dye production in 1847. The K.K. Factory at St. Joachimsthal started dye production in *1855* by taking over the old Schlick factory that had been used for smelting silver. The K and K stood for Imperial and Royal (*Kaiserlich und Königlich*). Even after the dissolution of the Holy Roman Empire in 1804/1806, the Hapsburg monarchy continued to reign as *Kaiser* over their Austrian territories and as the King (*König*) of Bohemia. The factory produced 10 t of uranium dyes a year for use in coloring ceramics and glassware—six to eight varieties of yellow, one orange, and one black dye [41–44]. Physicist Karl Ulrich from the Auer Co. transferred from Vienna to become the director of the new Joachimsthal radium factory. The radium factory was erected adjacent to the K. K. Urangelbfabrik, where the residues were being generated [45].

By 1907 they started processing the old pechblende residues which had not already been shipped to Vienna. Radium was first produced at Joachimsthal in 1909 (Table 4.2) [41–47]. The Haitinger-Ulrich process produced about 3.236 g of radium per 10,000 kg of residue [24]. Production was highest during the first 4 years when there was a large accumulation of residues. Thereafter, production only averaged about 2 g a year. In 1924, the dye business was closed and the radium factory took over all of the Urangelbfabrik. Radium production averaged 2.20 g a year from 1909 through 1936. Nazi Germany annexed the mines in 1938 and the old radium factory was demolished during 1938–1941 [41–44]. After the takeover of the mines by a German radium consortium, led by Auer Co., production increased to 3.3–3.5 g a year during World War II [38, 39].

The Svornost mine at Sankt Joachimsthal (SJT) dated back to 1530 when it was mined for silver. The silver ore was in the upper most layers of the earth. Deeper

Table 4.2 Bohemia uranium and radium production [43, 46, 47]

Uranium mined [43]		Production of radium [46–47]		Radium produced [46–47]	
Decades: Preradium	Metric tons	During	Grams	During	Grams
1854–1860	7.4	1909	Start up	1922	2.256
1861–1870	38.1	1909/1910	4.69	1923	1.521
1871–1880	39.9	1911	2.647	1924	1.453
1881–1890	44.5	1912	5.0	1925	0.120
1891–1900	53.7	1913	<2.0	1926	0
Postradium		1914	0.875	1927	3.0
1901–1910	104.4	1915	1.754	1928	0.792
1911–1920	50.1	1916–1919	WAR	1929	3.5
1921–1930	154.2	1920	2.257	1930	3.5
1931–1938	130.4	1921	3.221	*Average/year: 1909–1930*	*2.14*

down were found the ores with cobalt, nickel, and bismuth. Uranium was found in amorphous pechblende in the deepest parts of the mine, over 300 m or 12 levels down. In 1864, the Svornost became flooded when a strong current of spring water was encountered at a depth of 500 m. Heinrich Mache and Stefan Meyer detected radon in the mines of Joachimsthal in 1905. Their investigation was concerned more with the potential exploitation of the radioactivity for a Spa than with any health hazard [41–43].

After the identification of radon, a local entrepreneur began to collect the spring water in wooden buckets and offer healing baths in his bakery. Or one could bathe in a water barrel in which uranium ore residues were wrapped in a cloth bag. An official Spa/Bath House was established in 1906, when 43 patients were treated using two bath tubs. Horse power was used to haul buckets of water out of the mine until 1908 when a pipeline was installed to carry the water to bath cabins. In 1910, 373 patients were treated. The Radium Palace/Spa Hotel opened in 1912 and 2476 patients were treated in 1913. The radium factory was located across the street from the Radium Palace (Fig. 4.4). Today, there are four Springs and multiple new hotels servicing the world famous Health Spa [44, 45].

4.1.7 The Societe Annomyme des Traitments Chimiques

The Societe Annomyme des Traitments Chimiques was a Rothschild enterprise. In 1910, Henri de Rothschild set up this radium factory to supply the Paris Rothschild hospital. Albert Labordc, an ESPCI graduate and assistant to Pierre Curie, was in charge of purification and measurement of the radium. The factory was located in Isle-St.-Denis, *north* of Paris. Marie Curie also had a lab here. [8, 12] Ore was obtained from Portugal, the USA, and Madagascar [5–7, 48].

Fig. 4.4 Sankt Joachimsthal Radium Palace and Radium Factory, circa 1913 [45]. (Courtesy J. L. and V. R. Marshall, "Rediscovery of the Elements," 2010, ISBN 978-0-615-30795)

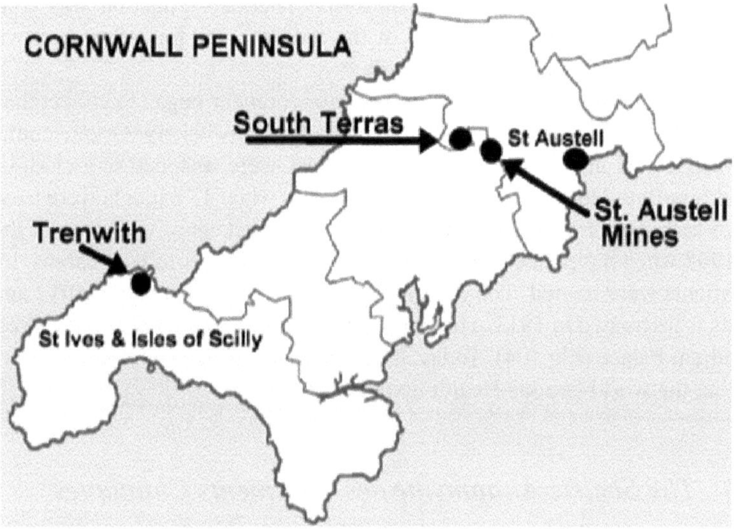

Fig. 4.5 Uranium mines on Cornwall peninsula. (Outline via public domain, labels by author)

4.1.8 The British Radium Corporation

The British Radium Corporation processed pechblende from St. Ives in Cornwall (Fig. 4.5). Cornwall County, located on the southwest peninsula of England, was reported to have over 70 sites of uranium ore, but almost all of them were originally considered to be economically unviable. Pechblende had been found among the copper ores at Wheal Trenwith, 2 mi. from St. Ives in 1843. In 1909, the National Mineral Corporation, through its subsidiary, St. Ives Consolidated Mines Ltd., became interested in the pechblende at the Wheal Trenwith mine. Nobel Chemist Sir William Ramsay (noble gases) was engaged to try and extract the radium from this ore. After this was accomplished, it was decided to form the British Radium Corporation in 1910 under the direction of Sir Ramsay.

The Trenwith mine produced 76 t of uranium ore in 1910, 67 t in 1911, and 42 t in 1912. The pechblende was shipped to a radium factory on Thomas St. at Baltic Wharf in Limehouse. The cornerstone was laid by Lady Ramsay on 18 October 1909. The plant was situated between a biscuit factory and a refining works for gold and silver. By the end of 1910, they had produced 500 mg of radium. This was the *first* commercial radium factory in Britain. By 1911 they had produced several grams of 10% pure radium bromide. Meanwhile a new Spa was opened in St. Ives for bathing in radon water. However, the British Radium Corporation went into receivership in 1918 and was dissolved in 1921 [10, 11, 13, 21].

4.1.9 The John Stewart Mac Arthur Company

The John Stewart Mac Arthur Company used American carnotite ore. J. S. Mac Arthur (1857–1920) was a Glasgow metallurgical chemist. He developed a continued interest in the mineral extraction business and studied the uranium ores in Cornwall, Portugal, Colorado, and Utah. In 1909, he became a Director of the International Vanadium Corporation of Liverpool. They planned to produce vanadium for use as an alloy in the steel industry. Their factory was located in the Garston district of Liverpool [13]. The other directors were all from the USA and they established an American subsidiary known as the General Vanadium Company in 1909/1910. The General Vanadium Company began acquiring carnotite-uranium ore, ($K_2(UO_2)_2(VO_4)_2 \times 3\ H_2O$), from the Paradox Valley in western Colorado, Chap. 5. By 1913, the General Vanadium Company had acquired 60 claims in the valley, mostly in the Long Park section. However, with the rising price of radium, the Vanadium Corporation of Liverpool developed another subsidiary, the Radium Extraction Company, in 1914 to extract radium from the carnotite ore, using the factory in Liverpool. The Atlantic transit proved unrealistic due to World War I and the Vanadium Corporation of Liverpool was dissolved in 1916 [13].

Meanwhile Mac Arthur had purchased his own claims for the carnotite-uranium ore in Utah. In 1911, he opened the *second* British radium factory. It was located it in the County of Cheshire and known as The Radium Works at Halton. Using carnotite ore he produced about 1500 mg of radium bromide by 1915. In that year he relocated the plant to Scotland in Dumbartonshire. It was known as The Loch Lomond Radium Works. The radium was supplied to the Admiralty for use in luminous paint.

Mac Arthur died in 1920 and his chief chemists, the Dempster brothers, continued the business until 1926, using ore from Portugal [13, 47].

4.1.10 The Societe Francaise du Radium Factory

The Societe Francaise du Radium factory was active during 1912–1930 using ore from South Terras, Cornwall (Fig. 4.5). The South Terras Mines had the largest collection of pechblende on that peninsula They were located between the villages of St. Stephen and Grampound Road on the banks of the River Fal, 5 mi. west of St. Austell. The Mines had been discovered in 1873 and they contained an average of 11 % U_3O_8. During 1889–1909, 600 and 89 t of it had been shipped to Germany for extraction and making green glass. After the discovery of radium in 1898, the ore was sent to the Buchler Company at Braunschweig for radium extraction [13, 49].

King Edward VII of England developed a skin cancer on his right naso-labial fold in 1906. There were no radium factories in the UK at the time. He was treated and cured with radium from Paris. The King then prevailed upon his friends Colonel Ashley and Lord Iveagh to donate funds to found a *London Radium Institute*. Lord Iveagh was heir to the Guiness brewing family coffers and Sir Ernest Cassel was a German-born financier. They purchased ore from South Terras to be refined at Buchler. They ordered enough ore for Buchler to produce 7.5 g of radium for the London Institute which began the treatment of patients in 1906/1907 [50].

Jacques Danne (1882–1919), had left the Curies and the Armet de Lisle radium company during 1907–1909 to set up his own commercial laboratory. This was the *Laboratoire d'Essais des substances Radioactives*, which contained a workshop for making instruments to measure radioactive substances. It was located at Coupieres, now called Gif-sur-Yvette, a suburb about 14 mi. southwest of Paris. In 1912, he was joined by his brother Gaston and by Marcel Pochon, both *Ecole de Physique et de Chimie Industrielles* graduates, in an attempt to start a Paris radium factory [11, 13, 49, 51]. They formed the **Societe Francaise du Radium factory as the** parent company for extracting the radium in Paris. Through a subsidiary, the *Societe Industrielle du Radium*, they acquired a 21-year lease on the South Terras Mines in Cornwall. There were four French directors, including Danne, and four British. The outbreak of War in 1914 led to the suspension of business [47].

A *British and General Radium Corporation* resumed operations in Cornwall during 1920–1922. Marcel Pochon moved to Cornwall to start a refinery at the reopened mine. Pochon had graduated from the *Ecole de Physique et de Chimie Industrielles* in 1907 and had trained under Pierre Curie [11, 13, 49, 51]. The Cornwall mines produced 736 t of ore before closing in 1929/1930. The mine director was a German named Straus, who suddenly disappeared in 1929 with several thousand dollars' worth of radium. Pochon stayed on-site in an attempt to refine what ore was left but the mine closed in 1930/1931. In 1932, Pochon and his family moved to Port Hope, Ontario, to run the radium refinery of the Eldorado Mining

Company. Pochon operated that refinery from 1933 to 1945 and Port Hope became a key component of the Manhattan Project, Chap. 8 [51].

4.1.11 The Oskar Knöfler Chemical Company

The Oskar Knöfler Chemical Company was the largest manufacturer of thorium in Berlin. Otto Hahn (1879–1968) had gone from organic chemistry to radio-chemistry in the 5 years since his Ph.D. at Marburg in 1901. Hahn returned to Germany after working with Rutherford in Montreal and continued his studies on radioactive thorium. Hahn contacted the Knöfler Company and they began supplying him with thorium residues in 1907, when he discovered another new radioactive isotope, mesothorium ($^{228}Ra_{88}$). Hahn quickly realized that the gamma ray emissions of mesothorium were as penetrating as those from radium. Hahn alerted Knöfler to its commercial significance as a radium substitute. They planned to market it commercially as Knöfler could produce it as a by-product of their thorium manufacture for Welsbach and others. Meanwhile, Fred Soddy patented the process for producing mesothorium. Knöfler then had to pay him 26,000 Marks (4.20 Marks to US$ 1) for use of the patent [52, 53].

It was 1913 before mesothorium ($^{228}Ra_{88}$) was identified as an isotope of radium, of which 35 have now been discovered with atomic weights of 201–235. None of the isotopes of radium are stable but the most common one and longest lived is radium-226. The gamma rays of significance for radium therapy come from the two decay products, lead-214 ($^{214}Pb_{82}$) and bismuth-214 ($^{214}Bi_{83}$), which are produced in the uranium decay series, Appendix V: Uranium Decay Emissions. The equally penetrating gamma rays from mesothorium came from the decay products actinium-228 ($^{228}Ac_{89}$) and thalium-208 ($^{208}Tl_{81}$), which originate from the decay of thorium [54, 55], Appendix VI: Beta and Gamma Decay Emissions.

4.1.12 The Auer Company of Berlin

The Auer Company of Berlin, which was manufacturing thorium for their gas mantles, decided that the Hahn publications made the Soddy patent moot. Auer's lawyers won and the two Berlin companies (Knöfler and Auer) started manufacturing mesothorium for medical use and then for luminous paint. Both companies were manufacturing thorium from the Brazilian *monazite* sand, which contained from 5 to 7% thorium and 0.3% uranium, Appendix III: Uranium Minerals. One metric ton of monazite yielded 2.0–2.5 mg of mesothorium. The mesothorium recovered contained 20–25% radium. By 1913 Auer had an annual production of 2 g of mesothorium, which sold for 200–300 Marks/mg or 200,000–300,000 Marks/g. This equaled US$71,428 a gram which was less than half the current price of radium, Appendix VII: The Price of Radium [56–58].

4.1.13 The Societe Miniere Industielle Franco-Bresilienne

The Societe Miniere Industielle Franco-Bresilienne was organized in 1909. It was the only mesothorium factory in France and was assisted by Marie Curie and Andre Debierne personally. The monazite ore was imported from Brazil. The Paris office was at No. 202 Quai de Clichy, Ave de Clichy [4–8]. World production of mesothorium was estimated at 7 g a year, with 5 g coming from Germany [59]. The other 2 g presumably came from France.

4.2 Curie-Therapy

Friedrich Otto Walkoff (1860–1934) was a dentist and amateur photographer in Braunschweig, Germany. He is credited with producing the first dental roentgenogram in January of 1896. He exposed his own teeth for 25 min to produce the image on a glass photographic plate that was wrapped in black paper (rather than shielded in a camera) [60]. By 1900 he was lecturing to his photography club about the "Invisible rays effective for photography." He explained the work of his fellow Germans: Plucker, Hittorf, Geissler, Lenard, and Roentgen as well as that of Sir Wm. Crookes. He described the discovery of the two new elements by "dem Franzosen Curie und seiner Frau" [61].

Walkoff revealed that his friend, Dr. Fritz Giesel of Braunschweig had prepared 2 g of radium from several thousand kilograms of *uranpecherz*. Giesel had then graciously provided 200 mg to Walkoff for physiological investigations. Walkoff reported that he made a repeated 20 min long application of the radium to his forearm. This produced a skin inflammation (*Hautenzundung*), of 2 weeks duration, which resembled that seen after long exposure to X-rays. His lecture was published in the October, 1900, issue of the Photographic Review, a magazine for the Friends of Photography [61].

In December of 1900 his friend, Fritz Giesel, confirmed Walkoff's report that *Radiumstrahlen* caused *Hautenzundung* similar to the effect from *Roentgenstrahlen*. Giesel had placed 270 mg of radium, which was enclosed in two celluloid capsules, on the inner aspect of his arm for 2 h. Giesel noted: "At first only a faint redness existed; after 2–3 weeks appeared an intense inflammation with pigmentation and finally, the effects lasted three times longer than had Walkhoff's, a sloughing of the epidermis, after which it almost completely healed" [62].

These reports prompted confirmation by Pierre Curie and Becquerel in 1901. Curie left a radium preparation, with an activity of 5000 times that of uranium, on his left arm for 10 h. It took over 42 days to heal. Becquerel used a preparation of 800,000 activity and left it in his waistcoat pocket for 6 h. The resultant burn healed in 49 days [11]. Curie consulted with the elderly dermatologist Ernest H. Besnier (1831–1909), the Director of the Hospital St. Louis in Paris, about treatment for his skin burn. Mindful of the use of X-rays for skin diseases, Besnier suggested that

Curie entrust some radium to dermatologist Henri Alexander Danlos (1844–1922), who had previously treated some lupus patients with X-rays [11].

Danlos agreed to some trials using radium for lupus vulgaris and for lupus erythematosus. At that time dermatologists had great difficulty distinguishing between the two. Lupus vulgaris was a superficial tuberculosis infection of the skin, commonly in the head and neck area. Lupus erythematosus is an autoimmune multisystem disease that is chronic, debilitating and essentially incurable. Danlos and his physicist Eugène Bloch (1878–1944) published their results in June of 1901 [63]. His first radium samples were described as a white powder in a little sack of 2–3 mm with activities of 5200 and 1000, which would be only 0.2 and 0.03 % pure radium, respectively, using the Boltwood-Rutherford ratio of $2.94 \times 10^6{:}1$ for the radium to uranium activity [21]. Not surprisingly his treatment times were gradually extended from 24 to 63 h [63]. In July of 1902 he described the treatment of five cases of "lupus" with samples that had activity of 19,000, 5200, and 2500, which would equal 0.6, 0.2 and 0.1 % pure radium [64].

The only patient with any significant follow up was #5, who endured treatment times of 24, 36, 72 (3 days), 96 (4 days), and 120 h (5 days). After four of his five lesions were treated and probably sensing the trend in therapy time, he ran away from the hospital. Six months later he consulted different dermatologists, M. Hallopeau and M. Gadaud, for a radium ulcer on the back of his hand [65, 66]. Danlos dutifully reported this as an overdose. No other clinical reports of treatment were published in 1902. Danlos abandoned radiation therapy in 1904 and returned to dermatology. He is now remembered mostly for his 1908 observations on the Ehlers-Danlos syndrome, a complex genetic connective tissue disease with redundant skin and loose joints.

During 1903 at least ten editorials discussing radium therapy appeared in the American literature [67]. It got so much press that a reader inquired of the Journal of the American Medical Association: "Can you tell me where I can get a small quantity…." He was informed: "Radium is hardly yet a commercial product, and samples are obtained more by favor than otherwise." "It is thus far more of a scientific curiosity than anything else" [68]. The Imperial Academy of Sciences in Vienna met in June of 1903. Both A. A. Exner and G. Holtzknecht reported on radium treatments. Holtzknecht reported on dermatological conditions like psoriasis and lupus. Exner, however, had treated an oral cavity cancer and a metastatic melanoma [69]. In October came the first claim of a cancer cure with radium. Drs. Semen Goldberg and Efin London in St. Petersburg reported the apparent cure of two patients with a rodent ulcer, a basal cell carcinoma of the skin [70].

In USA, the first physician to report on the actual treatment of a patient with radium was Margaret Abagail Cleaves (1848–1917) in September of 1903. Dr. Cleaves had a background of on-the-job training in psychiatry and gynecology. After her graduation from the Iowa Medical School in 1873 she had alternated between private practice and work at Insane Asylums in Iowa and Pennsylvania. However, she visited European asylums and wrote extensively on the merits of having female physicians to do pelvic exams, which were being deferred by male physicians. Electro-therapy was having its greatest application in neurology, psychiatry, and gynecology and Dr. Cleaves left Iowa to study this in New York City and Paris during 1887/1888 [71, 72].

Cleaves returned to New York City in 1890. In 1891, she was one of the founding members of the American Electro-Therapy Association, which also included physicians J. H. Kellogg of Battle Creek, R. H. Boggs of Pittsburg, W. J. Morton of New York City, M. K. Kassabian of Philadelphia, G. H. Stover of Denver and electrical engineer E. W. Hammer of New York City. She opened her own clinic, where she began her practice of Electro-Therapy and Physiological Therapy, in 1895, the year of the Roentgen discovery. The discovery led Cleaves and many in the American Electro-Therapy Association to begin using X-ray therapy in addition to Finsen ultraviolet therapy [73]. Apparently in September of 1903 she contacted the New York City radium vendors, Eimer and Amend, who arranged for her to briefly borrow some radium being ordered by chemistry professor Charles Baskerville at the University of North Carolina. He would transfer to the City College of New York in 1904. The loan was for a few days. The specimen was 1 g of 7000 activity, which was the equivalent of 2.38 mg of pure radium salt; $7000/2.94 \times 10^6 = 0.238\,\% \times 1000$ mg.

During September 15–16, 1903, she treated a sarcoma of the oral cavity and a squamous cell carcinoma of the cervix, stage IVA. She had been treating both patients unsuccessfully with X-rays and ultraviolet light for 5 weeks and 3 months, respectively. Both patients were inoperable and incurable with advanced disease. The sarcoma patient was given an exposure of 27 min in two fractions by holding the radium 0.75 of an inch from the skin. The treatments quickly relieved his intense pain and allowed him to open his mouth during 10 days of follow-up [74].

The cervical cancer patient was in partial remission after the X-ray treatments and remained at work. However, she still had cervical, bladder, and rectal involvement with rectal bleeding. In what was probably the first recorded *intra-vaginal* radium treatment, Cleaves inserted the glass water jacket of a vaginal X-ray tube into the vagina and placed the radium tube in the water jacket [71]. An exposure time of 15 min in two fractions was given, equally divided between anterior (bladder) and posterior (rectum) surfaces. The patient's rectal bleeding promptly ceased.

Both patients remained improved during the following week at which time Dr. Cleaves had to return the radium. Dr. Cleaves left for the annual American Electro-Therapy Association meeting in Atlantic City during September 22–24. She presented her novel treatments at the meeting and an abstract was published on 13 October in a Journal of the American Medical Association report of the meeting. She also submitted it to two other journals [74, 75]. Cleaves was using the equivalent of 2.38 mg of pure radium. The next two American therapists to report radium cases were also using nominal doses, 5.8, 5 and 1 mg. Because of the scarcity and cost, Dr. Cleaves was "one and done" with radium. However, after her reports in 1903, a flood of reports on radium began to appear in the literature, Appendix II B: Radiation Therapy Early Trials.

At the Colorado state medical convention on October 6–8, 1903, Dr. G.H. Stover of Denver mentioned three cases of cancer he was treating with radium. In January of 1904, Stover reported on his three patients. He was using 1 g of activity

17,000, which cost him US$ 200.00 and was equal to 5.8 mg of pure radium, 17/2940 = 0.58 % × 1000 mg. Stover was treating large cancers of the nasal cavity, oral cavity, and uterus with 1 h exposures and, not surprisingly, reported poor results [76]. Dr. Sam Tracy of New York City reported to the New York School of Physical Therapeutics on 23 October of 1903 on his treatment of two patients. For psoriasis he used a preparation of *1 g* with 15,000 activity that was equal to *5.0 mg* of pure radium, 15/2940 = 0.5 % × 1000 mg. The exposure rate was 20 min twice a week at a distance of 2.5 cm from the skin. For a squamous cell cancer on the neck he used 10 mg of 300,000 activity equal to 1.02 mg, 300/2940 = 10.2 % × 10 mg. The exposures were for 15–20 min three times a week at 2.5 cm from the lesion [77].

Dr. F.H. Williams (1852–1936) of Boston City Hospital was a graduate of the Massachusetts's Institute of Technology (MIT) and then Harvard Medical School in 1877. Working with physicists from MIT, Williams developed a Department of Radiology at Boston City between 1896 and 1905. His radiology text went through three editions: 1901–1903 and became a classic in the field. Already doing X-ray therapy as well as diagnostic work he acquired a small amount of weak radium in 1900. He acquired this from William Herbert Rollins (1852–1929), who was a Harvard Dental School (DDS) graduate in 1873, a medical (M. D.) graduate in 1877 and a resident Boston genius. Williams and Rollins had married sisters; Anna and Miriam Dunn. Dr. Rollins was an innovator and an inventor of all kinds of medical, radiology, and dental equipment. He was also reclusive and shunned people, meetings, and any patient contact [78].

Williams went to Paris, London, and Vienna in 1903 to observe treatments and to obtain a stronger radium source. He procured 100 mg of activity 1,800,000 or *61 mg* of pure radium, 1800/2940 = 61 % pure × 100 mg. He then discarded anything weaker. In February of 1904, he reported on 42 patients. Most of his patients had squamous cell cancer (#23) or basal cell cancer (#5) or lupus vulgaris (#4) though he did treat four women with breast cancer. The other six patients had the usual benign dermatologic conditions. His six page report gave practical instructions for the technique and the safety of the patient and the physician. His publications were a major influence in the professionalization of radium therapy, which had been dominated by amateur practitioners [79, 80].

Dr. Robert Abbey (1851–1928) was a graduate of the City College of New York and the Columbia Medical School (1871–1874). He practiced surgery in New York City and came to specialize in plastic surgery for inherited or acquired disfigurement. He was particularly aware of cancer deformities both before and after radical cancer surgery. Up to this point radium therapy had been largely in the hands of dermatologists or radiologists, but Robert and his nephew Truman Abbe introduced radium therapy to surgeons [81].

The Abbe family was well-to-do, well connected, well educated, well placed and well respected. Robert's father was merchant G. W. Abbe of New York City who married a Colgate toothpaste heiress and had two daughters and five sons. The eldest son Cleveland (1838–1916) was educated at the City College of New York and became the head meteorologist of the new National Weather Service in Washington D.C. In 1883, he convinced the railroads to adopt time zones, thereby earning the enmity of farmers everywhere [82]. Cleveland's son Truman (1873–1955) went to Harvard, then the Columbia Medical School (1895–1899), and the University of Berlin. He worked as a surgeon at Georgetown University and then George Washington University during 1902–1905 and was given an award in 1907 for his work on radium [66, 81].

On 7 March of 1903, Robert Abbe received from the Society Centrale de Produits Chimiques (SCPC) in Paris 150 mg of activity 300,000 equal to 15.3 mg of pure radium, 300/2940 = 0.102 × 150 mg [81]. He then purchased German radium (mesothorium) equivalent to 7.14 mg and to 61 mg of pure radium. Eight months later, in November, R. Abbe reported some of his findings to physicians in Trenton, NJ, and his nephew reported both their efforts at a meeting in the District of Columbia (D/C) [66]. Truman Abbe had obtained some weak radium from the Smithsonian after some support from the Weather Bureau. In *1904,* Robert published the 15-month results from his first 40 patients [83, 84]. By 1906 he had 125 cases and by 1910 over 500 [81]. He claimed that he *cured* only skin cancers and warts. For deep seated malignancies he favored surgery then post-operative radiation. Unlike the dermatologists, he was treating more surgical cases than benign skin conditions. By 1906, he had treated 40 "surgical" cancers, 9 sarcomas, 9 benign goiters, and 6 potential melanomas, as well as 30 skin cancers among his 125 patients [85]. Abbe pioneered preoperative radiation and interstitial radium therapy by surgically implanting the radium temporarily. And he anticipated the modern after-loading techniques by inserting plastic tubes into the tumor and inserting the radium postoperatively. He retired in 1924 and spent his last 4 years requiring constant blood transfusion because of bone marrow failure, presumably from radium toxicity [81].

Dermatologist Louis-Frederic Wickham (1861–1913) came from a Parisian industrialist family that had originated in Britain and never changed their surname (Fig. 4.6) [50]. After studying the Classics, Wickham began medical training in Paris in 1882 and finished his medical (M.D.) theses in 1890. He entered the maze of French specialty training in Dermatology-Syphilology. In 1895, he described what became known as "Wickham's striae" [86]. He became interested in radium in 1903.

On 5 March, 1905, he received eight radium sources from the Armet de Lisle radium company. They had been prepared by J. Danne [50]. Wickham planned an ambitious program of clinical therapy and pathology studies. The eight sources were calculated to have activities of 0.27, 0.34, 0.50, 1.6, 1.7, and 2.2 % that of pure radium. The samples weighed from 10 to 50 mg which were therefore equivalent to 0.034, 0.17, 0.65, 0.68, and 0.80 mg of pure radium: *all less than 1 mg.* However, within a year his results led him to start organizing a clinical and research center known as the *Laboratorie Biologique de Radium.* He recruited radium pharmacist A. Jaboin (chemical research); J. Danne (physics); Dr. Henri-Auguste Dominici, 1867–1919, (therapy and physiology); and Dr. Paul-Marie Degrais, 1874–1954, (surgical pathology). They opened the *Laboratorie Biologique de Radium* on 1 July 1906 with additional radium [50]. In October of 1906 he published his experiences of the preceding 18 months. His new inventory included ten specimens with less than 1 mg of pure radium and seven sources with more: 1.02, 3.37, 4.76, 6.12, 8.16, 16.3, and *39.5* mg [87].

Fig. 4.6 Louis-Frederick Wickham, 1923. (From the former library of the New York State Medical Association, public domain)

Wickham and Degrais published "Radium Therapy" in Paris in 1909. An English translation was published in 1910 [87]. It was 300 pages with the last 200 detailing their clinical results from 900 patients. The majority of the patients were dermatology cases but two chapters dealt with cancers usually treated only by surgery: known then as "surgical or deep cancers." He discussed the treatment of cancers of the oral cavity, the breast, lymph nodes, sarcomas, and lymphomas. In the English edition he reviewed the therapy for surgical cancers such as bladder, larynx, colon, and rectum. He also mentions interstitial and intraoperative therapy trials. In the chapter on gynecology and fibroid tumors, Wickham noted the life-saving treatment of hemorrhage and endometritis with radium. Blood transfusion and antibiotics were still decades in the future. The treatments of cervical cancer with the new techniques of intra-uterine radium were explained. Intra-uterine applicators were illustrated and preoperative and postoperative treatments were discussed [87].

Wickham credited Dr. Marie-Paul Oudin (1851–1923) with the first use of uterine *intra-cavitary* radium in 1906, 3 years after Dr. Cleaves used *intra-vaginal* radium. Oudin, Fig. 4.7, and Verchere successfully treated two women with chronic exsanguination from menometrorrhagia and large fibroid tumors [88]. Seven women with chronic gonorrhea were also successfully treated. Oudin had 27 mg of radium bromide of 37,500 activity, which was placed in a glass tube 25 mm long with a 2 mm diameter. This was placed in a hollow aluminum tube to block the alpha and beta rays. For the hemorrhaging patients they used four applications of 10–15 min each every 8 days. For the gonorrhea of the cervix and urethra they used two applications. The simplicity of this treatment and the avoidance of roentgen ray

Fig. 4.7 Marie-Paul Oudin, inventor of uterine intracavitary brachytherapy, 1923. (Public domain)

dermatitis of the pelvic skin led to its rapid acceptance in the treatment of cervical cancer as well.

Wickham's book was the *first* for radium therapists. It had such an impact that Sir Malcolm Morris of the British Radium Institute, noted in the introduction to the English edition that radium therapy was now separated into "before Wickham" and "after Wickham" [87]. Wickham set the stage for the next era: the treatment of deep seated or "surgical cancers" treated with "mass radium," which would come to mean 1 g or more of pure radium or radon. The International Radium Standards were accepted in 1912 and were disseminated internationally in 1913. This allowed radium therapy to become a more exact scientific discipline in contrast to roentgen therapy. That same year, Wickham died just as large quantities of American radium entered the market, Chap. 6. The *Laboratorie Biologique de Radium* was disbanded in 1914 as World War I started.

After Marie Curie's second Nobel Prize in 1911, the Pasteur Institute received a financial gift to begin constructing a Radium Institute, sometimes mistakenly called the Curie Institute. Construction started in 1912 but would not be finished until after World War I. Twin pavilions were erected: the Curie Pavilion and the Pasteur Pavilion. The former was for Marie Curie and her research. The Pasteur Pavilion was for clinical radiotherapy and the study of radio-physiology. It was placed under the direction of Claude Regaud (1870–1940), who was assisted by Antoine Lacassagne (1884–1971) and Andre Debierne (1874–1949). In 1923, two more Pavilions were added to the complex, the Rothschild and the Roentgen. The former was for outpatient clinics and pathology laboratories. Dr. Henri Coutard took over the Roentgen Pavilion for X-ray therapy and diagnostic studies. In 1932, the Regaud Pavilion was created for research and tele-curie therapy and in 1936 the Hospital Curie was added to the complex [89].

References

1. de Haën E (1899) Über eine radioactive substanz. Ann Phys Chem (Leipzig, Neue Folge) 68:902
2. Brucer M (1990) A chronology of nuclear medicine. Heritage, St. Louis
3. Adloff JP, MacCormick HJ (1995) The dawn of radiochemistry. Radiochim Acta 70/71:13–22
4. Roque X (1997) Marie Curie and the radium industry. Hist Technol (special issue) 13(4):267–281
5. Roque X (2001) Displacing radioactivity. In: Joerges B, Shinn T (eds) Instrumentation between science, state, and industry. Springer, Heidelberg, pp 51–68
6. Rentetzi M (2008) Biography of a trafficking material. In: Trafficking materials and gendered experimental practices. Columbia, New York. http://www.gutenberg-e.org/rentetzi/chapter01.html. Accessed 29 June 2014
7. Boudia S, Roque X (eds) (1997) Science, medicine and industry. The Curie and Joliot-Curie laboratories. Hist Technol (special issue) 13(4):241–343
8. Bovdia S (1998) Marie Curie: scientific entrepreneur. Phys World 11:35–39
9. Adloff JP, Kauffman GB (2007) The troubled story of polonium. Chem Educ 12:94–101
10. Landa E (1987) From buried treasure to buried waste. Colo Sch Mines Q 82(2):1–77
11. Mould RF (2007) Radium history mosaic. Nowotw J Oncol 57(Suppl 4):1–272
12. Rentetzi M (2008) The US radium industry. Minerva 46:437–462
13. Harvie DI (2005) Deadly sunshine: the history and fatal legacy of radium. Tempus, Stroud
14. Ramstetter H (1966) Carl Johann Eugen de Haën. Neue Dtsch Biogr 7:436–437. http://www.deutsche-biographie.de/sfz25156.html. Accessed 30 June 2014
15. SigmaAldrich (2014) Seelze plant history. Riedel-de Haën Chemicals—SigmaAldrich.com. http://www.sigmaaldrich.com/analytical-chromatography/fluka-analytical/about-fluka-and-riedel.html#Riedel%20History. Accessed 30 June 2014
16. Marshall JL, Marshall VL (2010) Braunschweig Germany. Rediscovery of the Elements DVD. http://www.jennymarshall.com/products.htm. Accessed 30 June 2014
17. Schneider T (1964) Giesel, Friedrich. Neue Dtsch Biogr 6:387. http://daten.digitale-sammlungen.de/0001/bsb00016322/images/index.html?seite=403. Accessed 30 June 2014
18. Giesel F (1899) Einiges uber das verhalten des radioaktiven baryts und uber polonium. Ann Phys Chem (Leipzig) 69(3):91–94
19. Giesel F (1899) Uber radium und polonium. Phys Z 1:16
20. Davis KS (1924) The history of radium. Radiology 2:334–342
21. Badash L (1979) Radioactivity in America. Johns Hopkins, Baltimore
22. Curran TFV (1911) Radium, uranium and vanadium. The mineral industry during 1910, vol 19. McGraw-Hill, New York, pp 597–599, 668–672
23. Le Radium (Paris) (1904–1919) Merged with Journal de physique théorique et appliquée to form Journal de physique et le radium
24. Moore RB, Kithil KL (1913) A preliminary report on uranium, radium, and vanadium Bulletin 70. Dept.of Interior, Bureau of Mines, Washington, DC
25. Calvert JB (2003) Rare Earths & A. von Welsbach. http://mysite.du.edu/~jcalvert/phys/rare.htm. Accessed 30 June 2014
26. Ortel S (2008) Dr. Carl Freiherr Auer von Welsbach und die Geschichte von Auergesellschaft und OSRAM, Berlin. Pressglas-Korrrespondenz, Marz 2008, Stand 11.06.2008. Seite 303-310. Available via http://www.pressglas-korrespondenz.de/aktuelles/pdf/pk-2008-2w-oertel-sg-auer-welsbach.pdf. Accessed 9 April 2013
27. Marshall JL, Marshall VR (2002) Althofen and Auer von Welsbach. Hexag Alpha Chi Sigma 93(1):8–11
28. Auer Museum (2014) Auer von Welsbach biography. http://www.althofen.at/AvW-Museum/Englisch/biographie_e.htm. Accessed 30 June 2014
29. Adunka R (2000) Carl Auer von Welsbach (auf Deutsch). http://pluslucis.univie.ac.at/PlusLucis/001/welsbach.pdf. Accessed 22 Aug 2014

30. Hedrick JB (2013) REE Handbook. http://www.reehandbook.com/definition.html. Accessed 22 Aug 2014
31. Kithil KL (1915) Monazite, thorium and mesothorium. Technical Paper 110. U.S. Bureau Of Mines, Washington, DC. http://archive.org/stream/monazitethoriumm00kithiala#page/32/mode/2up. Accessed 30 June 2014
32. Indo Auerlicht (2013) History of Auerlicht Co. http://www.auerlicht.com/. Accessed 9 April 2013
33. Allhofen Museum (2014) Auer von Welsbach Museum. http://www.althofen.at/welsbach.htm. Accessed 30 June 2014
34. Frame P (1999) Doramad radioactive toothpaste. Oak ridge associated universities. http://www.orau.org/ptp/collection/quackcures/toothpaste.htm. Accessed 30 June 2014
35. James J et al (2011) One hundred years at the intersection of chemistry & physics. DeGruyter, Berlin. books.google.com/books?isbn=311023954X. Accessed 22 Aug 2014
36. Fritz-Haber-Institute (2005) Historical review of the Fritz-Haber-Institute of the Max-Planck-Society. http://www.fhi-berlin.mpg.de/history/h1.epl. Accessed 22 Aug 2014
37. Friedrich B (2006) Fritz Haber (1868–1934) Angewandte Chemie (International Edition) 44, 3957 (2005) and 45, 4053 (2006). https://www.fhi-erlin.mpg.de/history/Friedrich_HaberArticle.pdf. Accessed 22 Aug 2014
38. Hayes P (2004) From cooperation to complicity, Degussa in the 3rd Reich. Cambridge University, New York
39. Owen ER (ca 1945) Production of luminous compounds at the works of Auer Gesellschaft A.G., German Radium Syndicate. http://www.cdvandt.org/bios-303.htm. Accessed 30 June 2013
40. Riehl N, Seitz F (1996) Stalin's captive. American Chemical Society/Chemical Heritage Foundation, Washington, DC (Diane, Darby)
41. Veselovsky F, Ondurs P, Kominek J (1997) History of the Jachymov ore district. J Czech Geol Soc 42(4):127–132
42. Zeman A, Benes P (1995) St. Joachimsthal mines and their importance in the early history of radioactivity. Radiochim Acta 70/71:23–29
43. Zeman Z, Karlash R (2008) Uranium matters. CEU Press, Budapest
44. Marshall JL (2011) Uranium mining & radium refining in St. Joachimsthal/Jáchymov. Nowotw J Oncol 61(1):11e–15e
45. Marshall JL, Marshall VL (2010) Jachymov CR. Rediscovery of the Elements DVD. http://www.jennymarshall.com/products.htm. Accessed 30 June 2014
46. Of C (ed) (1912–1913) The mineral industry during 1911–12, vol 20–21. McGraw-Hill, New York
47. Roush GA (ed) (1914–1938) The mineral industry during 1913–37, vol 22–46. McGraw-Hill, New York
48. Curie M, Curie I (1926) Radium. Encyclopedia Britannica. http://www.britannica.com/nobelprize/article-9435287. Accessed 30 June 2014
49. Smale CV (1992) South Terras mine. http://www.dangerouslaboratories.org/rcw3.html. Accessed 30 June 2014
50. Mould RF et al (2010) Louis Wickham: father of radium therapy. Nowotw J Oncol 60(4):79e–103e
51. Bothwell R (1984) Eldorado, Canada's national uranium company. University of Toronto, Toronto
52. Hahn O (1962) Otto Hahn: a scientific autobiography. Vieweg & Sohn, Braunschweig. (English edition: Hahn O (1966) (trans: Ley W), Scribners & Sons, New York)
53. Hahn O (1970) Otto Hahn: my life (trans: Kaiser E, Wilkins E), Herder & Herder, New York. (German edition: Hahn O (1968) Bruchmann, Munich)
54. Lapp RE, Andrews HL (1972) Nuclear radiation physics, 4th edn. Prentice Hall, Edgewood Cliffs
55. Schlundt H (1922) Mesothorium technical paper 265. U.S. Bureau of Mines, Technical Papers for 1921–1923. Department of the Interior, Bureau of Mines, Washington, DC

56. Lubinau JO, Mould RF (2008) The roller coaster price of radium. Nowotw J Oncol 58(4): 9e–14e
57. Hahne FJ (1989) Early uranium mining in the United States. World Nuclear Association. http://db.world-nuclear.org/reference/usumin.html. Accessed 28 June 2014
58. Landa ER (1982) Radium—the first nuclear industry. Sci Am 247(5):180–193
59. Helvoort TV (2001) Scapel or rays? Med Hist 45:33–60
60. Grigg ERN (1965) The trail of the invisible light. Thomas, Springfield
61. Walkoff FO (1900) Unsichtobare, photographisch wirksame strahlen. Photogr Rundsch 14:189–191
62. Giesel FO (1900) Ueber radioactive stoffe. Ber Dtsch Chemisch Gesellsch 33:3569–3571
63. Danlos HA, Block P (1901) Note sur le traitement du lupus erythematiux par des aplicaitions de radium. Bull Soc Fr Dermatol Syphiligr 12:438–440
64. Danlos HA (1902) Quatre cas de lupus traits par let radium et par d'autres procedes. Ann Dermatol 3:723–727
65. Cleaves MA (1904) Light energy. Rebman, New York
66. Abbe T (1904) Notes on the physiological and the therapeutic action of radium. Wash Med Ann II 6:363–367
67. Chase C (1906) American literature on radium and radium therapy prior to 1906. Am J Roentgenol 8:766–778
68. American Medical Association (1903) Letter to the editor. J Am Med Assoc XLI:324
69. American Medical Association (1903) Current medical literature. J Am Med Assoc XLI: 344–346
70. Goldberg SW, London ES (1903) Zur frage der beziehungen zwischen Becquerel-strahlen und hautaffectionen. Dermatol Z 10:457–462
71. Aronowitz JN et al (2007) Margaret Cleaves. Brachytherapy 6:293–297
72. Gagliardi RA, Wilson JF, Knight N (1996) Radiation oncology. In: A history of the radiological sciences. Radiology Centennial, Reston
73. Cleaves MA (1903) Methods of Roentgen-ray treatment of malignant diseases of the uterus, rectum and bladder. Med Electrol Radiol IV:213–218
74. Cleaves MA (1903) Radium. Med Rec 64(16):601–605
75. Cleaves MA (1903) Radium. J Adv Ther 21(11):667–682
76. Stover GH (1904) Radium. Colo Med I(3):92–97
77. Tracy SG (1903) Radium. J Adv Ther 21(12):758–767
78. del Regato JA (1993) Ch 2 FH Williams, Notes WH Rollins. In: Radiation oncologists. Radiology Centennial, Reston, pp 13–24, 261
79. Williams FH (1925) Reminiscences of a pioneer in roentgenology and radium therapy. Am J Roentgenol 13:23–59
80. Williams FH (1904) Some of the physical properties and medical uses of radium salts. Med News 84:241–246
81. Aronowitz JA (2012) Robert Abbe. Brachytherapy 11(6):1–8
82. Humphreys WJ (2006) NOAA history, Cleveland Abbe. http://www.history.noaa.gov/giants/abbe.html or http://www.history.noaa.gov/giants_index.html. Accessed 14 Feb 2014
83. Abbe R (1904) Radium and radioactivity. Yale Med J X(12):433–447
84. Abbe R (1904) Subtle power of radium. Trans Am Surg Assn 22:253–262
85. Abbe R (1906) Radium in surgery. J Am Med Assoc XLVII:183–185
86. Steffan C, Dupree ML (2004) Louis-Frederick Wickham and the Wickham striae. Dermatol Clin (Sept/Oct), 207–209. ISSN 15-40-9740
87. Wickham L, Degrais P (1910) Radiumtherapy. Funk & Wagnalls, New York
88. Oudin P, Verchere (1906) Du radium en gynecologie. C R Acad Sci 143:90–92
89. del Regato JA (1993) Chap. 7 C. Regaud, Chap. 11 A Lacassagne, Chap. 10 H Coutard. In: Radiation oncologists. Radiology Centennial, Reston, pp 53–64, 99–108, 87–98

Chapter 5
Colorado Uranium

5.1 Mexican Cession

After an 11-year struggle, the Republic of Mexico achieved independence from Spain in 1821. Mexico claimed the territory from the north 42nd parallel (the south border of the Oregon Territory) all the way south to the 16th parallel at Guatemala and the Bay of Tiajuanapec. The northern half of this Mexican Republic was sparsely populated and organized into three large Territories: Alta California, Nuevo Mexico, and Tejas-Coahuila. Administrative capitals were established in Monterrey, California, and Santa Fe, Nuevo Mexico. In 1836, American settlers in Mexican Tejas-Coahuila (Texas) staged a military separation from Mexico. Mexico could not suppress the rebellion and the Republic of Texas was formed. After 10 years, the Republic of Texas was bankrupt and wished to join the USA. When the republic was accepted in late 1845, Mexico was drawn into a disastrous war during 1846–1848. The victorious Americans now took the northern half of Mexico—Alta California, Nuevo Mexico, and Texas. This "Mexican Cession" was more accurately a confiscation, and it extended the western border of the USA to the Pacific. The Americans now claimed everything to the west of the Rocky Mountains. This confiscated Mexican land was thinly populated with random isolated missions and rancheros as well as hostile Indians [1–9].

The Louisiana Purchase had been considered "the Great American Desert"; a relative desert compared to the fertile eastern USA, because of the lack of trees and water. The "Mexican Cession" between the Rocky Mountains and the Pacific Ocean was more like a real desert. Technically, a desert gets less than 6–10 in. of rain a year. The 100th meridian runs north from Monterey, Mexico, along the eastern border of the Texas panhandle through Abilene, Texas; Dodge City, Kansas; Pierre, South Dakota; and Bismarck, North Dakota [10, 11]. It divides the USA into an eastern side which gets 20 or more inchcs of rain a year and a western side which gets less (Fig. 5.1).

R. F. Robison, *Mining and Selling Radium and Uranium*,
DOI 10.1007/978-3-319-11830-7_5

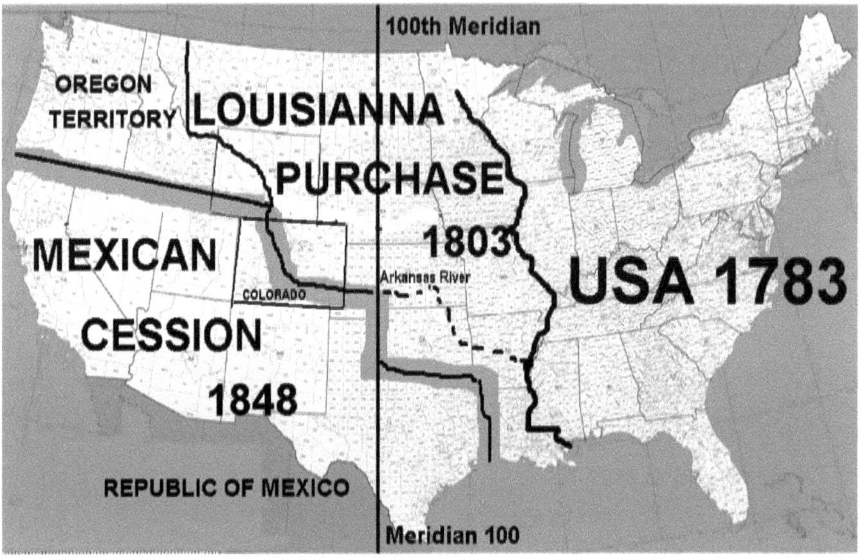

Fig. 5.1 The 100th meridian and the Mexican Cession of 1848. (By author)

How big a desert was it? The Mormons, having been expelled from Illinois, had just arrived at the Great Salt Lake in 1849. They quickly recognized that they were in a desert where no one was likely to bother them about their peculiar practice of polygamy. They proposed a provisional state of *Deseret* to be formed by them out of the Mexican Cessation [12–14]. This lasted, at least on paper, for 2 years before being rejected by the USA (Fig. 5.2).

Fig. 5.2 The provisional mormon State of Deseret: 1849–1851. (By author)

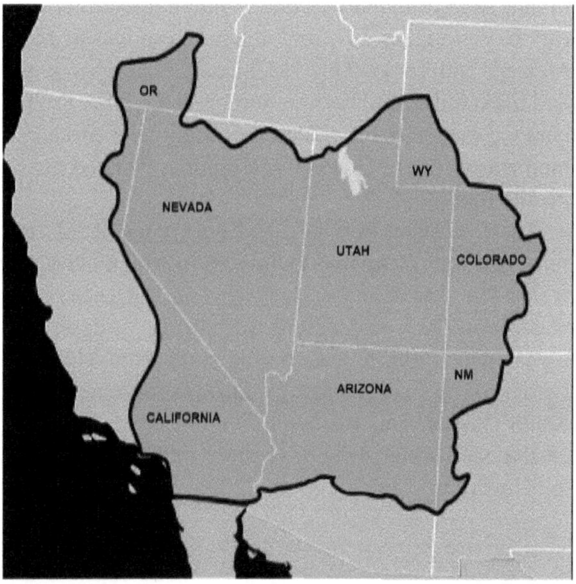

In spite of all the mineral riches found in their colonies of New Spain, New Granada, Peru, and Rio de la Platta (Argentina), the Spanish had never found precious metals west or north of the Rio Bravo del Norte (Rio Grande) River, north latitude 30–32. *Ironically, the ink was hardly dry on the* Mexican Cession treaty of 1848 when the California Gold Rush of 1849/1850 commenced. Gold was discovered on 23 January 1848 in the foothills of the Sierra Nevada Mountains of Alta California; 40 miles west of what is now Sacramento. The peace treaty with Mexico was signed 10 days later on 2 February and ratified by the Senate on 10 March. The eastern newspapers noted the gold strike in August and the California Gold Rush started in December of 1848. During 1849, there were 70,000 "forty-niners" looking for gold in California. In 1850, 45,000 more joined them, using placer (superficial or surface) mining for gold. By 1853, hydraulic mining was necessary, and by 1855 the gold rush was all over [6–9].

In order to get to California, the miner who could not afford to go by way of the Panama Isthmus had to walk across the Continental Divide; the Rocky Mountains. The three major Emigrant Trails—the Mormon (1847–1869), the Oregon (1843–1869), and the California (1843–1869)—all followed the Great Platte River Road westward from the Missouri River. Emigrants left the Missouri River from established ports at either Independence, Missouri, or Council Bluffs, Iowa. Once they crossed to the west bank of the Missouri, the emigrants were on their own in the Unorganized Territory as they walked west to the Platte River Valley. French explorers had named it the Platte (Flat) because it was shallow, only 3–4 ft deep, very wide, and in a flat valley. The Americans described it as "a foot deep, a mile wide, too muddy to drink and too thick to plow." However, it provided a route through the Divide and onto California and Oregon [6–9, 15].

As the emigrants approached the Rockies, the Platte River forked into north and south branches. The North Platte was the main emigrant trail to the West (Fig. 5.3). The North Platte originates in northern Colorado and goes north into central Wyoming where it is joined by the Sweetwater River, which flows east from the

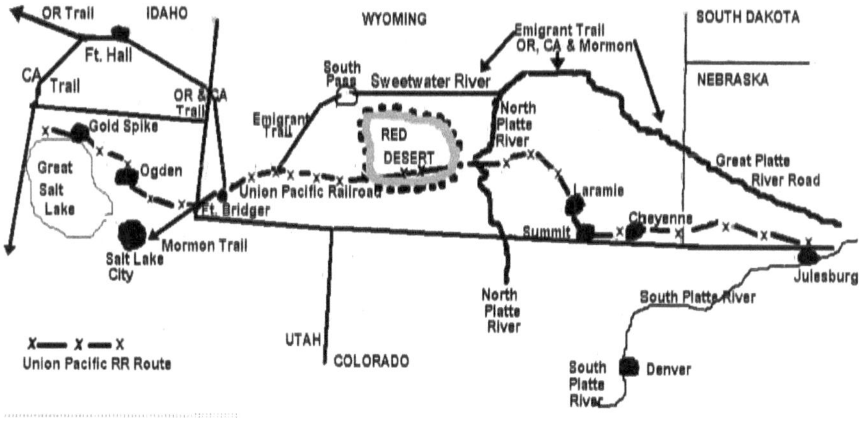

Fig. 5.3 The emigrant trails and the Union Pacific Railroad route. (By author)

Divide. That junction is only 100 miles away from the Divide. After being joined by the Sweetwater, the North Platte flows due east to the Missouri. The westward-bound emigrants followed the North Platte west through Fort Laramie and Fort Caspar to that junction with the gentle shallow Sweetwater. The Sweetwater route was a gradual incline to the venerable *South Pass* through the Divide, at *7400 ft* altitude and 800 miles west of the Missouri River [15].

On the west side of the pass, the Mormon Trail went south through Fort Bridger to the Great Salt Lake. The Oregon and California Trails continued west to the Snake River, where they diverged. The Oregon emigrants followed the Snake River west to the Columbia River. The California Trail went south to follow the Humboldt River through Nevada to the Sierra Nevada Mountains of eastern California. After the California Gold Rush was over in *1855,* many of the gold seekers (Argonauts) returned home the same way they had come. Some lucky ones found gold and silver on the eastern slopes of the Rockies during 1859–1864, the Pikes Peak Gold Rush.

The discovery of gold in Alta California in 1848 and California statehood in 1850 spurred the American Congress toward establishing railroad connections to the West Coast. By 1854, Secretary of War Jefferson Davis of Mississippi had secured surveys of five different routes to California from the Mississippi River (Table 5.1) [16, 17]. The army did the surveys in 1853/1854 and faced hostile Indians everywhere. Routes A, B, D, and E eventually acquired standard gauge railroads after the Civil War. The Union Pacific, Central Pacific, and subsequent transcontinental railroads avoided both the central Rockies with their high mountains and the high deserts of the Colorado Plateau. The rugged route C was more suited for a narrow-gauge rail line. The Denver and Rio Grande Western Railroad was eventually able to connect Denver with Salt Lake and Ogden in 1883 [18–20].

The most favorable route was the Great Platte (River) Valley route along the 41st and 42nd parallel. This route had been used by buffalo herds, Indians, mountain men, fur traders, Pony Express, and emigrants on the Oregon, California, and Mormon Trails. In the future, it would accommodate the Lincoln Highway and Interstate 80

Table 5.1 Pacific railroad surveys of 1853–1854 [16–17] and construction era [21–23]

Routes	Parallels	a-Army (Army) routes surveyed	a-Current routes
		b-RR built	b-RR construction
A	47–49	Army: St. Paul–Spokane–Ft. Vancouver	I-94, I-90
		RR: Great Northern and Northern Pacific	1884–1893
B	41–42	Army: Omaha–Wyoming–Sacramento, CA	I-80
		RR: Central Pacific and Union Pacific	1862–1869
C	38	Army: Ft. Leavenworth–Pikes Peak–CA	I-70
		RR: Narrow gauge DRGW	1870–1880
D	35	Army: Memphis–Santa Fe–Los Angeles	US 66
		RR: Atchison, Topeka and Santa Fe	1883–1888
E	32	Army: New Orleans–El Paso–San Diego	I-10
		RR: Southern Pacific	1883

RR railroad, *DRGW* Denver and Rio Grand Western

as well. The 41/42 parallel was favored by Abe Lincoln and the nonslave states, but not the South. However, after the cessation of the Confederate States in 1861, there were no Southern objections. The Yankees passed the Pacific Railroad Act in May 1862, which confirmed the route along the Platte River, and it was signed by President Lincoln in July 1862 [16, 17].

The California-based Central Pacific Railroad—organized by Stanford, Crocker, Huntington, and Hopkins—started building east into the Sierra Nevada immediately. California was too far away to be bothered with significant wartime restrictions on labor and material. The Union Pacific Railroad was financed by eastern capital but the organization and engineering would come from the Union Army Generals when the Civil War ended in April 1865. Colonel Grenville Dodge was put in charge of building the Union Pacific road and finding a suitable pass through the Rocky Mountains.

The Union Pacific planned to lay their track west along the flat valley of the Platte River, to the town of North Platte, Nebraska, where the river forked north and south. The Union Pacific had hoped to follow the South Platte into Denver and then west to the gold and silver mines in Gilpin County. The Union Pacific did not want to go north through the desolate South Pass at 7400 ft above sea level. However, west of Denver, the lowest pass through the Rockies was at *11,000* ft altitude and would require a grade of over 2.2 %, which was not allowed. Colonel Dodge, the Union Pacific chief engineer, and Jim Evans, the surveyor, searched along the North Platte River for an easier pass through the Rockies. In 1865, they found a natural pass, at *8200* ft elevation, some 28 miles west of Cheyenne, Wyoming. It was about 550 miles from Omaha, and they named it Evan's Pass. The summit was named after Dodge's longtime comrade, General W. T. Sherman [16, 17]. It was higher than South Pass but easier to ascend; less than a 2.2 % grade.

The Union Pacific tracks were laid on a gradual ascending slope for the entire 522 miles from Omaha, at sea level, to Cheyenne at 6060 ft altitude by the winter of 1887/1888. In spring, they laid a track 18 miles west to reach a solid rock roadbed at Granite Canyon, 7300 ft altitude. The Union Pacific had only to cover 10 miles while ascending 900 ft of fairly rolling slope to crest the Sherman summit *at 8200* ft on 5 April 1868. From Evan's Pass westward the Union Pacific was essentially going downhill all the way to Ogden, Utah.

Laramie is 24 miles west of the Pass at an altitude of 7171 ft and Rawlings is at 6700 ft. West of Rawlings, the Continental Divide in southern Wyoming forms a basin instead of a mountain summit. The Union Pacific was able to cross the southern edge of this circular basin at two locations, which are now shared with Interstate 80 [15–17]. On the east side of the Great Divide Basin, at exit 208 on Interstate 80, the elevation is only 6750 ft. On the west side of the basin, between exits 154 and 156, the elevation is about 7500 ft. The basin itself is occupied by the Red Desert, 9000 sq. ft of desert and sagebrush (Fig. 5.3).

The Union Pacific bypassed the Salt Lake City Basin and its surrounding mountains to connect to Ogden, at 4300 ft elevation. Ogden is 40 miles north and 10 miles east of the Great Salt Lake. The Union Pacific and the Central Pacific received huge government contracts/subsidies that were paid by the miles of track they laid. The

Central Pacific had laid *691 miles* from Sacramento, and the Union Pacific had laid *1108 miles* of track from Omaha when they met at Promontory Summit, 66 miles northwest of Salt Lake, on 10 May 1869. During the celebration, at least one gold spike was (briefly) implanted where the tracks met. The rail line was known as the Overland Route. The Express train was known as the Overland Flyer and after 1895, as the Overland Limited [16, 17].

5.2 Colorado Minerals

Colorado was organized into a territory in 1861 and was created as a true rectangle; 380 miles wide and 280 miles long [12, 13]. The crest of the Rocky Mountains forms the Continental Divide as it extends the 280 miles length of Colorado, while essentially partitioning off the western third. Eight major rivers arise in the Rockies, four on each side. The eastern slopes of the Rockies, as they extend south from Casper, Wyoming, to US 50, are known as the Front Range. The eastern slopes give rise to the headwaters of the Rio Bravo del Norte, the Arkansas, the South Platte, and the North Platte. The Arkansas arises near Leadville and travels east for 1469 miles to enter the Mississippi. From 1803 to 1848, the Arkansas was the arbitrary and disputed southern border of the USA (Fig. 5.1). After 1848, the Rio Grande River (Rio Bravo del Norte) became the southern US border [12, 13].

The South Platte starts in the grassland basin known as South Park, just south of Fairplay, Colorado, and goes southeast toward Cripple Creek and Pikes Peak before heading north through what is now Denver (Fig. 5.3). It flows north through the Littleton and Englewood suburbs of Denver, where it is joined by Little Dry Creek, then Cherry Creek, and then Clear Creek among many others. Then, it runs north to Greeley before heading east to join the North Platte in western Nebraska. Gold-seeking Argonauts on the California Trail during *1848–1855* traversed both the North and South Platte Rivers when going west and again when coming home afterwards. In 1857, the returnees were searching for gold along the South Platte near the confluence of Clear Creek. In May *1858,* at the junction of the South Platte and Cherry Creek "some color was raised." About 7 miles south of that junction, the Little Dry Creek joins the South Platte in what is now Englewood; near the intersection of US 285 and US 85. In the first significant strike in the Rocky Mountains, 622 troy grains of gold were found there in July 1858. Both of these sites are now in downtown Denver [18, 19].

That did it. Word got out and thousands of Argonauts embarked for Pikes Peak (or Bust) during the winter of 1858/1859. Few Easterners knew any Colorado geography but they all had seen pictures of Pikes Peak, and it was visible from miles away. Eventually, they found their way to the Front Range. However, the South Platte Gold Rush was forever mislabeled the Pikes Peak Gold Rush. In November 1858, a promoter selling real estate convinced the prospectors to name their settlement after the territorial governor of Kansas, in whose jurisdiction they were encamped. Ironically, Governor James W. Denver had just left office. The

Colorado Territory was organized out of the Kansas and Utah Territories in 1861 but Statehood was delayed until 1876 [12, 13].

In January of 1859, a miner with California experience prospected up Clear Creek toward the steam arising from a visible hot spring, now Idaho Springs. He made his strike there in Clear Creek County. Later that spring, another California miner followed Clear Creek up its north fork into what is now *Gilpin County*. He made the first major strike of rich placer (surface) deposits. More strikes followed in Gilpin County. Additional discoveries occurred in Summit County, Parke County, and Lake County. Then Aspen, Gunnison, and the San Juan communities of Telluride, Ouray, and Silverton found silver and then gold. Between 1870 and 1890, the value of the discoveries increased almost seventimes; from US$ 2.9 million to 19.5 million [18, 19].

The mining communities were to be connected by narrow-gauge railroads that had rails only 3 ft apart [20, 24]. The standard gauge rail lines had track that was 4 ft 8.5 in. apart, which was not suitable for the sharp curves and steep grades of the Central Rockies. The narrow gauge locomotives, cars, rails and everything else was smaller, lighter, and cheaper and thus easier to finance. During 1870–1883, the Denver & Rio Grande Western (DRGW) Railroad laid 1500 miles of narrow-gauge track between Denver and Ogden, Utah; Table 5.2. The DRGW was the brainchild of General William Jackson Palmer (1836–1909), a Philadelphia quaker, who had led his regiment of the 15th Pennsylvania Volunteer Calvary during the War of Southern Rebellion and had earned the Medal of Honor. Dr. William Bell (1840–1921), a recent London medical graduate was his business partner and lifelong friend. Their connections in both Philadelphia and the UK provided their necessary capital [20, 24].

Table 5.2 Denver & Rio Grande Western Rail lines: 1870–1883 [20, 24, 25]

I	Denver & Rio Grande out of Denver
I A	East Colorado
1871–1874	Denver south to Colorado Springs
1874–1876	Colorado Springs south to Pueblo and Trinidad
1877–1880	Pueblo West via Canon City to Salida
1880	Salida northwest to Leadville
I B	Over the divide
1877–1882	Walsenburg and Alamosa west via Cumbres pass to Durango to Silverton
1881	Salida west via Marshall pass to Gunnison
1883	Gunnison via Cerro Summit to Montrose to grand junction
1887	Leadville west via Tennessee pass to Glenwood Springs to grand junction
II	Rio Grande western routes
1883	Grand junction to Salt Lake city
1883	Grand junction to Ogden and overland route of Union Pacific–Central Pacific

Palmer and Bell, respectively, developed the sister cities of Colorado Springs and Manitou Springs as health resorts [26–28]. They recruited retired military men, titled remittance men, and wealthy invalids, such as those with tuberculosis. High-altitude therapy was then in vogue [26]. So many English immigrants were recruited that the two springs came to be known as Little London [26–28]. Bell brought his English bride to Manitou in 1872 and built both a hotel and his Victorian home, Briarhurst Manor. The General was known as a stern, aloof teetotaler and no liquor was served in his Colorado Springs until 1933; but it flowed freely in Manitou Springs and at Dr. Bell's manor. The population at the springs soared from 2 to 11,200 by 1890 to 20,000 by 1900. The Bells were blessed with a happy marriage and five children. They entertained frequently at Briarhurst, including the family of Her Majesty Queen Victoria's vice-consul in Denver, Richard Pearce (1837–1927). Pearce's oldest son, Harold, married the eldest Bell daughter, Cara Rowena in 1895.

Pearce was the first to discover uranium in the USA in 1871. At 18, he had graduated from the mining school at Camborne and then studied metallurgy in London and at Freiburg in Saxony. He was the first to discover pitchblende ore in Cornwall, 1859–1863. He became a manager at the Roachdale Mining Company in Swansea, South Wales. Roachdale sent Pearce to *Gilpin County* on the Colorado Front Range. He was to inspect their deteriorating (played-out) gold and silver properties, which they had been mining since 1859. Hydraulic-placer techniques had been used to obtain the precious metals. After the gold and silver had been washed away, the residue (the tailings) had been discarded as worthless. Pearce recognized the mineral pitchblende in these tailings and informed the owners that it was worth US$ 2000.00 a ton in London. He presented his uranium discovery at the Royal Geological Society in Cornwall [29, 30].

Initially, uranium had value only as a coloring agent for (a) glass, (b) ceramics, (c) paint, or (d) for producing a purplish tint to photographs. However, after the Franco-Prussian War of 1870–1871, there was an increased interest in steel alloys like ferro-uranium and ferro-vanadium for the European armaments industry [31]. The European powers were involved in military adventures across the globe and preparing for war at home. Scientists continued to experiment with using both uranium and vanadium to strengthen steel but ferro-uranium was preferred for armaments. Uranium steel alloys were used to make heavy cannons that would sustain the heat and stress of rapid firing. The European demand for uranium paralleled their increasing expenditures for armaments [29–32].

In *1872,* Pearce moved his family to Colorado, where he personally leased a Front Range claim to produce uranium. Other vendors also secured Front Range uranium for export. Uranium from mines in Gilpin County was shipped to Europe; mostly between 1897 and 1906 and mostly to Great Britain, France, and Germany. The Europeans were interested only in the uranium until 1900, after which both uranium and radium were extracted abroad. [29, 30, 32–37]. By 1873, Pearce had become the Superintendent for the Boston and Colorado Gold and Silver Smelting Co. at Black Hawk in Gilpin County. Pearce managed the Front Range smelter during the mining boom in Colorado [34]. He returned to Cornwall in 1902 and brought some radium crystals with him. Pearce (1837–1927) added a few crystals to everything he drank; believing it was a stimulating tonic [29, 30].

Fig. 5.4 General W. J. Palmer and Queen, 1890 [38]. (Public domain)

In contrast to the Bells, General Palmer had a disastrous marriage. In 1870, at age 32, he had married Mary L. "Queen" Mellon, the 20-year-old daughter of a prominent New York businessman (Fig. 5.4). They honeymooned in the UK, where the General sought investors. Upon their return, Queen went home to New York, while the General started construction of his railroad in Denver [20, 25] and a 22-room estate, Glen Eyrie, at the springs. Queen came West in October of 1871, became pregnant, and left 5 months later to live back East. As she was leaving, her father arrived with his pregnant and fertile third wife, aged 35, and their six children. He was broke and in failing health, and they became permanent houseguests. He died 2 years later aged 58, leaving his family to the care of the General.

Queen could not tolerate the altitude in Colorado and was subsequently diagnosed with severe heart disease. Somehow, they managed three daughters, who were born between 1872 and 1881, at which point Queen moved to England. The General paid all the bills and visited them twice a year until Queen's death at age 44 in December of 1894. He brought the three young girls to Glen Eyrie the next summer. His life was transformed as he devoted the rest of his life to his daughters. He retired from the railroad business in 1901 and spent his last decade in extensive philanthropy at the springs—funding colleges, libraries, schools, parks, and lakes. A freak riding accident in 1906 resulted in a crushed cervical vertebrae and he became paraplegic. He got around in a wheelchair and electric and steam automobiles; he hated the smell of gasoline.

He had one last hurrah in 1907. Not able to travel to Philadelphia for the 35th reunion of his former 15th Pennsylvania Volunteer Cavalry regiment, he paid for all 280 men, aged 59–82 from seven states, to come to the springs for 1 week.

He arranged ten Pullman cars to transport them from Philadelphia. They were boarded, wined, and dined at his Antlers Hotel. Led by the General in his White Steamer, the regiment paraded down Pikes Peak Boulevard, four abreast, with regimental banners and captured colors flying, through decorated streets, and bouquets of flowers. And at the farewell banquet, the Quaker General did not spare the Mumms. When he died in March 1909, it appeared that the springs had lost their greatest benefactor [18–20, 27, 28, 30, 34].

By 1890, it had appeared that the Colorado gold and silver discoveries were beginning to wane. The precious metals were thought to be confined to a circumscribed Mineral Belt that extended along the Continental Divide from Boulder County in the northeast through Breckenridge, Leadville, Aspen, Gunnison, and Ouray to San Juan County in the southwest [18, 19]. However, just east of the Mineral Belt, lay Cripple Creek. At first, nobody believed there could be gold there: 70 miles south of Denver and 7–12 miles west from Pikes Peak. Some locals were trying to raise cattle there, and the uneven terrain caused at least one cow to become crippled [39, 40]. It was a desolate country and only received 14 in. of precipitation a year. It was located in an extinct volcano crater that had formed a 10,000-acre bowl of volcanic rock at an elevation of 8500–10,500 ft, above sea level. Only one man was crazy enough to look for gold there; Crazy Bob Womack. He found it in October of *1890* after 12 years of digging. He then sold the claim for US$ 300.00! Within the month, there were 200 prospectors and 100 claims. In 1891, W. S. Stratton, became the first of 30 millionaires in Cripple Creek [39, 40].

However, Cripple Creek was not a solo miner's camp, as this was not placer or pan gold. This required hard rock mining, which the individual prospector could not finance. Corporations supplied the capital investments and miners were employed by a daily wage. The big money was made on trading stocks at the mine exchanges and on providing services in saloons, barbershops, general stores, saw mills, newspapers, and real estate. Taking advantage of this situation were the Penrose brothers from Philadelphia. The Penrose family had immigrated to Philadelphia in 1700 and had been prominent and prosperous ever since [41]. Brothers Dick (1863–1931) and Spencer Penrose (1865–1939) were roommates at Harvard. Dick graduated summa cum laude in chemistry in 1884 while Spencer finished last in his class. Dick excelled in sports, especially crew, while Spencer took an oar to the head and ended up in the infirmary. In 1886, Dick finished a Masters and a Ph.D. in *geology*. During 1891–1892, Dick was evaluating potential gold, silver, and copper mines in the Southwest. Dick teamed up with Spencer and a German banker in New Mexico to invest in a gold mine in Southwest Arizona. Within 18 months, they were able to share a million dollar profit. The autumn of 1892, Dick took up a job at the brand new University of Chicago as a professor of economic geology. When Dick left for Chicago, he suggested that his younger brother might try Colorado Springs for employment opportunities as gold had been discovered in nearby Cripple Creek [42].

Spencer arrived in Colorado Springs in December of 1892. He was welcomed by an old Philadelphia neighbor, Charlie Tutt, who had become a successful businessman at the springs. Tutt offered Spencer a partnership in his real estate venture in Cripple Creek, on the other side (west) of Pikes Peak. Charles MacNeill, who actually knew something about the chemistry of ore reduction, joined the team in 1893. Spencer spent the period between 1893 and 1896 in Cripple Creek and the partners invested in mining properties as well as other real estate. Dick Penrose took a leave of absence from Chicago in 1897 to help investigate the gold finds in Cripple Creek [39, 40, 42]. Cripple Creek became the biggest gold field in the Americas and second only to South Africa. Colorado produced US$ 37 million worth of gold in 1898, 37 million in 1899, and 41 million in 1900, which was the peak year. After a decade of making big money, the Penrose–Tutt–MacNeill group, with help from Dick Penrose decided to form a multimillion dollar trust to control all the mill and refining facilities and associated transport. Capitalized at US$ 13 million in 1901/1902, they took over all the mills and transport in Cripple Creek. Then there was copper.

In 1903, the Cripple Creek Trust, composed of the three partners, plus Dick Penrose and most of the Penrose family, created the Utah Copper Company and took control of the copper mines in Bingham Canyon, Utah. By 1906, the copper production had increased by tenfold. As the profits soared, many investors sold out and Spencer became the major stockholder. In 1906, at age 41, Spencer married a local wealthy widow, Julie V. L. McMillan (1870–1957) and the copper company was sold to Kennecott [27, 28, 42, 43]. Spencer became the major stockholder in Kennecott Copper and had an annual income of US$ 1 million in 1909. After the death of General Palmer in 1909, Spencer and Julie Penrose gradually took over as the major benefactors and boosters for Colorado Springs.

By 1908, Dick changed from geology fieldwork to mining investment and economic geology. In addition to advising Spencer, he had also invested in the Cripple Creek mines and Utah Copper. In 1927, he established the Penrose Medal of the Geological Society of America, of which he had been a member since 1889. He had remained a bachelor and left an estate of nearly *US$ 8 million in 1931*. It was divided equally between a scientific society in Philadelphia and the Geological Society, of which he had been president in 1930. The Geological Society of America was transformed by the *US$ 3.9 million* endowment into a dynamic world leader in global resources and the environment, which it remains today [44, 45].

Spencer had a long history of alcohol and tobacco abuse, and it was not surprising when he developed cancer of the laryngopharynx in 1932. He was referred to Henri Coutard (1876–1950) at the Roentgen Pavilion of the Curie Radium Institute in Paris, the one place in the world where, at that time, he could be cured without mutilating surgery. Six years later, he developed cancer of the esophagus, which could not be cured. He brought Coutard, an advanced 400-kV X-ray machine, and physicist Dale Trout (1922–1977) to the springs for his unsuccessful treatment. Before his death in 1939, Spencer and Julie created a Trust, the El Pomar Foundation, to fund local projects and a Cancer Pavilion at the local hospital in Colorado Springs, to which he donated the machine and Coutard as well. This developed into the Penrose Cancer Hospital from 1949 to 1974. Under the guidance of radiation oncology innovator, Juan A. Del Regato (1909–1999), who had trained in Paris, this became the first training program in the USA for radiation oncologists [46].

5.3 Colorado Plateau

The Union Pacific, Central Pacific, and subsequent standard-gauge railroads had avoided both the central Rockies with their high mountains and the high deserts of the Colorado Plateau. While rail lines were being built across the West, two government agencies were sent to make scientific explorations of the western territories. Both the War Department and the Department of the Interior each sent survey teams between 1867 and 1879 to explore, map, and assess the geology of the land from eastern Colorado into western California. The Department of the Interior civilian survey team was led by Ferdinand V. Hayden and Clarence King. The US Army survey team was led by General John W. Powell and Lt. George Wheeler. Powell and Wheeler concentrated on the area surrounding the Colorado River. Geologist G. K. Gilbert, primarily with the Powell Survey, is credited along with Powell for identifying the Colorado Plateau. Gilbert recognized the features that separated the plateau from the Rocky Mountains on the east and the strange Basin and Range Province formations farther to the west [21–23, 47] (Fig. 5.5).

On the western slopes of the Rockies, the mighty Colorado River, known before 1930 as the Grand, begins to form as it is joined by its major tributaries—the Green, the Gunnison, the Rio Dolores, the San Juan, and the Little Colorado. The Colorado Plateau is centered over the Colorado River and its tributaries (Fig. 5.6). The plateau has four basic characteristics. (1) It is a flat (smooth and un-deformed) plateau situated *between* two mountainous (highly deformed) regions: the Rocky Mountains and the Basin–Range Province. (2) The Colorado Plateau is a series of plateaus, mesas, and buttes, which have an average elevation of 5300–6000 ft altitude, *but there are no mountain ranges!* (3) Most of the surface rocks are sedimentary; from disintegrating older rocks and other mineral wastes. At higher elevations, the rocks

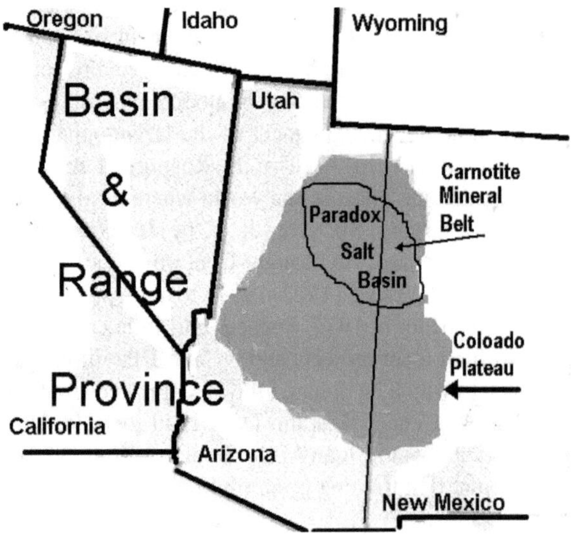

Fig. 5.5 The Colorado Plateau and the Basin–Range Province. (By author)

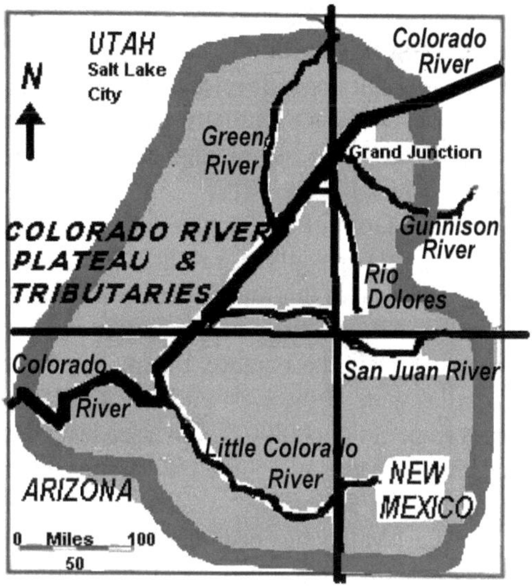

Fig. 5.6 The Colorado River Plateau and tributaries. (By author)

are igneous: from molten rock. (4) There is little permanent standing water in this environment of sparsely vegetated plateaus, mesas, deep canyons, and barren badlands. The average precipitation is only 10 in./year and quickly drains into the five Colorado River tributaries: the Green, Gunnison, Dolores, San Juan, and Little Colorado.

The Colorado Plateau begins at the western edges of the Uinta Mountains in Utah and the Rocky Mountains of Colorado and New Mexico. It extends southwest to the Mogollon Rim in northern Arizona, near Las Vegas, Nevada. The Rim is an escarpment which defines the southwestern edge of the Colorado Plateau. The name Mogollon comes from Don Juan Ignacio Flores Mogollón, the Spanish governor of New Mexico from 1712 to 1715. The plateau is prone to massive erosion and the Colorado River system has cut deep through the rock, while seeking sea level, as seen at the Grand Canyon and at Monument Valley, Utah. The plateau is high and dry, flat and large, barren and eroded but scenic; it encompasses 10 national parks and 17 national monuments. It encompasses 130,000–140,000 sq. miles [47–49].

The western border of the plateau, from Salt Lake City to Las Vegas, Nevada, abuts the much larger Basin and Ranges Province, which encompasses 300,000 sq. miles. The province is largely occupied by the Great Basin and extends from the Oregon–Nevada border south to the Mexican border and west to the Sierra Nevada Mountains. Between Reno and Salt Lake, it is 500 miles wide. The province is characterized by abrupt changes in elevation which alternate between narrow mountain ranges and flat deserts and basins. This is well seen at the western end where Death Valley is at 282 ft below sea level and the adjacent Panamint Mountains peak at 11,000 ft above sea level [49].

The northern half of the *Colorado Plateau* is a large salt basin: the Paradox (Salt) Basin (Fig. 5.5). Geologists estimate that 320 million years ago, the Paradox was under water and without drainage, which led to the eventual evaporation of the ocean water with its 3.5 % salt content. Over time, several 100 ft of salt were deposited in this basin area. In one area of this salt basin, the Paradox Valley of Colorado, a salt dome developed that was 10,000 ft in thickness. Being soluble, the salt dome roof was eventually eroded away creating the Paradox Valley, which is located at the northeastern edge of the Colorado Plateau, on the far western slopes of the Rockies [47,48, 50].

The Rio Dolores bisects the valley into east and west sections by cutting north through its steep walls at a right angle instead of flowing lengthwise through the valley. Because of this anomalous route, the geologists named it Paradox Valley. As the valley was the first salt dome site to be discovered, the geologists named the whole salt basin the Paradox Basin. The Paradox Valley is about 25 miles long. It is 5 miles wide at the west end and 3 miles wide at the east end. The valley floor varies from 5000 to 5800 ft above sea level. Rocky cliff-like walls of sandstone and shale, with heights of 1500–2000 ft, surround the valley on three sides; north, south and west (Fig. 5.7).

The East Valley has been described as an enormous prairie dog town in a desert of sand, cactus, greasewood, and sagebrush. Contact with civilization was usually made by going east 56–75 miles through low mountain passes, to Placerville, Colorado [50, 51]. The Rio Grande Southern narrow-gauge Railroad was constructed during 1890–1892 and connected with Placerville [52]. The West Valley abuts against the foothills of the nearby La Sal (French: salt) Mountains of Utah. This way in and out, by a steep rocky trail, was favored only by "outlaws." Water from the La Sal Mountains enters the West Valley. The early French explorers had found and named Roc Creek and La Sal Creek, which descend from the La Sal Mountains and flow into the Rio Dolores on the north and south sides of the West Valley. The Creeks are only 10 miles apart "as the crow flies" (Fig. 5.8).

Fig. 5.7 The Paradox Valley: 2012. (Author's collection)

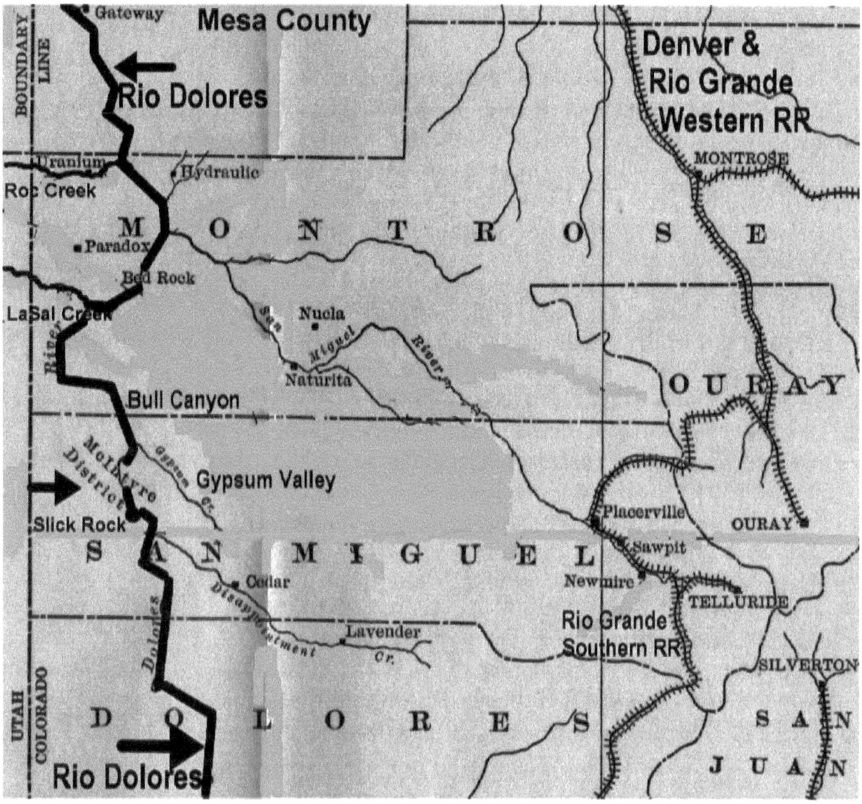

Fig. 5.8 The Carnotite Mineral Belt: 1913 [37]. (Public domain)

The first Americans to enter the valley came in *1875* as part of the US Geological Survey. It was then occupied by the friendly Ute Indians. Colorado statehood came in *1876* and the first settlers trickled in during 1877–1880. By 1881, mail service had been initiated and the Utes were removed. Cattlemen and ranchers moved in and began a bitter and violent competition to obtain water for their part of the desert valley. The West Valley accumulated about 200 inhabitants and two population centers: Paradox and Bedrock. Paradox became the residential center for families, and Bedrock became the commercial center of the valley [50, 51]. However, by 1885, it was estimated that 50 % of the Bedrock residents were outlaws; passing through or stopping for rest, food, and fresh horses. Between 1884 and 1898, both Bedrock and Paradox got a school and a store, but the law was 140 miles away at Montrose, the county seat. The Pioneers had to develop their own "sagebrush court," which was based on local circumstances. In many cases, the victim was a "dangerous member of the community and needed to be done away with" [51]. After 1885, the killings in the Paradox were documented:

- 1885 Virgil Prentiss kills John Prentiss at Bedrock; found guilty.
- 1889 Tom Kelly kills J. Z. Stephens at Paradox; found guilty.
- 1889 Tom McCarty & Butch Cassidy rob Telluride bank.
- 1893 Bill and Fred Mc Carty, father and son, die in robbery attempt at Delta CO bank. Tom McCarty escapes to Sew 'Em Mesa then Oregon
- 1895 Tom Peppar (a hired gun) kills Kan Young in west Paradox Valley; acquitted self-defense.
- 1897 Bart Young kills Wm. Skelton at Red Mountain Pass between Silverton and Ouray CO; Young then escapes to Idaho.
- 1905 Bill Young and Dick Williams survive ambush in west Paradox Valley.
- 1907 Bill Young kills Charlie Wilkinson at Rio Dolores Bridge, charged with manslaughter.
- 1907 John Neftus, Wilkinson's accomplice, found dead in Valley; no suspects.
- 1907 Sheepherder Pete Swanson killed by cattlemen; no suspects.
- 1910 Peg Leg Foster kills Axel Peterson in west Valley; Jury agreed he deserved it.
- 1916 Cleve Unk Estes kills Dick Granath for teasing him.
- 1921 Bedrock gang decapitates Slim Hecox, Cashin mine watchman.

In 1885, beloved schoolteacher John Prentiss was killed with a hammer by his alleged brother. John was later identified as Roland Wilson, an ex-convict hiding from the law. In 1889, Tom McCarty joined Butch Cassidy and Matt Warner in robbing the San Miguel Valley Bank in Telluride, Colorado. In 1893, Tom, Bill, and Fred McCarty robbed the Farmers and Merchants Bank in Delta, Colorado. Tom lost his brother Bill and nephew Fred in the robbery. Tom survived the shoot out and escaped to Sew-Em-Up Mesa and then to Oregon. The Mesa was so named because it was where local cattle rustlers surgically removed brands from stolen cattle and then sewed their hide back together [50, 51]. During 1895–1898, copper was found in the valley and miners moved in. The miners were single, rough, and tumble and given to drinking, gambling, and violence. Bedrock became the headquarters for the miners. In 1897, killer Bart Young escaped to Idaho where he "struck it rich" with a gold mine discovery. He then self-destructed into a dissolute lifestyle in San Francisco. He killed his wife, and then he was murdered by her lover.

Bill Young suspected Charlie Wilkinson and John Neftus of trying to kill him in 1905, so he murdered both of them. Young was released from prison in 1920. Unk Estes, in Disappointment Valley, was known to be mentally impaired but armed. No one was thought to be stupid enough to tease him [51]. The foreman at the Cashin mine was decapitated by a Bedrock gang in 1921. The owner of the Cashin mine lived in Chicago, and she sent three Pinkerton detectives to Paradox to find the gang that decapitated Slim Hecox. The gang was promptly rounded up, and Slim's head was retrieved. Slim was dug up to be reunited with his head. Unfortunately, the pine box had been made only big enough for his headless body. They had to place his head in the crook of his left arm so he would fit in the coffin for the reburial [51].

5.4 Carnotite Mineral Belt

The indigenous Ute Indians of the Paradox Valley had been noted to use a yellow pigment for body paint and to color buckskin. Around *1880,* a sample of the yellow ore from Roc Creek was sent to Leadville for analysis. Since it contained no significant gold or silver, there was no interest in it. However, in *1896* specimens were sent to the Smithsonian, where it was determined that the mineral contained uranium [29, 30, 53, 54]. Both the Colorado Plateau and Paradox Basin contained no significant gold or silver to attract prospectors. However, when the precious metals began to wane after 1890, miners began looking for other minerals of commercial value. The Paradox Basin was eventually found to have copper, uranium, and vanadium [30, 35, 37]. The Copper Prince mine had opened on Roc Creek in *1895* when *copper* was discovered on the north side of the West Valley. The *Cashin copper* mine opened in *1898* on La Sal Creek, on the south side of the West Valley [50, 51] (Fig. 5.8).

Charles Poulot arrived in Colorado from France in *1896*. He was a recent graduate of the University of Paris (Sorbonne) School of Mines. There he had studied under Professor M. M. Charles Friedel (1832–1899). Poulot had been hired by a French syndicate to identify and purchase uranium deposits in Colorado. In December of 1897, he was purchasing pitchblende ore in Gilpin County. Poulot and a partner, Charles Voilleque, founded the Rare Metal Mining & Manufacturing Company (RMMC) in 1897/1898 in an attempt to extract *uranium* from Front Range pitchblende, but the enterprise failed [29, 30, 53–55].

Then, in *1898,* he was hired by a Michigan copper company to erect and supervise the smelting plant at the *Cashin* copper mine on La Sal Creek; on the south side of the Paradox Valley. The local copper miners questioned the expert Poulot about the identity of the mysterious, bright yellow, powdery deposits found on both *La Sal and Roc Creeks*. Poulot thought the mineral was possibly (a) autunite; Ca $(UO_2)_2(PO_4)_2 \times 10\text{-}12\ H_2O$; or (b) uranochre; a yellow impure uranium oxide [29]. However, he sent samples from Roc Creek to his mentor, Professor *Friedel,* at the Sorbonne. Friedel (1832–1899) was both an organic chemist and a mineralogist. At the Sorbonne, he had become professor of mineralogy in 1876 and then professor of organic chemistry in 1884 [56].

Friedel and his associate, E. Comenge, identified the mineral as a new compound of *uranium;* $K_2\ (UO_2)_2(VO_4) \times 2\text{-}3H_2O$ [57]. They named it "carnotite" in honor of Marie Adolphe Carnot (1839–1920). He was a mining engineer and chemist, who was the French inspector of mines and the brother of the president of the French Republic (1887–1894), Marie Francois Sadi Carnot (1837–1894). This M. F. S. Carnot was the fourth president of the third republic. Their uncle had been N. L. S. Carnot (1796–1832), a physicist and pioneer of thermodynamics. Thus was this American radium ore christened with an illustrious but foreign eponym [56, 57]. The identification of carnotite was published in Paris in February of 1899, just 2 months after the discovery of radium. Friedel and Cumenge reported that the Curies had confirmed the radioactivity of the new mineral and its various components—uranium, radium, and vanadium [57]. Friedel died 2 months later.

The eastern part of the Paradox Basin that lies within Colorado would become known as the Carnotite Mineral Belt—the Colorado counties of Mesa, Montrose, and San Miguel. The belt was centered over the Rio Dolores and its tributaries and contained the mining districts of Hydraulic, Uravan, Paradox Valley, and McIntyre. The northern border was the Rio San Miguel and the southern border was Disappointment Creek in San Miguel County. The Carnotite Belt would supply half the world's radium from 1910 to 1922, while producing uranium and vanadium as byproducts [30, 37]. After 1952, it was known as the Uravan Belt (Fig. 5.8) [30, 37]. After the discovery of carnotite in 1898/1899, prospectors swarmed into the area; exploring (a) Roc Creek and La Sal Creek in the Paradox Valley, (b) the old Hydraulic mining area just north of the Valley, and (c) the McIntyre District in San Miguel County, south of the Valley. Even though carnotite had first been found in Montrose County, the initial attempts to start a factory near the mines for mineral extraction took place along the Rio Dolores in *San Miguel County.*

In 1900, Poulot and Voilleque and their company, the RMMC, opened a pilot plant for the extraction and concentration of uranium and vanadium from *carnotite* ore in the McIntyre District. Their plant was built south of the Paradox Valley in San Miguel County near Stevens Camp, which is now the hamlet of Slick Rock. They obtained enough carnotite ore to get a concentrate with 50 % uranium oxide and 16 % vanadium oxide for sale to Europe. Encouraged, they built a new three-story plant. Due to diminishing grades of available ore, the concentration process proved unprofitable. The plant closed in 1901 and Poulot returned to France. This was the first American mill to attempt the extraction of uranium and vanadium from carnotite [29, 30, 32–36]. There had been no attempt to extract radium (Table 5.3).

Chemistry professor Herman Fleck and Charles Carpenter took over the former Poulot-Voilleque factory at Steven's Camp during 1902–1903 and tried an extraction process developed by Fleck and metallurgist W. G. Haldane. When that failed, Carpenter tried again with chemistry professor Justin Haynes using the Haynes–Engle process and operating as the Western Refining Company. After that failed, Haynes and Engle tried again in 1905 when they build the Dolores Refining Company on the Rio Dolores about a mile upstream from the former plants. After 4 years they gave

Table 5.3 Carnotite factories in San Miguel County: 1900–1913 [29, 30, 32–36]

Year	Companies	Principles	Product
1900–1901	Rare Metal Mining & Manufacturing Company	Poulot and Voilleque	UR, VN
1902–1903	RMMC factory plant	Fleck and Carpenter	UR, VN
1903–1904	Western Refining Company	Carpenter and Haynes	UR, VN
1905–1909	Dolores Refining Company	Haynes and Engle	UR, VN
1912–1913	American Rare Metals Company	Fleck and Haldane	UR, VN, RA

UR uranium, *VN* vanadium, *RA* radium

it up. None of these carnotite plants in the McIntyre District attempted to recover radium; it was left in the tailings after the removal of the uranium and vanadium.

The first attempt in the USA to extract radium, as a byproduct, was made by Steven T. Lockwood (1875–1971), an attorney from Buffalo, New York. Near Richardson, Utah, carnotite had been found in 1898 by miners Welsh and Loftus. In 1900, Lockwood contacted the discoverers and convinced them to join him in establishing the Welsh-Loftus Uranium and Rare Metals Co. Lockwood served as president and general manager. Mining work began in Utah, while Lockwood planned an extraction plant just south of Buffalo, in Lackawanna, a steel town. The Buffalo–Niagara–Lackawanna area was becoming a steel production center and it is assumed that Lockwood's major interest was in using vanadium as a steel alloy [29, 30, 32, 36].

During 1902–1904, Lockwood worked on an extraction process. He turned to chemistry and physics professors at Columbia and at his alma mater, Princeton, for help. He also asked Pierre Curie for advice on radium extraction. The Welsh-Loftus Co. became the Rare Metals Reduction Company in 1903, when a reduction plant was erected to extract (a) radium, (b) uranium and (c) vanadium from Utah carnotite. The first ore arrived from Utah in April 1904. The Rare Metals Reduction Company was able to produce U_3O_8 and iron vanadate salts. The vanadium was converted to ferro-vanadium alloy in the electric furnaces of the new Niagara Research Laboratories during 1904–1907. The Niagara Research Laboratories were also involved with the American Vanadium Company of Pittsburgh in producing ferro-vanadium. The Laboratories prospered and in 1906, merged with Union Carbide. However, the vanadium sales of the Rare Metals Reduction Company were not enough to overcome the lack of profitability of the other two minerals. There was little market for the uranium, and their process failed to extract enough radium to cover expenses. The radium was of such low purity that it was not worth the cost of fractional crystallization. This radium could not be sold but was donated to research institutions.

By early 1908, the Rare Metals Reduction Company could not obtain enough carnotite to keep going as prospectors sought more profitable ores. The company failure was thus credited to (a) the low demand for uranium, (b) poor methods of radium recovery, and consequently (c) an inadequate ore supply due to prospector disinterest [29, 30, 32]. Lockwood returned to his law practice and in 1915 was appointed the US attorney for the Western District of New York (1915–1922) by his old Princeton professor, Woodrow Wilson [32].

The second attempt in the USA to extract radium, as a byproduct, was made in San Miguel County, Colorado, during *1912–1913*. Professors Fleck and Haldane tried again with carnotite. They purchased the carnotite claims of the old Western Refining Company on the Rio Dolores and reorganized as the American Rare Metals Company. They sought to separately extract the three carnotite minerals, and they were able to produce 100 mg of radium before closing in the spring of 1913 [29]. Various carnotite claims in the McIntyre District were later acquired in 1930 by the Shattuck Chemical Company in Denver which operated them from *1930* to *1942* and produced small amounts of radium [29]. Today, the Rio Dolores and its picturesque canyons around Slick Rock have become a tourist recreation area for bicycling and river rafting.

The US Mineral Industry began to issue annual reports in 1892. They reported that uranium and vanadium ores were being shipped to France in 1901 and 1902 [58]. In 1907, they reported that "so far but few uses have been found for uranium" [59]. They noted in 1910 that radium has no noticeable luminescence nor produces any chemical changes and that "Uranium-radium minerals are likewise unable to produce any known physiological effects" [60]. Also, the Colorado Bureau of Mines and the Colorado School of Mines did not think the low-grade Colorado and Utah carnotite uranium deposits were economically viable for uranium, vanadium, or radium. Amidst this wealth of information, the US Bureau of Mines was organized in late 1910. Chemist Charles L. Parsons (1867–1954) was hired as the chief of mineral technology at the Bureau and Richard B. Moore (1871–1931) was retained as the chief of the Bureau's Denver laboratory. In 1912, Moore, a physical chemist, and K. L. Kithil, a mineral technologist, made the first accurate survey of the carnotite ore in Colorado and Utah. Their findings were published in 1913 in a lengthy comprehensive government document known since as Bulletin 70 [36, 37].

The Bureau found that the quantity of radium obtained from carnotite was nearly four times that from pitchblende. The pitchblende had about 50 % UO_2 and the carnotite only 2–3 %. However, the pitchblende was widely scattered in hard granite rock. The carnotite was in soft sandstone and near the ground surface. It was, therefore, easier to mine and there was much more of it—100 sq. miles more. They stated that the "richest known radium-bearing ore in the world is that of the Paradox Valley" but that the mining area extended from the Hydraulic Area north of the valley to the McIntyre District on the south [37] (Fig. 5.8).

Parsons stated that "Seemingly the country has been quite unaware of the extent to which uranium ores have been sent abroad." He noted that during the year 1912 carnotite ores carrying 28.8 t of uranium oxide were obtained from carnotite ores and practically the entire amount was exported. The exported American ores were purchased for their radium content. Parsons noted: "This means that 8.8 g of radium chloride, or an equivalent 11.43 g of radium bromide, will be obtained from the ore shipped from this country in 1912." "It can probably be definitely stated that in 1912 there was obtained from American ores nearly two and one-half times as much radium as from all other sources combined." "Uranium was being shipped abroad in almost equal quantities in 1911" [36, 37, 61–63].

The Bureau noted with particular interest the General Vanadium Company of Baltimore, which was really a subsidiary of the International Vanadium Company of Liverpool, England. This subsidiary had purchased about 60 carnotite claims in the Paradox Valley since 1909, to allegedly mine vanadium. Starting in 1911, they began to ship the radium-bearing ore to the *Radium Extraction Company,* another one of their subsidiaries in Liverpool [37]. In addition to the General Vanadium Company, a number of independent operators were mining and shipping Paradox Valley carnotite to Europe. One of the main suppliers was Thomas F. V. Curran, of Placerville, Colorado, and New York City. The Curran brothers and their associates had 81 carnotite claims in the Paradox Valley. O. Barlow Willmarth of Telluride, Colorado, had started shipping carnotite to Europe in 1911, and he stated before Congress in 1914 that every pound of his output for the last 3 years had been sold to Paris, London, or Berlin. W. L. Cummings of Placerville, Colorado, sold ore containing 7.2 g of radium salt abroad while selling 15.2 g to American companies in 1914 [29, 32].

By 1913, it was estimated that the European refineries had produced a total of almost *30 g of radium* during the 15 years from 1899 to 1913. It was estimated that the USA had furnished at least 10 g of radium ore (a third), if not more, to Europe from 1899 through 1913 [36, 37, 61–63]. Meanwhile, between 1909 and 1913, the price of radium had doubled; from the range of US$ 75,000–90,000 to 160,000–180,000 per gram [33, 64, 65]. See Appendix VII.

References

1. Nugent W (2008) Habits of Empire. Knopf, New York
2. Bailey TA (1950) A diplomatic history of the American people, 4th edn. Appleton-Century-Crofts, New York
3. McGuire E, Portwood TB (1961) Our free nation, 2nd edn. Macmillan, New York
4. Nevins A, Commager HS (1967) A pocket history of the United States. Washington Square, New York
5. Todd LP, Curti M (1972) Rise of the American nation, 3rd edn. Harcourt Brace Jovanovich, New York
6. Hofstadter R, Miller W, Aaron D, Jordon WD (1976) The United States: conquering a continent, part one, 4th edn. Prentice Hall, New Jersey
7. National Park Service (1982) Exploring the American west; 1803–1879, handbook #116. US Dept. of Interior, Washington DC
8. Davis WC (2009) The American frontier: 1800–1899. University of Oklahoma Press, Norman
9. Hayes D (2009) Historical atlas of the American west. University of California Press, Berkeley
10. Reisner M (1993) Cadillac desert. Penguin, New York
11. Rosenberg M (2014) 100th Meridian. Geography of the Earth. http://www.geography.about.com/od/learnabouttheearth/a/100thmeridian.htm. Accessed 29 June 2014
12. Stein M (2008) How the states got their shapes. Smithsonian & Harper Collins, New York
13. Stein M (2011) The people behind the border-lines. Smithsonian Books, Washington DC
14. Poll RD (2014) State of deseret. Utah History Encyclopedia. http://www.historytogo.utah.gov/utah_chapters/pioneers_and_cowboys/deseret.html. Accessed 22 Aug 2014
15. Vance JE (1990) Capturing the horizon. Johns Hopkins, Baltimore
16. Bain DH (1999) Empire express. Viking Penguin, New York
17. Ambrose SE (2000) Nothing like it in the world. Simon & Schuster, New York
18. Brown RL (1965) An empire of silver. Caxton, Caldwell ID
19. Smith DA (2009) The trail of gold and silver: mining in Colorado. University of Colorado Press, Boulder
20. Athearn R (1962) The Denver and Rio Grande western railroad. University of Nebraska Press, Lincoln
21. Goetzmann WH (1959) Army exploration in the American west: 1803–63. Yale University Press, New Haven
22. Bartlett RA (1962) Great surveys of the American west. University of Oklahoma Press, Norman
23. Goetzmann WH (1966) Exploration & empire. Knopf, New York
24. Beebe L, Clegg C (1958) Narrow gauge in the rockies. Howell-North, Berkeley
25. Chappell G, Hauck CW (1977) Scenic line of the world and black canyon revisited. Colorado Railroad Historical Foundation, Golden

26. Clapesattle H (1984) Dr. Webb of Colorado springs. Colorado Associated University Press, Boulder
27. Ellis AM (1954) The Colorado springs story. Dentan, Colorado Springs
28. Sprague M (1987) Newport in the Rockies, 4th edn. Ohio University Press, Athens
29. Landa ER (1987) Buried treasure to buried waste-the rise and fall of the radium industry. Colorado school of mines quarterly, vol 82. Colorado School of Mines Press, Colorado, pp 1–77
30. Bruyn K (1955) Uranium country. University of Colorado Press, Boulder
31. Craig GA (1961) Europe since 1815. Holt, Rinehart & Winston, New York
32. Badash L (1979) Radioactivity in America. Johns Hopkins, Baltimore
33. Hahne FJ (1989) Early uranium mining in the united states. World Nuclear Association. http://www.db.world-nuclear.org/reference/usumin.html. Accessed 28 June 2014
34. Bueler GR (1981) Colorado's colorful characters. Pruett, Boulder
35. Hart SS (1999) Atomic age narrow gauge. RGS Technical Page, vol 2, No. 4-December, 1999. http://www.riograndesouthern.com/RGSTechPages/_bdwhite/atomic.html. Accessed 29 June 2014
36. Parsons CL (1913) Our radium resources. J Ind Eng Chem 5:943–946
37. Moore RB, Kithil KL (1913) A preliminary report on uranium, radium, and vanadium, Bulletin 70. Dept. of Interior, Washington DC
38. Blevins T, Daily D et al (eds) (2009) Legends, labors & loves of W.J. Palmer. Pikes Peak Library District History, Colorado Springs
39. Taylor RG (1966) Cripple creek; geographic monograph series, vol 1. Indiana University Press, Bloomington IN
40. Feitz L (1969) Cripple creek. Golden Bell, Denver
41. Leach JG, Penrose GH (1903) History of the Penrose family (1903). Biddle, Philadelphia
42. Eckel EB (1982) The geological society of America; history, #155. GSA Boulder, Colorado. http://www.books.google.com/books?id=y5CESCKPlpoC&printsec=frontcover#v=onepage&q&f=false. Accessed 22 Aug 2014
43. Stone WF (1918) Spencer penrose. History of Colorado vol III. http://www.books.google.com/books?id=TS6U22oRZXYC&dq=Spencer+Penrose+and+Colorado+banking&source=gbs_navlinks_s. Accessed 22 Aug 2014
44. Hiegert C (1993) Spencer and Julie Penrose. http://www.w3.d11.org/Bristol_Wall/1900/front_1900_penrose.html. Accessed 8 Aug 2011
45. Aber JS (2013) History of geology; Penrose & GSA. Emporia State University. http://www.academic.emporia.edu/aberjame/histgeol/penrose/penrose.html. Accessed 29 June 2014
46. Wilson JF, Chahbazian CM (1988) Penrose cancer hospital 1949–74. IJROBP 15:1475–83
47. Foutz DR (1994) Geology of Colorado illustrated. DR Foutz, Grand Junction
48. Wyckoff J (1976) The story of geology. Golden, New York
49. American Automobile Associavation (2013) AAA tour book and map of NV. AAA, Heathrow FLA
50. Deets LE (1934) Paradox valley-an historical interpretation of its structure and changes. Colo Mag 11(5):194–196
51. Greager HE (2000) The hell that was paradox, 3rd edn. B&B, Gunnison
52. Crum JM (1961) The rio grande southern railroad. Hamilton, Durango
53. McKee TM (1955) Early discovery of uranium ore in Colorado. Colo Mag 32(3):194–196
54. Curran TFV (1911) Carnotite in paradox valley. Colo Eng Min J 92:1287–88
55. Curran TFV (1913) Carnotite-I. Eng Min J 96:1165–67
56. Badash L (1966) (1966) Carnotite, what's in a name. Chem Br 2:240–241
57. Friedel C, Cumenge E (1899) Sur un noveau minerai d'urane, la carnotite. Compt Rend 128:532–34
58. Struthers J (ed) (1902–03) The mineral industry during 1901–02, vol 10, 11. Sci Publ, New York
59. Ingalls WR (ed) (1906–10) The mineral industry during 1905–09, vol 14–18. McGraw-Hill, New York

60. Curran TFV (1911) Radium uranium & vanadium. In: Fay AH (ed) The mineral industry during 1910, vol 19. McGraw-Hill, New York, pp 597–599, 68–72
61. Of C (ed) (1912) Radium, uranium & vanadium. In: Of C (ed) The mineral industry during 1911, vol 20. McGraw-Hill, New York, pp 655–658, 734–741
62. Of C (ed) (1913) Radium, uranium and vanadium. In: Of C (ed) The mineral industry during 1912, vol 21. McGraw-Hill, New York, pp. 752–754, 852–860
63. Fischer S (1914) Radium, uranium & vanadium. In: Roush GA (ed) The mineral industry during 1913, vol 22. McGraw-Hill, New York, pp 657–660, 773–777
64. Landa ER (1982) Radium—the first nuclear industry. Sci Am 247(5):180–193
65. Lubinau JO, Mould RF (2008) The roller coaster price of radium. Nowotw J Oncol 58(4):9–14e

Chapter 6
US Radium and Uranium Sales

6.1 American Vanadium Company

James J. and Joseph M. Flannery were colorful and wildly successful Pittsburgh businessmen. James J. (1855–1920) opened a series of funeral parlors in 1874. Joe (1867–1920) joined him after graduation from the Catholic College of the Holy Ghost. The Flannerys were lifelong supporters of the new Catholic University, founded in 1878. In 1911, it was renamed for the Marquis du Quesne (1700–1778), a former governor of New France (1534–1763) when Pittsburgh was Fort Duquesne. In 1904, the Flannerys obtained the patents on a stay bolt specifically designed for railroad locomotive fireboxes: the Tate flexible bolt. They formed the Flannery Bolt Company in Bridgeville, PA, USA, 13 miles southwest of Pittsburgh, for its manufacture and sale to the railroads [1–5]. The younger brother John S. Flannery complained that his brothers became prominent citizens of Pittsburgh while he was stuck with the bodies (mortuary) [6].

The Flannerys entered into a study of steel alloys and learned of the ferrovanadium alloy being used in Europe. During 1904–1905, the Flannerys began to obtain claims and options on the vanadium ores, carnotite and roscoelite, in the western Colorado counties of Mesa, Montrose, and San Miguel [7]. However, it was thought that the American vanadium deposits were inadequate and not commercially viable, as had been suggested to Lockwood by the Niagara Research Lab in 1904–1907 [8, 9]. Vanadium had been discovered twice: in 1801 in Mexico and in Sweden in 1831. A second vanadium mineral, roscoelite, was discovered by the English chemist Henry Roscoe (1833–1915), and it was found near Placerville, Colorado, in 1905. Carnotite was the only vanadium mineral to also contain uranium (Table 6.1).

In the fall of 1905, the Flannerys were approached by two Spaniards wishing to sell them an option on a vanadium mine in the Andes Mountains of Peru. Mining engineer and geologist J. Foster Hewitt and an A. J. Thompson were dispatched the 3600 miles to Peru to check it out. It was found to be worthless, but while preparing to depart from the port of Callao they heard rumors of a new rich vanadium discovery. This mine, Minas Ragra, was owned by Don Eulogio E. Fernandini of Lima and was 220 miles from Lima, in the central Andes highlands at 16,000 ft. altitude. It was however, within 30 miles of the railroad at Cerro de Pasco, a mining

R. F. Robison, *Mining and Selling Radium and Uranium*,
DOI 10.1007/978-3-319-11830-7_6

Table 6.1 Vanadium minerals 1905 [10]

Mineral—discovered	Chemistry	Location	% vanadium
Vanadinite—1801, 1831	$Pb_5Cl(VO_4)_3$	AZ, NM, Spain	1–3
Roscoelite—1876	$K(V^{3+}, Al, Mg)_2AlSi_3O_{10}(OH)_2$	CO	1–3
Carnotite—1897	$K_2O(2U_3O_3)(V_3O_5)$	CO, UT	1–3
Patronite—1905	VS_4	Peru	20

center for various minerals ever since the Spanish found silver there in 1630. At an altitude of 14,000 ft. in the Andes, Cerro de Pasco is one of the highest cities in the world. Cerro is Spanish for hill, spine, or backbone.

Rather than return to Pittsburgh empty-handed, Hewitt and Thompson trekked up to the site. It took them a week, which was just as well because of the altitude acclimatization needed [9, 10]. The Ferrocarril Centro Andino rail line went 138 miles from sea level at Callao over the Andes summit at 15,673 ft. elevation to La Oroya at 12,000 ft. altitude. It had been completed in 1893 and was 138 miles long. Connecting at La Oroya was a railway operated by the Cerro de Pasco Mining Company to serve ore mining in their district, 82 miles away. These lines are still operating over this 220-mile route from Callao to La Oroya and on to Cerro de Pasco at 14,000 ft. altitude [11]. Hewitt and Thompson arrived at the Minas Ragra, which was about 30 miles from a railroad shipping point at Ricran on the Cerro de Pasco Railroad. The new vanadium mineral there was vanadium sulfate (VS_4), which had just been discovered by Antenor Riza Patron of Lima and named for him as "patronite." Patron was a mining engineer in the employ of Fernandini, who owned several mining properties in the Pasco Region (State). Fernandini had a coal smelter in this area but the surrounding coal deposits were heavily contaminated with sulfur, which prevented their commercial development. Earlier in 1905, Patron had the sulfurous material analyzed and found vanadium sulfate. It was 20% vanadium, which made it the richest vanadium ore known ([10], Table 6.1).

Hewitt tried to arrange an immediate option on the property. Fernandini refused, saying that he was negotiating with an Armstrong company, in England, that made guns. However, he was persuaded to delay the sale to the English until the Flannerys could get down there to see him. Hewitt caught the next boat and gave the extremely enthusiastic report to the Flannerys. Joe caught the next boat back and met with Fernandini. He argued that the British were just experimenting with vanadium but that his company was actually working with the alloy and making steel products. Joe had only US$ 20,000 in cash when he left Pittsburgh, but somewhere along the way, he had it converted into gold coins. Legend has it that he carried the gold in an old carpetbag, and for dramatic effect, he dumped the lot, perhaps 1000 American gold double eagles (US$ 20), onto Don Fernandini's office desk. That did it. In addition to the cash, Don Fernandini was to get 25% of the stock in the new company and was to be paid a nice salary to run the mine. The deal was signed in November of 1905 [2, 9].

The Flannerys knew the ore was rich, 20% vanadium, but they did not know the extent of the deposits. Immediately, geologists were sent to probe the site to determine its size. On Christmas Eve of 1905, they got the news that the ore was wide and deep. On this basis, the Flannerys were able to get friends in Pittsburgh to invest in their new American Vanadium Company, which was organized in early 1906. The stock sold originally for US$ 50–55 a share, paid healthy dividends for years. It sold for US$ 1200 a share 10 years later (1916). The American Vanadium Company, with Peruvian patronite and with carnotite and roscoelite from Colorado and Utah, would control 92 % of the world's vanadium for several years [2–5]. The Minas Ragra, at 16,000 ft. elevation in the Andes, would provide 80 % of the world's vanadium. At the mine, the ore was loaded into sacks and placed on the backs of llamas. The llamas were herded 7.5 miles to the shore of Lake Pun Run at 14,000 ft. altitude. The sacks were then loaded on a barge for a transit of 8.0 miles across the lake, where they were reloaded on llamas for a 15.5-mile trek to the Cerro de Pasco railhead at Ricran. After the scenic 220-mile train trip to Callao, the sacks were loaded on steam ships for the USA [9, 10].

In 1906, the Flannerys hired the Niagara Research Lab to extract the vanadium from the Peruvian patronite. They recruited outside experts in metallurgy such as *J. Kent Smith* from Sheffield, England. By 1907, the American Vanadium Company had erected their extraction plant at Bridgeville, some 13 miles southwest from Pittsburgh. B. D. Sakiatwalla from Calcutta, India, was established at the Bridgeville plant. From June 1906 to January 1909, the Peruvian mines shipped 1800 t of patronite with 20 % vanadium oxide to the American Vanadium Company. Later, by roasting the sulfurous ore at the mine they burned off enough sulfur to increase the vanadium content to 50 % [9, 10].

Joe Flannery marketed ferrovanadium to the railroads, Henry Ford, and the Panama Canal. The Flannerys were already in the railroad business with the Flannery Bolt, made of ferrovanadium. Locomotives had continued to get bigger, and the side frames could not handle the stress, resulting in frequent breakdowns. The Flannerys introduced ferrovanadium side frames, which proved greatly superior. Shortly the American Vanadium Company was selling the alloy for practically all the most powerful locomotives under construction. Eventually, even the rails were made of ferrovanadium steel.

The Flannerys suspected that the auto industry could also create a constant demand for vanadium steel. In 1896, the US auto industry had emerged when the Duryea brothers in the Chicopee-Springfield, MA, area produced and sold 13 identical cars that year. In 1902, R. E. Olds (1864–1950) sold 2500 gas buggies, costing only US$ 650 each. Between 1902 and 1906, the gasoline engine won out over steam and electric for automobiles [12]. The Henry Ford Motor Company had started in 1903 with cars made by the Dodge brothers, John and Horace, who originally owned 10 % of the company. In 1905, legend has it that Henry Ford (1863–1947) examined a French racing car that had crashed during a race in Michigan. He noted the intact ferrovanadium parts amidst the wreckage. In 1906, Ford was known to be planning a new model, which would be known as the Model T (1908–1928), and which would make history as the first automobile that a working man could afford. Starting in 1906, the Flannerys were working with Ford on using ferrovanadium for the Model N and then the Model T. Figure 6.1 shows H. Ford on his visit to the United Steel Company in Canton, OH, where he met with the Flannery brothers and *J. K. Smith* in July of 1906 to celebrate the first batch of vanadium steel with cigars

Fig. 6.1 The Flannerys and H. Ford, 1906 (purchased from the collections of Henry Ford [13]). From the *left:* J. Kent Smith (*JKS*), 2nd; Joe M. Flannery (*JMF*), 4th; James J. Flannery (*JJF*), 5th; Henry Ford (*HF*), 7th

for everyone [13]. Ford was convinced that the alloy would allow him to safely reduce the weight of the Model T to one half the weight of the existing competition. He launched the Model T in 1908 with heavy advertising of its vanadium steel construction. The newly burgeoning auto industry embraced ferrovanadium as did the railroads.

The Panama Canal construction (1904–1914) had started under civilian engineer John F. Stevens. In 1907, US Army General George Washington Goethals took over as chief engineer and received a visit from Joe Flannery. After more than a year of study, Goethals became convinced of the superiority of ferrovanadium. In early 1910, he placed the order for ferrovanadium steel for the hinges and other components of the locks (locking gates) of which there are three on each coast. Five thousand tons of vanadium steel were used for the order [2–6].

Joe had concentrated on sales and marketing while James took charge of the American Vanadium Company and the Bolt Company. To handle their financial resources, they created the Oakland Savings and Trust Company, in suburban Pittsburgh, with James as president. This Oakland residential suburb is now home to the University of Pittsburgh Medical Center. In 1911, the Flannerys built the five-story Vanadium Building in Oakland at 3530 Forbes Ave, on the corner with Meyran (Fig. 6.2). They commissioned a 9-by-12-ft. stained glass window,

Fig. 6.2 The vanadium building, 2012. (Author)

depicting the Norse goddess Freyja Vanadis, to be mounted on a rear wall facing the main entrance [14, 15]. Translating "Vanadis" had been difficult for Norse scholars but Nils Gabriel Sefström, the Swedish chemist, named the element vanadium after Freyja in 1831 when he became the second chemist to discover it. The window has been preserved at the Heinz History Center, which has beautiful colored pictures of it "on line" [15] as does an article by J. Lubenau, the Standard Chemical historian [14].

6.2 Standard Chemical Company

It was during 1910 that the Flannery brothers lost their sister, Eleanor Flannery Murphy (1856–1910), to uterine cancer, which was a leading cause of death in the USA until the 1940s. When she was diagnosed in 1908, she was inoperable, as were 50 % of the cases then seen. Even the operable cases had only a 20 % survival. Radium was being tried experimentally for the inoperable cases that faced a certain and painful death. However, it would be 1915 before physicians in Baltimore and New York acquired the large amounts of radium necessary to begin successful therapy (see Chap. 7). In 1908/1909, the Flannerys became aware of the possible medical benefits of radium and sought to purchase it in Europe. However, the European radium factories could not meet the demand, even at the price of US$ 170–180,000 a gram (Appendix VII), and there were no American radium producers (see Chap. 5). During the 2 years of their sister's painful illness, 1908–1910, the Flannerys became interested in the radium business and its connection with vanadium and uranium in carnotite ore.

In 1909, Joe Flannery decided to go into the radium business and create an American supplier using American carnotite ore deposits. He faced two large problems: American physicians were divided about the effectiveness of radium therapy and the USA had a shortage of radiochemists who could manage the necessary large commercial production of radium [2]. The American Vanadium Company already owned some carnotite claims in Montrose, Mesa, and San Miguel Counties of southwest Colorado. These claims had never been extensively developed by the American Vanadium Company because of their high-grade Peruvian vanadium. As they had done with vanadium, the Flannerys would first acquire the necessary mining claims, and then hire the chemical experts to extract the radium. The Flannerys identified the Paradox Valley area and the nearby confluence of the Rio Dolores and Rio San Miguel in Montrose County, as the best source of carnotite–uranium ore. The ore would need to be transported back east for extraction as that was easier than getting the chemicals and the scientists to the remote Paradox Basin. As early as 1909, they purchased a 19-acre industrial site some 20 miles southwest of Pittsburgh and about 7 miles from their Bridgeville vanadium plant. This was on the west end of Canonsburg, Pennsylvania, and it had good railroad, light rail (interurban), water, and highway connections.

The Paradox deposits were rather well defined, occupying an area 20 miles wide by 40 miles long [7]. The La Sal Mountains of Utah formed the western border and the eastern limit was about 6 miles west of the town of Naturita (Fig. 6.3). Access to the Paradox Valley was best accomplished by wagon road from Naturita at the east end. The nearest railroad, the Rio Grande Southern was 40 miles east of Naturita. Four miles to the west of Naturita were old coal plant ovens and a water supply furnished by natural springs at Coke Ovens. In 1909, the American Vanadium Company acquired the Coke Ovens, which served as the initial headquarters for their planned new radium mining company. They built a small hydroelectric power station, concentrating mill, and ore house there during 1910. Between 1909 and 1913, they purchased 90–100 carnotite claims. In May of 1911, the Flannerys were encouraged enough to incorporate a new company, *Standard Chemical,* to manage their radium business. Joe assumed the presidency.

Between 1913 and 1930, several other American radium companies became operational (Table 6.2). They all began to acquire mining claims in the Paradox Valley or nearby. The field was *dominated* by the three largest companies: Standard, the Radium Company of Colorado, and the US Radium Corporation. The General Vanadium Company of Liverpool (60 claims) and local independent contractors like the Curran brothers (80 claims), O. B. Willmarth, W. L. Cummings, and David Taylor also began acquiring Paradox Valley claims. The Flannerys eventually

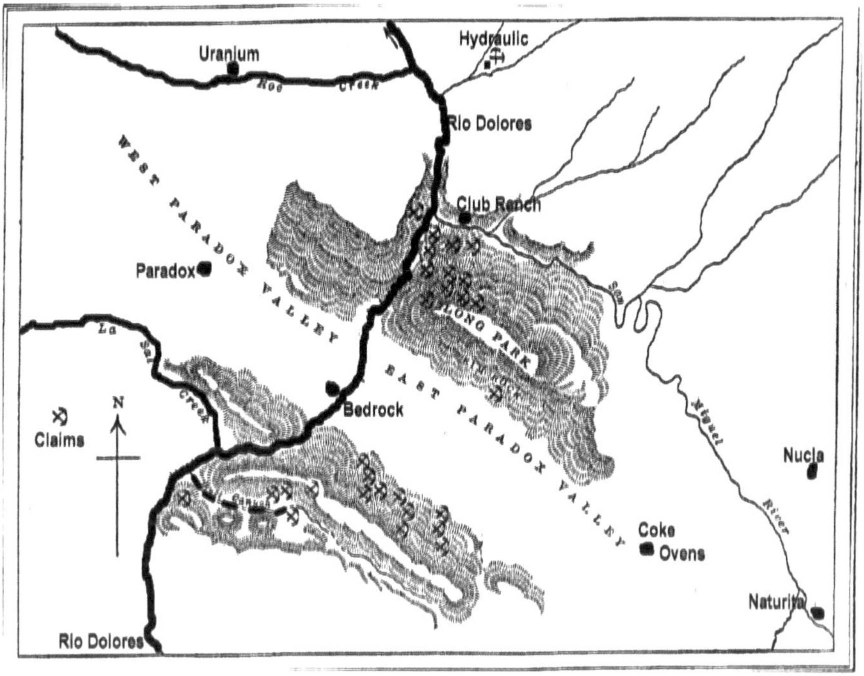

Fig. 6.3 Paradox Valley, 1913, [7]. (Public domain)

acquired 360 claims in 23 groups comprising four major camps: the *Club Ranch* and the *Dolores Camp* in the northeast valley, and the *Joe Dandy* and the *Monogram* camps in the south valley [1, 7, 16–22]. The Paradox Valley mining claims of 1916 are listed in Table 6.3 and their locations are illustrated in Fig. 6.4 [16].

Standard sent their ore by mule to Coke Ovens. There, the ore was mechanically concentrated from 1–2 U_3O_8 to 3–4 %. Low-grade ore < 1/% went to a dump and high-grade ore > 3–4 % went directly to Canonsburg. The concentrated ore was sent by wagon to Naturita and then another 40 miles to the narrow gauge Rio Grande Southern Railroad at Placerville. That railroad connected mountainous Placerville to Salida by narrow gauge rail. At Salida, the Rio Grande Southern narrow gauge line connected to the Denver and Rio Grande Western, a standard gauge line, based in Denver. From the Denver railroad hub, the ore was shipped back east to Canonsburg [23, 24].

Radium experts from Europe and America were recruited by the Flannerys to develop the Canonsburg extraction factory, which had started receiving ore in January 1911. W. A. Schlesinger came from Heidelberg in 1911 and Otto Brill, D. Sc. (Vienna), came from the world famous Austrian mines at Joachimsthal, Bohemia, in 1912 [6, 17–19]. American universities were just beginning to study radiochemistry after the discoveries in Europe during 1895–1900. The American graduates found immediate employment in the new American radium businesses.

Table 6.2 American radium companies [1–26]

	Principals	Operational
1. Rare Metals Reduction Co.: Lackawanna, NY	S Lockwood	1904–1908
2. American Rare Metals Company: Colorado	H Fleck, WG Haldane	1912–1913
3a. Standard Chemical Co.: Canonsburg/Pittsburgh	J Flannery	1911–1933
3b. Radium Chemical Co.: Manhattan, NYC	C Viol, WH Cameron	1913–1987
3c. Radium Luminous Corporation: Manhattan, NYC		1917
3d. Radium Dial: Brooklyn, NYC		1918–1922
3e. Radium Dial: Ottawa, IL	J Kelly Sr	1922–1934
3f. Luminous Processes: Ottawa, IL	J Kelly Jr	1934–1978
4. National Radium Institute: New York and Baltimore	H Kelly, J Douglas	1913–1915
5. Radium Co. of America: Sellersville, PA	S. Kohn	1913–1917
6a. Schlesinger Radium Co. also known as 6b	WA Schlesinger	1915–1917
6b. Radium Company Colorado: Denver	W and H Koenig	1917–1924
6c. Cold Light Manufacturing Co	C Whittlemore	1916
6d. Tungsten Products Merger	K Kithil	1920
7. Carnotite Reduction Co.: Chicago	H Mc Coy	1915–1920
8. Chemical Products Co.: Denver	H Baker	1916–1920
9a. Radium Luminous Material Corp.: Newark, NJ	S Sochocky	1917–1921
9b. US Radium Corp.: Orange, NJ	V Hess, H Baker	1921–1980
9c. Safety Light Corp.		1980–2007
10. W.L. Cummings Chemical Co.: Lansdowne, PA	Philadelphia	1920–1922
11. Keystone Metal Reduction Co.: Cheswick, PA	Pittsburg	1921–1922
12. Shattuck Chemical Co.: Denver	SW Shattuck	1930–1941
13. Vitro Manufacturing Co.: Canonsburg, PA		1930–1942

NYC New York City

Two of the more prominent Americans working in radioactivity were chemists Bertram B. Boltwood (1870–1927) at Yale and Herbert N. McCoy (1870–1945) at the University of Chicago until 1917 when he left to begin a full-time career in industry. During 1905–1908, Boltwood and McCoy, along with E. Rutherford, were responsible for establishing the genetic relationship of uranium and radium and working out the correct ratios. Both also worked as consultants in the radium business and the thorium-rare earth gas mantle industry. In 1910, Boltwood became the first professor of radiochemistry at Yale. McCoy held a patent on radium extraction and was working with the Carnotite Reduction Company at their extraction plant in Chicago from 1915 to 1920, the past 3 years as president. After Boltwood's suicide in 1927, McCoy semiretired to Southern California [19].

McCoy, who had a BS from Purdue in 1893 and Ph.D. from Chicago in 1898, had worked up to full professor at Chicago during 1901–1910. He trained several graduate students in radioactivity including Charles H. Viol (1886–1928). Viol had

Table 6.3 Paradox Valley mining claims [1–26]

	East valley	South slopes site	Owners and claims
1	Monogram Camp	12 miles west of Naturita	STD = 9; Misc. = 7
2	Joe Dandy	2 miles east of Monogram	STD; GVC = 4; Misc. = 3
3	Thunderbolt Camp	2 miles east of Jo Dandy	GVC = 2; Hequembourg = 1
4	Bull Canyon	5–8 miles south of Monogram	GVC = 2; W.L. Cummings = 2; O. B. Willmarth = 1 Misc. = 16
	East valley	*North slopes site*	
1	Hydraulic District	North of valley	GVC; RCC = 2
2	Dolores Camp	North on Rio San Miguel	STD group
3	Cliff Camp	East of Dolores Camp	RCC = 4
4	Club Ranch Camp	Confluence of Rio D. and S. M.	STD group and mill
5	Long Park Camp	21 miles west of Naturita	[a]Currans = 13; Primos = 7; RCC = 22 Crucible steel = 7; STD ≥4
6	Liverpool IVC	North of Long Park	GVC
7	Saucer Basin	3 miles south of Club Ranch	USR; Wilson = 4
	West valley	*North side*	
1	Roc Creek	Copper Prince (Rajah) mine	RCC = 13
2	Carpenter Flats		RCC = 10
	West valley	*South side*	
1	La Sal Creek	Yellow Bird Camp 6–7 miles above Cashin Copper mine	USR
2	McIntyre Canyon	8 miles south of valley	Carnotite factories (Chap. 5)

STD Standard, *GVC* General Vanadium Corporation, *RCC* Radium Co. Colorado, *USR* US Radium

[a] Several of these claims had originally belonged to Tom Curran and his brothers. The Curran family was also from Pittsburgh and had been neighbors of the Flannerys. Their best claim was called the Maggie C. in honor of their sister [10]

a B.S. from Purdue in 1907 and a Ph.D. in radiochemistry from the University of Chicago in 1912. When Brill was recalled to Austria in 1913, Viol replaced him as director of research at *Standard.* Glenn D. Kammer (1888–1927) finished his Ph.D. at the University of Pittsburgh in 1912 and became Viol's assistant director in 1913. They both died of radiation toxicity: radium martyrs. Arthur R. Miller (B. S. Purdue, 1914) became an assistant to Kammer and other staff members were hired from the Universities of Pittsburgh, Cornell, and Michigan [6, 19].

A pilot program was in place at Canonsburg during 1911–1913, during which time *Standard* produced 2.1 g of radium while modifying the Curie extraction process for industrial scale. They were also able to secure vanadium and uranium from the carnotite ore for use as ferroalloys. Uranium still had a small market but Standard organized the Standard Alloy Company in Pittsburgh to become the only American producer of ferro-uranium. It was used to make airplane propellers [5, 6]. After 1913, *Standard* could recover 1000 mg (1 g) of radium from *500 t* of concentrated carnotite. It would take four separate stages over 6 months' time to process the carnotite from 5000 lbs of rock into *1 g* (1000 mg) of marketable radium.

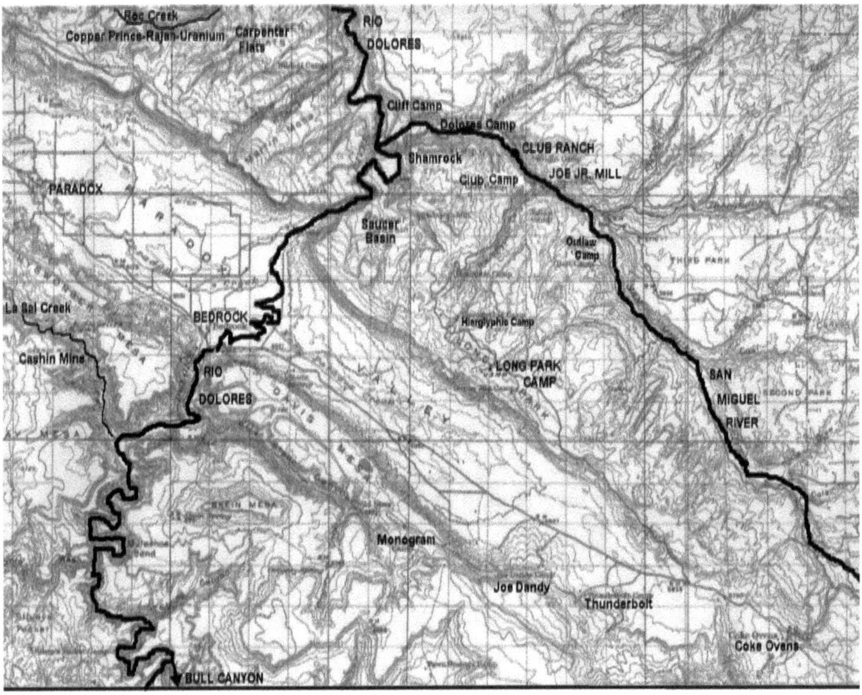

Fig. 6.4 Paradox Valley carnotite claims and camps, 1916. (Public domain)

a. Two hundred miners hand-sorted through 2500 t of ore to select *500 t* of at least 1 % U_3O_8. This was sent to the concentrating mill at Coke Ovens. After 1914, it went to the Joe Jr. Camp, at Club Ranch Camp on the San Miguel River about 4.5 miles east of the junction with the Rio Dolores. This site occupied five acres and contained the concentration mill, a laboratory, dormitory housing, a commissary, and a hydroelectric generator powered by the Rio San Miguel [1, 6].
b. At the mill, the 500 t of ore was concentrated to 3–4 %, which reduced the volume to 125 t. This was transported by mule and wagon to the railhead at Placerville. The narrow gauge Rio Grande Southern connected to the standard gauge Denver and Rio Grande Western at Salida. The later connected to the transcontinental Central Pacific rail stub at Denver. Various rail connections were then available to Chicago and the Pennsylvania Railroad route to Canonsburg.
c. At Canonsburg, 150 men in eight departments processed the 125 t of concentrate by using 500 t of chemicals, 1000 t of coal, and 10,000 t of distilled water. This yielded 220 lbs or 100,000 g of a radium–barium–calcium chloride (*Ra–Ba–CaCl*) solution. The initial fractional crystallization was started using porcelain bathtubs. The solutions were dried in steam pans and the powder placed in large glass bottles with corks. These bottles were placed inside galvanized steel cans, which had hinged locking tops and metal handles (bails). Two staff men, with a can in each hand, then boarded the interurban light rail train

Fig. 6.5 Flannery Bldg. fifth floor radium refinery, 1921, [25]. (Public domain)

which ran from Washington, Pennsylvania, through Canonsburg to downtown Pittsburgh. There they transferred to the regular Forbes Ave. streetcar, which ran east to the Vanadium Building, which was renamed the Flannery Building in 1920.

d. Standard Chemical used the fifth floor of the Vanadium Building for the final series of fractional crystallizations. Large open porcelain evaporating dishes were heated over gas burners in the unvented rooms (Fig. 6.5). Thus, the Oakland suburb, now home to the University of Pittsburgh Medical Center, was the site of a radium refinery from 1911 to 1922 [6, 25]. The fifth floor continued as a laboratory until at least 1937 and Radium Dial painting had an office there until 1938 [2]. The building went through multiple owners until the Parkvale Bank & Savings Association bought the building in the 1980s. They could not sell it because of widespread radium contamination, including the heating and plumbing systems, so it was vacated in 1999. The building was decontaminated between 2000 and 2003 [26]. The fifth floor and the roof were among the areas with the greatest radioactivity. It is now a general office building.

Standard Chemical, in May 1913, became the first American company to put radium on the market and they became the largest American producer. However, at first there were few American buyers [2, 27]. During 1913–1915, Standard produced a total of 13.4 g of radium and most of it went to Europe. American physicians and scientists had imported only about 500 mg before 1913 [2, 17]. Therefore, Standard formed a subsidiary, the Radium Chemical Company of New York City, in 1913 to handle marketing and sales. The Radium Chemical subsidiary encouraged medical sales by producing a monthly journal called *Radium,* from 1913 to 1925 [28, 29]. The uranium and vanadium by-products were also being sold. The editors were C. H. Viol, the radiochemist, and W. Herron Cameron (1879–1944), a physician. Viol and Cameron were also instrumental in the organization of the American Radium Society in 1916 [30]. Radium Chemical also manufactured radium-sealed sources and brachytherapy (intracavitary) devices for cancer therapy. These were made at their headquarters in midtown Manhattan.

World War I (1914–1919) essentially ended the European demand for medical radium but created a new market for radioluminescent paint to be used in military vehicles and on wristwatches.

Phosphorescence has been known since the seventeenth century and luminous paint had been used for decades. This usually involved crystalline zinc sulfide, known as Sidot's blend, which is phosphorescent but requires light for activation (see Chap. 2). Adding small amounts of radium supplies an almost infinite supply of alpha particles, which serve to activate the Sidot's Blend in the absence of light. This phenomenon had been observed by W. Crookes, J. Elster, and H. Geitel in 1903 and was used by E. Rutherford to count alpha particles (see Chap. 3). F. Giesel had marketed radioluminescent paint in 1906. It became a novelty in Europe but it was not exploited commercially until World War I. Then, the combatants began using radium paint on all military vehicle instruments and on the dials of the new military necessity, the wristwatch [31–34].

In the USA, three different radium companies began marketing radioluminous paint: US Radium in 1914, the Radium Company of Colorado in 1916, and Standard in 1917. Their corresponding paint products were known as Undark, Marvelite, and Luna respectively [1, 17, 29].

Beginning in 1916, radium paint began to account for more and more of the radium market. The Standard product, Luna, was produced by their subsidiary, the Radium Luminous Corporation in New York City. Another subsidiary was the Radium Dial Company, which applied the radium paint to timepieces and was located in Long Island City in Queens, New York. By 1917, the use of radium in paint had created a market as large as that for medical use. In the year 1918, it was reported that 95 % of the radium produced in the USA that year was used for military paint and only 5 % for medical therapy. The 1918 price of radium was fluctuating between US$ 100 and 125 per mg [31–33]. The paint industry was consuming about 500 mg a month or 6 g a year of radium. The USA produced about 60 g from 1914 through 1918 and about 30 g was used for paint [17]. By Armistice Day (11-11-18), every American soldier had a radium-dial wristwatch. After the war, consumer demand for these skyrocketed. After sales in 1913 of only 8500 radioluminescent watches and clocks, 2.2 million were sold in 1919 [34]. In 1920, four million luminous timepieces as well as other radium novelties including sex aids such as rectal suppositories (Vita), scrotal pads (Radi-endocrinator), and condoms (Nutrex) were sold [35].

The year 1920 was marred by the deaths of both James, age 65, and Joe Flannery, age 53 [2, 3]. It also brought (a) the beginning of a postwar business recession that decimated the automobile industry, (b) a return of the deadly Spanish Flu, and (c)

Table 6.4 US radium production in grams [17, 37]

Year	Standard production		All US annual production	Total US production
	Annual	Total		
1913	02.1	02.1	02.1	02.1
1914	09.6	11.7	10.1	12.3
1915	01.7	13.4	04.7	17
1916	05.0	18.4	08.2	25.2
1917	07.0	25.4	13.8	39
1918	13.6	39	22.8	61.8
1919	11.8	50.8	28.6	90.4
1920	18.5	69.3	32.5	122.9
1921	04.7	74	35.7	158.6
1922	0	74	24.2	182.8
1923	0	74	12.2 (USR)	195
1924	0	74	03.4 (USR)	198.4
1925	0	74	03.0 (USR)	201.4
1926	0	74	01.7 (USR)	203.1
1927–1930	0	74	0	
1931–1942	Shattuck and Vitro production		15.0	*218*

USR US radium

the deaths of the hard-drinking Dodge brothers; John at age 56 from the flu and his brother Horace of liver cirrhosis, at 52. Many of the smaller radium companies, such as Carnotite Reduction, Chemical Products, W. L. Cummings, and Keystone, all went out of business during the economic recession from January 1920 to July of 1921 [17, 18, 36]. However, in 1921 the three major US radium companies reached *their zenith,* producing 35.7 g of radium for sale. The American factories had produced an average of 18 g a year and by 1921 the USA was producing 80% of the world's radium [17, 36, 37]. From 1913 to 1922, the USA was the world leader in radium production and sales. America produced 182.7 g, of which 74 g came from Standard (Table 6.4).

6.3 The Denver and Canonsburg Legacies

Willy A. Schlesinger (Heidelberg) left Standard before 1914 and teamed up with Henry and Walter Koenig, from the University of Pittsburgh, to study radium chemistry at Princeton University. In 1915, they established the *Schlesinger Radium Company* plant in Denver. Walter Koenig was plant manager and Henry worked on radium paint. C. F. Whittemore, from the US Bureau of Mines, later joined the company as chief physicist. Mineralogist Karl L. Kithil, also from the Bureau of Mines, became the vice president. Financial backing was provided by Dr. V. C. Thorne of New York City. Schlesinger acquired 90 Paradox Valley claims including mines at Long Park, Hydraulic and Roc Creek, where the Copper Prince was renamed the Rajah (Fig. 6.4).

In 1916, they organized a subsidiary, the Cold Light Manufacturing Company, to produce their brand of radium paint: Marvelite. Their Cold Light paint subsidiary was located at *2001 East Colfax,* just north of Cheesman Park. Factory tailings were donated to the city to supplement the asphalt paving of the streets around the park. The Schlesinger Company had their downtown office at the Radium Building. Their radium factory was 0.5 miles south of downtown and occupied the area between 11th and 12th Avenue between Umatilla and Quivas. The property adjoined the Denver and Rio Grande Western Railroad tracks at *1045 Tejon* [17, 18, 22, 24, 38, 39]. Schlesinger was reorganized in 1917 as the *Radium Company of Colorado.* They published an instructional book for physicians, the *Radium Therapist.* Their branch offices were impeccably located: Market St. in San Francisco, Madison Ave. in New York City, South Michigan Ave., across the street from the Art Institute, in Chicago, and the Ave Des Champs Elysees in Paris [27, 29]. Because of financial problems, the *Radium Company of Colorado* later became a subsidiary of Metals Exploration Company. The two companies shared office space at the Radium Building, *1429 18th St.,* in downtown Denver. In 1921, the *Radium Company of Colorado* was consolidated with Tungsten Products, which had rich carnotite deposits near Moab, Utah, and in Mesa County, Colorado. The *Radium Company of Colorado* was second only to Standard in radium production.

The *National Radium Institute* was privately owned and had no commercial sales during its brief existence from 1913 to 1915 [17, 18, 24, 40]. Dr. Howard A. Kelly of Johns Hopkins and industrialist/mining engineer James Douglas, the president of Phelps Dodge Mining, formed a joint venture with the US Bureau of Mines in 1913 to mine and produce radium (see Chap. 7). The *National Radium Institute* leased claims in the Paradox Valley. Their main extraction factory was located 2.0 miles south of downtown at *500 S. Santa Fe Drive,* adjacent to the Denver and Rio Grande Western Railroad tracks as the ore was shipped into Denver by that rail line. To aid the *National Radium Institute,* the US Bureau of Mines operated a laboratory in downtown Denver at the Foster Building, 910 16th St., from early 1912 to 1916. The final fractionation separations and measurements of radium were done there [40].

The S. W. Shattuck Company was started in 1921 by an Illinois dentist, who came to Denver for his health and died. He started a small plant downtown for extracting molybdenum. His widow and the plant foreman expanded the business and moved it to South Denver. The extraction plant acquired some carnotite from broker David Taylor and began processing it for uranium and radium. In 1930, the Shattuck Chemical Company began working the tailings left at the old mines in the McIntyre District. The tailings were sent to their extraction facility at *1805 S. Bannock* St., 4.0 miles south of downtown [17, 18, 39].

In 1979, the Environmental Protection Agency (EPA) became aware of 35 former Colorado properties where radium had been processed. Thirty-one (31/35) of these were in metropolitan *Denver* and involved 65 sites where the soil was contaminated by radium, thorium, and uranium, leading to the production of radon and its polonium daughters. The areas of Denver that were identified in 1978/1979 as having radioactive contamination that required remedial action [38, 39] are identified in

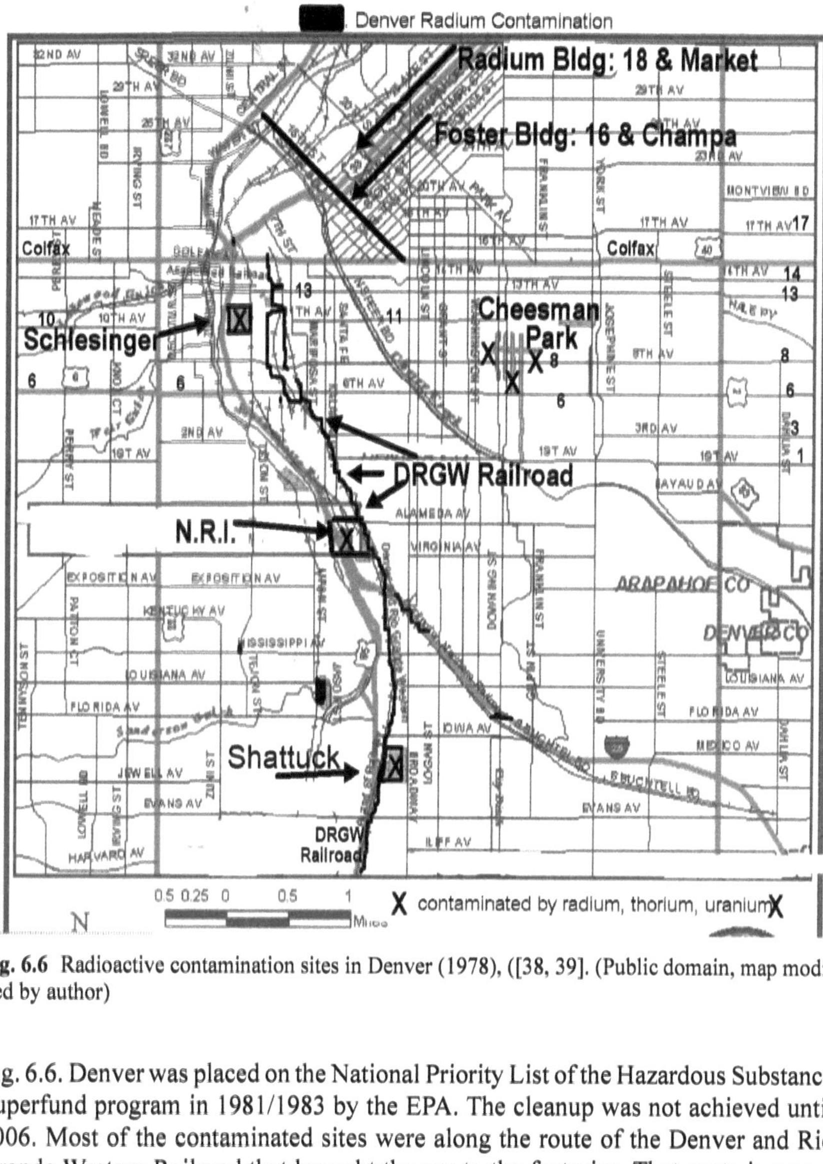

Fig. 6.6 Radioactive contamination sites in Denver (1978), ([38, 39]. (Public domain, map modified by author)

Fig. 6.6. Denver was placed on the National Priority List of the Hazardous Substance Superfund program in 1981/1983 by the EPA. The cleanup was not achieved until 2006. Most of the contaminated sites were along the route of the Denver and Rio Grande Western Railroad that brought the ore to the factories. That route is now a light rail commuter line for the city of Denver.

The old *National Radium Institute* plant was donated to the government by January of 1917 and went through several owners but never resumed radium production. The site was identified by the EPA to have the most radioactive contamination in Denver. After a 6-ft. thick cap of cement was placed on the permanently radioactive site in 1995, it is now occupied by a Home Depot store at 500 S. Santa Fe Drive, between the Interstate and the rail line [41]. The site is accessed, with difficulty, from exit 207-B of Interstate-25 or by exiting the light rail line at Alameda or Broadway.

High-grade, > 50 % U_3O_8, pechblende had been discovered in the Belgium Congo during World War I by the Union Miniere du Haut Katanga, which was a mining company jointly owned by the British and Belgians. Copper, rubber, and ivory were more important during the war and the pechblende was ignored until after the war (see Chap. 8). The first Congo pechblende arrived in Belgium in December of 1921 for radium extraction at Olen. The Union Miniere created a sales and marketing division, Radium Belge, in 1922 [42, 43]. They convinced the two largest American radium producers, Standard and the Radium Company of Colorado, that they could not compete against the vast and rich reserves of the Congo ore, which was mined by forced or slave labor. Radium Belge cut the price from US$ 120 to 70 per mg to prove it. The Belgians produced their first commercial radium in December 1922 and started sales in 1923. Standard and the Radium Company of Colorado were offered and accepted a 5-year contract, 1922/1923–1926/1927, to gradually exit the radium business. Standard had already suspended radium production in mid-1921 as Canonsburg had an oversupply or ore. Standard remained as a Radium Belge sales representative until 1927 [17, 42, 43].

By 1923, the American radium mining business was over, and many former radium companies, mill owners, and miners now turned their attention to extracting vanadium from Colorado and Utah roscoelite and carnotite. The Radium Company of Colorado closed their mines at Naturita in 1923 and went out of business in 1924. However, their sister company, Metals Exploration, went into the vanadium business and later into vanadium-uranium extraction [24]. The US Radium Corporation was primarily in the paint business and not in radium sales. They were not intimidated by the Belgian takeover and continued to produce their own radium from Long Park during 1917–1926. US Radium was the only American producer during 1923–1926 (Table 6.4).

Radium Belge refused to renew the American contracts in 1927 and Standard divested all its remaining mining operations between 1928 and 1933. The Canonsburg plant had been sold in 1923 and burned down in 1925. The site was purchased in 1929 by Vitro Manufacturing, later Vitro Corporation of America, which made uranium for glass and ceramic coloring. Reworking the tailings at Canonsburg, Vitro was able to continue producing some radium during 1930–1942. During World War II, the US government hired Vitro to process uranium that came from the Belgium Congo. After the war, Vitro was supplied with various tailings, sludge, sweepings, and waste products from multiple Atomic Energy Commission sites across the country. Vitro closed in 1957 but remained a storage facility for the Atomic Energy Commission [1, 44].

In 1975, the Nuclear Regulatory Commission (NRC) replaced the Atomic Energy Commission and was given jurisdiction for the tailings from uranium ore mills. Congress funded the cleanup for 22 former uranium ore millsites. The only one east of Denver was *Canonsburg*. In 1983, the NRC found that 34 acres in *Canonsburg* were contaminated to a depth of 16 ft. The buildings were razed and the residual radioactive waste was consolidated into a 6-acre "disposal cell." The "cell" is lined by a 6-ft. thick cap. At the bottom, the cell has 3 ft. of soil and clay as a radon barrier, then a riprap layer of stones 2 ft. thick, overlaid with a top layer of 1 ft. of soil

Fig. 6.7 Canonsburg property for rent, 2012. (Author)

planted with local grasses. The surrounding 18.6-acre site is now owned by the state of Pennsylvania. It is enclosed by fences, posted for radioactive danger, and periodically inspected by the appropriate state and federal agencies [1, 17, 44]. However, there is property for lease across the street (Fig. 6.7).

6.4 The Uravan Mineral Belt

Prior to World War I, there were at least six companies in America mining vanadium and five major producers of ferrovanadium, led by the Flannery American Vanadium Company [45, 46]:

- American Vanadium Company at Bridgeport, Pennsylvania
- US Vanadium Company of Telluride, Colorado
- Dolores Refining Company (1905–1909) of Cedar (Slick Rock), Colorado
- Vanadium Alloys of Newmire, Colorado
- General Vanadium Company of Liverpool, UK
- Primos Chemical Company of Primos (SW Philadelphia), Pennsylvania

Demand for vanadium had increased at the start of World War I and was initially met by roscoelite deposits near Rifle and Placerville, Colorado. During that war, two new conglomerates were formed: the Vanadium Corporation of America in 1916 [31] and the Union Carbide and Carbon Corporation, a chemical company, in 1917 [32, 34, 47]. These two companies would come to monopolize the American vanadium and uranium businesses and were eventually taken to court, in the 1960s, for collusion, price fixing, and other monopolistic practices [48]. The Vanadium Corporation of America absorbed both the American Vanadium Company and the

Primos Company by 1919 [34, 36]. By 1923, carnotite ore was no longer competitive for either radium or vanadium and carnotite mines began to close. However, after 1930, carnotite began to replace roscoelite, and by 1935 most of the carnotite mines were reopened. In 1934–1935, Vanadium Corporation of America took over the assets of the defunct Radium Company of Colorado and the former mines of the US Radium Corporation. They acquired the Radium Company of Colorado vanadium mill at Naturita and established the town of Vancorum to the west of it [18, 24].

Union Carbide and Carbon Corporation entered the mining business in 1926 when they acquired US Vanadium as a subsidiary. The latter purchased the former Standard Chemical Company refinery and their roscoelite claims at Rifle, Colorado, in 1928 and expanded the mill in 1928 and 1939 [24, 47–50]. Also in 1928, US Vanadium purchased the Joe Jr. Mill, which Standard Chemical had closed in 1921, along with 14 km of carnotite claims. In 1936, US Vanadium created the new town of *Uravan* at the site of the old Joe Jr. Mill and opened a refinery for carnotite. In addition to housing, the company town had stores, a post office, fire station, schools, a health clinic, swimming pool, sports field, and a recreation hall [47–50].

The old Carnotite Mineral Belt, along the Rio Dolores, came under study by the US Geology Service during 1939. Explorations continued during World War II for wartime strategic minerals and continued after 1947 at the request of the Atomic Energy Commission. The carnotite sandstone deposits were found to average 2 % V_2O_5 and 0.25 % U_3O_8. The largest known deposits of carnotite were found at *Uravan,* Colorado [51]. By 1952, the old Carnotite Belt was being identified as the *Uravan* Mineral Belt because those deposits were of higher grade, of larger size, and occurred closer together than in surrounding areas.

During 1942–1944, the US government increased their purchase price for vanadium. New refineries were built by the Vanadium Corporation of America and US Vanadium in Utah and Colorado. In 1944, the USA stopped buying vanadium and began buying the tailings which were then processed for uranium at Grand Junction, Colorado. At the end of World War II, Colorado Plateau mining included uranium as well as vanadium. In 1947, the Atomic Energy Commission contracted with the Vanadium Corporation of America to refine uranium at their Naturita and Durango, Colorado, mills. US Vanadium was contracted for uranium processing at their Rifle mill in 1946 and at Uravan in 1949/1950. The Grand Junction mill began to refine uranium ore instead of tailings in 1951. From 1948 to 1958, the USA sustained uranium mining by price supports and discovery bonuses. The zenith of uranium production was reached in 1960 when 2100 t were produced [18, 24, 51]. During 1947–1960, there would be ten uranium refineries in Colorado alone: *Naturita, Uravan,* Durango, Grand Junction, Rifle, Slick Rock, Rifle #2, Maybell, Gunnison, and Cannon City [24].

The Atomic Energy Commission uranium procurement program ended on 31 December of 1970 after 27,000 t of U_3O_8 had been produced. The Uravan Mineral Belt had produced 85 % of all the uranium and the Uravan area accounted for 80 % of all the vanadium [51]. The Uravan mill processed uranium until 1963 [52] and produced 11 % of all the Cold War uranium [51]. Uravan was evacuated in 1984 and razed while 680 acres of ground was decontaminated by the US Department

Fig. 6.8 US Dept. of energy disposal cell at Uravan, CO [53–55]. (Public domain)

of Energy (DOE) during 1986–2001 [52, 53]. Radioactive soil and building debris were deposited in three disposal cells on a 27-acre site created above Uravan on Club Mesa adjacent to Hieroglyphic Canyon [54] (Fig. 6.4). During 1993–1999, a fourth disposal cell was created on 10 of the 27 acres for relocating radioactive waste from Naturita. The Naturita cell contains 79 g (Ci) of radium-226 in an old sandstone quarry lined by 2 ft. of clay [55]. It is covered by four protective layers: (a) a 36-in. radon barrier, (b) a 66-in. frost protection layer of compacted soil, (c) a 6-in. bedding layer of sand and gravel, and (d) 12 in. of rock (Fig. 6.8).

After the end of the Cold War uranium frenzy, the populations of Paradox and Bedrock dwindled, and the Paradox Valley reverted to an open cattle range (Fig. 6.9). The peaceful and picturesque valley has now acquired a new reputation as an outdoor recreation Mecca and an international tourism destination. Paradox now has at least two bed-and-breakfast inns and the valley is on a popular bicycle route from Gateway south along the Rio Dolores to Dove Creek in San Miguel County [56].

Fig. 6.9 Downtown Bedrock, 2012. (Author)

A Gateway Canyons resort had been developed by a founder of the Discovery Channel. Rafting trips are arranged for 3-day adventures from Slick Rock to Bedrock and/or from Bedrock to Gateway [57]. However, the former carnotite mines may again see activity. A Canadian company received approval to build a new uranium mill to meet the increasing demand in China and India for nuclear reactor fuel. The mill would be located on Pinon Ridge, about halfway between Bedrock and Naturita, on Highway 90. If the price of uranium warrants it, they are planning on using ore from 41 nearby carnotite mines [58, 59].

6.5 US Radium Corporation

In 1914, just prior to World War I, the first American company to produce radium paint for watch dials was established. Physicist Sabin A. von Sochocky (1883–1928), attended the University of Lvov, then in the Austro-Hungarian Empire, before obtaining an M.D. in Moscow. He earned a Ph.D. in physics while studying in Vienna, Prague, Dresden, and Paris. After immigrating to America about 1906, he developed a formula for radium paint. In 1914/1915, he partnered with George S. Willis M.D. (1876–1924) to establish the *Radium Luminous Material Corporation* in Newark, New Jersey, to paint dials for the Ingersoll Watch Company and others. Their radium paint was marketed as *Undark.* All the radium paint companies sought to glamorize the work of dial painting by labeling the workplaces as "studios" and having the girls wear smocks while "painting" [60–62]. With the entry of the USA into the European war in April 1917, the Radium Luminous Material Corporation had to expand to 250 dial painters and move to larger facilities in Orange, New Jersey. The demand was such that they decided to mine and refine their own radium and purchased carnotite claims. The latter were obtained in Long Park, on La Sal Creek, and on Saucer Basin overlooking the Rio Dolores. The Radium Luminous Material Corporation mining headquarters were located at Naturita, Colorado (Fig. 6.3).

In Orange, they built a two-story factory at Alden and High Street. Radium extraction was done on the first floor while dial painting was on the second. From 1917 to 1926, they daily processed 0.5–2.0 t of carnotite ore which had 2–4 % UO_2, and which yielded 5–7 mg of radium per ton. Their process could remove only 80 % of the radium and the tailings were discarded on site.

In May 1919, they started using cheaper mesothorium, ^{228}radium, as well as ^{226}radium [34]. Also in 1919, they began selling the Undark radium paint directly to clock companies or consumers, who could then do their own dial painting [60]. Connecticut had several clock companies, but the Waterbury Clock Company, later consolidated into Timex, was producing five million cheap watches a year and, through consolidations, became dominant.

By 1920, the Radium Luminous Material Corporation began to sell their excess radium. They sold 2250 mg to New York State that year for use at the State Institute for Malignant Disease in Buffalo [2, 34, 36]. In 1921, both Willis and Sochocky became symptomatic from the chronic radium poisoning that would eventually kill them in 1924 and 1928, respectively. The Radium Luminous Material Corporation was taken over in 1921 and was reorganized as the *US Radium Corporation*. In 1925, they were supplying radium or mesothorium paint to 110 of the 120 timepiece and novelty companies that utilized it [60–62]. *US Radium Corporation* ceased their radium mining operations in 1926; however, they would continue to produce radium paint products until 1978/1979 when they switched to tritium, ^{3}H.

Replacing S.A. von Sochocky as chief physicist was Austrian Victor F. Hess (1883–1964). After physics degrees at the University of Graz, he studied with Franz Exner at the University of Vienna and was then first assistant to Stefan Myer at the Institute of Radium Research in Vienna from 1910 to 1920. He then became professor of physics at the University of Graz, where he married a Jewish widow, 15 years his senior, who had two children. In 1921, the new family man was lured to New Jersey to work for the *US Radium Corporation* as chief physicist and director of research. He remained there until the 1925/1926 collapse of their radium mining [63]. Hess returned to the University of Graz. He later shared the 1936 Nobel in physics for his work on cosmic rays. This economic windfall was then confiscated by Nazi Austria after the Anschluss in 1938, at which point the Hess family returned to America. Hess resumed at the *US Radium Corporation* as a consultant while serving as professor of physics at Fordham University from 1938 to 1956. Among his 128 publications were papers on dial-painting hazards as well as radium toxicity and hazards; he lost a thumb from radium damage in 1933 [63].

During 1922–1924, the first deaths, from radium poisoning, among of the *US Radium Corporation* dial-painting employees in New Jersey began to appear. One of the many ploys to avoid legal responsibility was to relocate to a different state and they moved from Orange to New York City in 1925. There, they continued in business providing radium paint as well as radium brachytherapy devices. By 1926, two chemists and six dial painters had died. It was estimated that of the 800 dial painters working between 1917 and 1925 at the Orange, New Jersey, plant, 5.4% (43/800) died of radium poisoning. The deceptive and fraudulent efforts of the radium companies, aided by their devious academic consultants, to avoid their obligations to these victims of industrial toxicity have been well chronicled over the years. Dr. H. S. Martland (1883–1954) reported this new occupational disease during 1925–1929 [64]. Radium toxicity killed the dial painters by (a) aplastic anemia, (b) mandible and facial bone necrosis, and (c) osteosarcoma [64]. Out-of-court settlements were made by the *US Radium Corporation* to victims and/or their families in 1928. However, in court, the *US Radium Corporation* was exonerated in 1935 after a decade of legal battles.

By World War II (1939–1945), the military insisted on safety standards for the radium paint industry, the Manhattan Project workers, and the uranium miners. The proper tolerance levels continued to be a source of debate. In 1943, radium safety expert Robley Evans at the Massachusetts Institute of Technology reported that 43 of 288 (15%) of the dial painters in nine different plants on the East Coast had exceeded the tolerance dose for radium exposure [60–62].

The Orange, New Jersey, site vacated by the *US Radium Corporation* in 1925 was leased out until it was sold in 1943. Radioactive tailings had been left at the site, which changed owners several times in the era before NRC regulations developed. In 1980, however, the owner could not sell the property until it had been decontaminated of radioactive materials. This led to legal action against the *US Radium Corporation* during 1981–1991 over the complex EPA laws regarding responsibility of distant owners for radium-contaminated property: "toxic tort litigation." In Orange, in addition to the 2-acre factory site at Alden and High, 250 noncontiguous residential and commercial properties in Orange, West Orange, and South Orange were contaminated as well. Evidently, in addition to the factory painters, over 100 workers were doing piecework at home. Removal of the original factory buildings and excavation and disposal of radium-contaminated soil and debris was started by the NRC in 1997 and was completed in 2005 [65, 66].

In New York City, the *US Radium Corporation* had multiple offices and factories, especially in lower Manhattan. They listed 30 Church St. and 58 Pine St. as business addresses. From 1924 to 1954, they used five floors of a ten-story building on Pearl St. for dial painting and radium processing. During World War II, the *US Radium Corporation* was employing 1000 dial painters, but by 1945, only 135. In 1947, hand painting was replaced by a machine. The *US Radium Corporation* also used a penthouse on Lafayette St. for preparing the radium paint and for reclaiming radium from abandoned older dials and from the new dial-painting machine. It was also the site for manufacturing radium–beryllium neutron sources used in borehole analyses of oil and gas well exploration. The New York State Department of Labor started monitoring the *US Radium Corporation* operations during World War II. There were radon breath surveys of the workers and external gamma irradiation measurements in the buildings. Between 1948 and 1954, the company moved its radium business out of New York City. The Pearl St. building was then sold and was scheduled for demolition in 1958 so that a federal office building could be constructed there! However, in 1961, the new owner could not afford to decontaminate it and vacated it. In 1962, it had to be decontaminated by the Atomic Energy Commission and New York State before being demolished [66].

The *US Radium Corporation* moved its operations to a 10-acre site in a valley 5 miles southeast of Bloomsburg, Pennsylvania, which was an isolated small town in the northern mountains of the state. The site was shared with two other *US Radium Corporation* companies. They continued supplying medical radium sources to physicians but their main activity was manufacturing radioluminous dials, gauges, deck markers, paint, and light sources for military and civil defense needs. In addition to ^{226}radium, they used ^{137}Cs, ^{210}Po, ^{90}Sr, ^{85}Kr, ^{63}Ni, and ^{3}H. Radioactive waste was deposited in an abandoned canal, two lagoons, and two new underground silos. Inspections by the Atomic Energy Commission in 1953 and 1959 noted that radioactive waste was leaking into the groundwater and then into the adjacent Susquehanna River [26, 67–69].

Between 1969 and 1975, the Atomic Energy Commission was replaced by the NRC, which ordered Decontamination Action for the Bloomsburg plant in 1977. The use of radium paint was discontinued in 1979 and replaced by tritium, ^{3}H.

Using a familiar legal move, the *US Radium Corporation* underwent a corporate reorganization in 1980, which would limit its liability, protect corporate assets, and keep an active NRC license. The *US Radium Corporation* was terminated and replaced by US Radium Industries Inc., which transferred all the non-NRC license activities to several new subsidiaries: US Radium Chemical Products, US Radium Lighting Products, US Radium Metals, US Natural Resources, the GAF Corporation and Mitsubishi Chemical Industries [26, 67–69].

The NRC-licensed operations went to a new corporation, Safety Light. Safety Light continued to manufacture radioluminescent signs containing tritium, such as exit road signs and aircraft-airport signs, while stalling decontamination efforts at the Bloomsburg site. Attempts by the NRC to cancel their license were repeatedly overturned in court. In 2005, there was a court-approved settlement agreement among the NRC, Safety Light, and the state of Pennsylvania for an orderly shutdown and monthly payments into a decontamination trust fund. The Safety Light license finally expired on 31 December 2007 [26, 68, 69].

6.6 Radium Chemical Company

From 1928 to 1933, Standard Chemical was gradually being liquidated. At the final dissolution of the company, 90% of the stock was held by Thomas J. Gearing, a Flannery in-law. The second largest stockholder was Joe Kelly, Senior (1884–1965), a son-in-law of the late Joe Flannery. It was Gearing and Kelly who had been denied the renewal of the Radium Belge contract in 1927. At the dissolution of Standard Chemical in 1933, two other companies absorbed their assets: (1) Radium Services and (2) *Radium Chemical*; the officers of the companies were essentially all the same people. The new president of Radium Services was Joe Kelly, Sr. The *Radium Chemical Company*, located in Manhattan, had several subsidiaries throughout the country engaged in radium dial painting which included (a) Radium Dial and (b) Radium Luminous, a paint manufacturer [1, 2, 60–62].

Radium Dial had been a subsidiary of *Radium Chemical* since 1917 and was one of the two largest radium dial-painting companies in the USA. They were located in Brooklyn, New York. In 1920, Radium Dial opened another "studio" in Chicago to be closer to the Western Clock Manufacturing Company (Westclox) of Peru, Illinois. Radium Dial, in 1922, moved to Ottawa, Illinois, to be within 15 miles of Peru. They took over an old school building and the women employees worked at the school desks. From 1918 to 1924, they employed about 1000 dial painters. In 1925, they were painting 4300 dials a day or 1,075,000 a year: the most prolific radium dial business in the USA. The US Radium lawsuits started in New Jersey that same year [60–62]. Knowing of the problems in New Jersey, Radium Dial began testing their employees in Ottawa for radium toxicity in 1925, but never informed them about the results! Worse yet, when they found toxic levels in an employee, they quit testing her! And, if a dial painter began to show signs of bone disease, she was fired so as not to alarm the other workers! Between 1927 and 1931, five Ottawa girls died of radium poisoning. After the sixth death in 1934, the legal battles began. It is estimated that eventually 35 Ottawa dial painters died of radium poisoning.

In 1934, to try and avoid legal responsibility, Radium Dial quit the dial-painting business. Their assets, which included the desks and the workers, were then shifted a few blocks away to a new company: Luminous Processes. This new company was run by Joe Kelly, Junior [60–62]. In 1936, Radium Chemical was again reorganized; this time into four companies: Radium Services, Radium Chemical, Luminous Processes, and Radium Dial. The later was left as a separate entity, an orphan with no assets, although they maintained a corner office on the fifth floor of the Flannery building in Pittsburg [2]. The lawsuits went all the way to the US Supreme Court, which ruled Radium Dial liable in October 1939. By that time, many of the plaintiffs had died and Radium Dial was still without assets.

Luminous Processes continued radium dial painting for watch dials and instruments through World War II. In 1975, they failed an NRC inspection and switched to tritium (^3H). They were shut down in 1978 when the NRC found 1666 times their allowable level of radiation. Luminous Processes generously tried to donate their old work desks to a local Catholic school until their contamination was revealed! When they left the town, the building was demolished. Decontamination of the site was paid for by the taxpayers of Illinois [60–62].

Radium Chemical also manufactured radium-sealed sources and brachytherapy (intracavitary) devices for physicians. After the demise of Standard in 1933, *Radium Chemical* continued the manufacture of these radium therapy devices. From 1938 to 1954, they operated in midtown Manhattan in a two-story brick building, where they employed 11 chemists in the production of radium needles and the operation of a radon plant for making 24-carat gold radon seeds. Services included the reconditioning of older needles that needed straightening, sharpening, and/or polishing.

In 1955, *Radium Chemical* moved from Manhattan to a one-story brick building at 60–06, 27th Ave., Woodside, Queens, New York City [70, 71]. By 1975, the business had significantly diminished due to the availability of newer, safer medical brachytherapy sources. In 1981, the New York State Department of Labor ordered a shutdown and disposal of the 25-year-old radon plant. An inventory of the gold radon seeds was also ordered as radioactive gold jewelry had been detected in New York City. It was suspected that the gold radon seeds had been sold to gold recyclers and then used in making jewelry [72]. After Radium Chemical failed to produce an inventory and also failed waste disposal and safety inspections, the Department of Labor ordered a suspension of all activities with radioactive materials. The disposal and decontamination of the premises was ordered in 1983. *Radium Chemical* appealed but the license was finally terminated in 1986/1987. They were unable to find a buyer or to fund the cleanup. Instead, they agreed to abandon the premises and its contents and to allow New York State to request the US EPA to take remedial action under the Federal Superfund program to clean up abandoned hazardous waste sites [70–72].

In August of 1988, the EPA found 10,000 radium needles in a storage vault. Source removal began in 1989 with 120 Ci of radium being sent to Beatty, Nevada. It was estimated that this 120 g of radium represented 10 % of all the radium in the Western Hemisphere. Additional low-activity waste of 189 Ci was placed in 150 steel drums and boxes that were removed to Richland, Washington, near the infamous Hanover plutonium plant (1943–1971), which remains America's largest radioactive waste site. Ninety-two tons of radioactive hazardous waste and 197 gallons of nonradioactive hazardous waste were also removed. Between 1990 and 1994, the remaining building site was decontaminated, dismantled, and cleared of 812 t of radioactive soil and debris. The New York City site was finally removed from a National Priority List in *1995,* after a cost of about US$ 25,000,000.00 to taxpayers. Thus ended 96 years of American involvement in the radium business, 1899–1995 [1, 62–64, 70–72]

However, in May 2013, the US Food and Drug Agency approved intravenous radium-223 for the treatment of metastatic prostate cancer. Radium is bone seeking and produces significant DNA damage to the cancer cells by its alpha particle emission. This radium isotope has a half-life of only 11.4 days instead of 1600 years and comes in a shielded screw top vial. It decays so rapidly that it can eventually be disposed of in ordinary trash. It is produced in a cyclotron by the international Bayer Pharmaceutical HealthCare Company [73].

References

1. Lubenau JO (2005) Standard chemical company, Marie Curie and Canonsburg. http://www.canonsburgboro.com/MarieCurie/MCurie&StdChemical.htm. Accessed 31 Jan 2006
2. Lounsbury JE (1938) Famous Pittsburgh industries, the standard chemical co. Crucib 22:86–89, 109–113, 134–137
3. Viol CH, Cameron WH (eds) (1920) A sketch of the life and work of Joseph Flannery. Radium (Pittsburgh) 14(6):98–111
4. Fleming R (1922) Joseph M. Flannery. History of Pittsburgh and environs, vol 3. American historical society, New York, pp 804–808
5. Flannery JJ (1913) Book of prominent Pennsylvanians. Leader Publ. Co., Pittsburgh, p 77
6. Miller AL (1968) Personal reminiscences of the early history of the radium extraction industry. ANL Rep, July, ANL-7461, US A.E.C. Argonne National Laboratory, Chicago, pp 91–100
7. Moore RB, Kithil KL (1913) A preliminary report on uranium, radium, and vanadium, Bulletin 70. Department of Interior, Bureau of Mines, Washington Govt. Printing Office, Washington, DC
8. Ingalls WR (ed) (1909) Vanadium. The mineral industry, its statistics, technology, and trade during 1908, vol 17. McGraw-Hill, New York, pp 836–839
9. Fleming R (1933, July 27) Steel girders from the roof of the earth: how vanadium came to Pittsburgh. World of today section, Pittsburgh, pp. 1–2
10. Ingalls WR (ed) (1906–1910) The mineral industry; its statistics, technology, and trade during 1905–1909, vol XIV–IIXX. McGraw-Hill, New York
11. Galessio E (2014) Brief history of the railroads of Peru. http://www.pearcedale-conservation-park.com.au/c&b/peruhistory.htm. Accessed 22 Aug 2014
12. Georgano GN (1985) Cars: early and vintage, 1886–1930. Grange-Universal, London
13. The Henry Ford (2014) Benson Ford research center. http://www.thehenryford.org/research/. Accessed 28 June 2014
14. Lubenau JO, Mould RF (2008) The vanadium window. Nowotw J Oncol 58(4):187e–192e (Warsaw)

15. Heinz History Center (2013) Vanadium window. http://www.flickr.com/photos/casadeyork/9729614620/in/set-72157635483539823/. Accessed 10 Jan 2014
16. Kithil, KL, Davis JA (1917) Mining and concentrates of carnotite ore, bulletin 103. Department of Interior, Bureau of Mines, Washington Govt. Printing Office, Washington, DC
17. Landa ER (1987) Buried treasure to buried waste-the rise and fall of the radium industry. Colo Sch Mines Q 82:1–77
18. Bruyn K (1955) Uranium country. University of Colorado Press, Boulder
19. Badash L (1979) Radioactivity in America. Johns Hopkins, Baltimore
20. Curran TFV (1911) Carnotite in paradox valley. Colo Eng Min J 92:1287–1288
21. Curran TFV (1913) Carnotite-I. Eng Min J 96:1165–1167
22. Hahne FJ (1989) Early uranium mining in the united states. World nuclear association. db.world-nuclear.org/reference/usumin.html. Accessed 28 June 2014
23. Athearn R (1962) The Denver & Rio Grande Western Railroad. University of New Brunswick Press, Lincoln
24. Hart SS (1999) Atomic age narrow gauge. RGS Tech 2(4). http://www.riograndesouthern.com/RGSTechPages/_bdwhite/atomic.htm. Accessed 4 May 2009
25. Viol CH (1921) The story of Madame Curies gram of radium. Radium 17(3):37–52
26. Commonwealth of Pennsylvania (2014) Pennsylvania Bureau of radiation protection decommissioning site summaries. http://www.dep.state.pa.us/brp/Decom_and_Env_Sur/Environmental_Monitoring.htm. Accessed 18 Jan 2014
27. Mould RF (2007) Radium history Mosaic. Nowotw J Oncol 57(Suppl 4) (Warsaw)
28. Rowland RE (1995) Radium in humans: a review of US studies. http://www.osti.gov/accomplishments/documents/fullText/ACC0029.pdf. Accessed 28 June 2014
29. Grigg ERN (1965) The trail of the invisible light. Thomas, Springfield
30. Viol CH, Cameron WH (eds) (1916) American radium society. Radium (Pittsburgh) 8(3):58–59
31. Keeney RM (1917) Radium, uranium & vanadium. The mineral industry during 1916, vol 25. McGraw-Hill, New York, pp 641–643, 744–747
32. Keeney RM (1918) Radium, uranium & vanadium. The mineral industry during 1917, vol 26. McGraw-Hill, New York, pp 616–618, 720–725
33. Keeney RM (1919) Radium, uranium & vanadium. The mineral industry during 1918, vol 27. McGraw-Hill, New York, pp 639–643, 737–742
34. Keeney RM (1920) Radium, uranium & vanadium. The mineral industry during 1919, vol 28. McGraw-Hill, New York, pp 622–623, 710–718
35. Frame P (1998) The historical instrumentation collection. Oak ridge associated universities. http://www.orau/ptp/collection/quackcures/quackcures.htm. Accessed 26 Dec 2010
36. Moore RB (1921) Radium, uranium & vanadium. The mineral industry during 1920, vol 29. McGraw-Hill, New York, pp 615–620, 705–712
37. Sayers RR (1933) Radium in medical use in the United States. Radiology 20:305–309
38. U. S. EPA (2013) Denver radium NPL. http://www2.epa.gov/sites/…/n…. Accessed 28 June 2014
39. U. S. EPA (2013) Denver Radium Site | Region 8 | US EPA. http://www2.epa.gov/region8/denver-radium-site. Accessed 19 May 2014
40. Parsons CL, Moore RB, Lind SC, Schaeffer OC (1915) Extraction and recovery of radium, uranium, and vanadium from carnotite. National Radium Institute Cooperative Agreement, Bulletin 104. Department of Interior Bureau of Mines, Washington Govt. Printing Office, Washington, DC
41. U. S. EPA (1999) Denver radium study, superfund redevelopment program. http://www.epa.gov/superfund/programs/recycle/live/casestudy_denvradium.html. Accessed 3 Jan 2011
42. Radium Belge (ca 1931) Radium, production, general properties, therapeutic applications, apparatus. Union Miniere du Haut Katanga Brussels
43. Adams A (1993) The origin and the early development of the Belgium radium industry. Environ Int 19:491–501
44. PA Dept. Environmental Protection (2012) Pennsylvania decommissioning site summaries, Canonsburg. http://predoc.org/docs/index-4850.html?page=11. Accessed 28 June 2014

45. Of C (ed) (1912) Radium, uranium & vanadium. The mineral industry during 1911, vol 20. McGraw-Hill, New York, pp 655–658, 734–741
46. Fischer S (1914) Radium, uranium & vanadium. The mineral industry…during 1913, vol 22. McGraw-Hill, New York, pp 657–660, 773–777
47. Dow Chemical (2012) Union Carbide history. http://www.unioncarbide.com/history. Accessed 22 Aug 2014
48. Federal Reporter (1962) Union Carbide and Carbon Corporation and VCA. Federal Reporter, Third Series vol 300 US Court of Appeals. http://law.justia.com. Federal Reporter, Second Series, Vol. 300. Accessed 28 June 2014
49. U. S. EPA (2014) Uravan Uranium site. http://web.archive.org/web/20040908170056/ http://www.epa.gov/region08/sf/sites/co/uravan.html. Accessed 22 Aug 2014
50. Chenoweth WL (1987) Paradox valley Colorado. http://ulpeis.anl.gov/documents/dpeis/references/pdfs/Chenoweth_1987.pdf. Accessed 22 Aug 2014
51. Chenoweth WL (1981) The uranium-vanadium deposits of the uravan mineral belt. http://ulpeis.anl.gov/documents/dpeis/references/pdfs/Chenoweth_1981.pdf. Accessed 22 May 2014
52. Center for Land Use (2014) Uravan disposal cell. http://www.clui.org/section/uravan-disposal-cell. Accessed 11 Jan 2014
53. US Department of Energy (2014) Naturita (Uravan) CO disposal site. http://energy.gov/sites/prod/files/2014/03/f8/TITLE1.pdf. Accessed 19 May 2014
54. US Department of Energy (1999) Long-term surveillance plan for the upper Burbank (Naturita) disposal cell at Uravan, Colorado. http://www.lm.doe.gov/Naturita/ltsp_nat.pdf. Accessed 19 May 2014
55. DOE Office of Legacy Management (2014) Naturita, Colorado, Disposal Site. http://www.lm.doe.gov. LM Sites, Colorado. Accessed 19 May 2014
56. Solomon C (2010, August 1) Four corners, two wheels, a bicycle tour through SW canyon country. Times Travel Section, New York pp 6–7
57. Wilderness Aware Rafting (2014) Dolores River 3 Day Slick Rock Canyon. http://www.inaraft.com/dolores-river-slickrock-3day.php. Accessed 22 Aug 2014
58. Sullivan S (2013) Pinon ridge uranium mill granted radioactive materials license. http://www.postindependent.com/news/grandjunction/6289738-113/mill-energy-fuels-uranium. Accessed 2 Dec 2013
59. Harmon G (2011) State oks uranium mill near Naturita. https://groups.google.com/d/topic/west-slope/i919H4×9VQc. Accessed 28 June 2014
60. Rowland RE (1995) Radium in humans, a review of US studies. Argonne National Lab, Argonne. http://www.osti.gov/accomplishments/documents/fullText/ACC0029.pdf. Accessed 28 June 2014
61. Clark C (1997) Radium girls. University of North Carolina Press, Chapel Hill
62. Mullner R (1999) Deadly glow: the radium dial worker tragedy. American Public Health Association, Washington, DC
63. Seeger K (1993) Editorial; Victor F. Hess. Environ Int 19:421–423
64. Martland HS, Humphries RE (1987) Osteogenic sarcoma in dial painters using luminous paint. In: Holleb AI, Randers-Pehrson MB (eds) Classics in oncolocgy. American Cancer Society, New York, pp 247–255
65. U. S. EPA (2011) US radium corp. http://www.epa.gov/Region2/superfund/npl/0200772c.pdf. Accessed 18 Jan 2011
66. Pratt RM (1993) Review of radium hazards and regulation of radium in industry. Environ Int 19:475–489
67. US NRC (2013) Safety light corporation. http://www.nrc.gov/info-finder/decommissioning/complex/safety-light-corporation-slc-.html. Accessed 28 June 2014
68. Commonwealth of Pennsylvania (2014) Pennsylvania bureau of radiation protection decommissioning site summaries. Safety Light Corporation Superfund Site; Bloomsburg, PA DEP License PA-0166, NRC licenses 37-00030-02 & 37-00030-08. http://www.dep.state.pa.us/brp/Decom_and_Env_Sur/PA_Decommissioning_Site_Summaries.htm. Accessed 28 June 2014

69. Riker Danzig (1991) Case Decision: T & E Industries Inc. v. Safety Light Corp at Riker Danzig. http://www.riker.com/case-decisions/te-industries-inc.-v.-safety-light-corporation. Accessed 22 Aug 2014
70. U. S. EPA (1989) NPL site narrative for radium chemical company Inc., New York city. United States Environmental Protection Agency. http://www.epa.gov/superfund/sites/npl/nar1287.htm. Accessed 28 June 2014
71. U. S. EPA (2012) Radium chemical company, Inc. New York. http://www.epa.gov/region2/superfund/npl/0202850c.pdf. Accessed 28 June 2014
72. Lubenau JO (1999) A century's challenges, historical overview of radiation sources in the USA. IAEA Bull 41:49–53. http://www.iaea.org/publications/magazine/bulletin/Bull413/41305094954.pdf. Accessed 10 Jan 2011
73. Parker C et al (2013) Alpha emitter radium-223 and survival in metastatic prostate cancer. New Engl J Med 369(3):213–223, 276–278

Chapter 7
US Radium Oncology

7.1 Cancer Care in the Nineteenth Century

Until the later part of the nineteenth century, the diagnosis of cancer was tantamount to a death sentence. Anesthesia and antisepsis were introduced in midcentury, but both were used infrequently until the 1880s. General anesthetics, nitrous oxide and ether, were introduced during 1842–1846 by the American dentists H. Wells and W. T. Morton as well as by the physician C. Long. Considering the woeful state of medical education in the USA compared to Europe, this monumental discovery was a minor miracle. Chloroform was introduced in Europe in 1847 but was eventually considered too dangerous. Queen Victoria used chloroform for her eighth and ninth deliveries, her last, in 1853 and 1856. However, her physician John Snow (1813–1858) was condemned for it as well as for his theory that germs caused cholera. Unfortunately, the use of general anesthesia increased the surgical death rate from infection as more invasive surgery was now possible by bacterial-laden surgeons. O. W. Holmes in 1843 and Ignaz Semmelweis during 1847–1849 could not convince practitioners that "childbed fever" was due to germs they could not see [1–5].

Florence Nightingale (1820–1910) got around this disbelief by inventing epidemiology and biostatistics. She demonstrated during 1854–1858 that proper sewage and waste disposal, ventilation, sanitation, and nurses could save the lives of hospital patients. Surgeon Joseph Lister did his own epidemiological studies of the ubiquitous hospital infections and their fatality during 1867–1870. He began using carbolic acid dressings and aerosols in his operating room and the postoperative death rate plummeted along with the infection rate. The German language countries accepted Listerism, antisepsis, in 1870–1876 while the English and French did not [1–5].

The French chemist L. Pasteur (1822–1895) developed the germ theory to explain fermentation and beverage contamination, followed by the concept that microorganisms were causing disease in valuable commercial livestock. The germ theory of human contamination followed eventually. German physician Robert

R. F. Robison, *Mining and Selling Radium and Uranium*,
DOI 10.1007/978-3-319-11830-7_7

Koch (1843–1910) linked six different bacteria to specific surgical wound infections in 1878. He identified the bacteria causing anthrax in 1876, tuberculosis in 1882, and cholera in 1883. He was awarded the 1905 Nobel in Medicine, mostly for the epic discovery of the infectious nature of the "white plague" (tuberculosis), which has been omnipresent since at least 4000 BC. Pasteur developed a vaccine for anthrax in 1881 and for rabies in 1885. Germ prevention changed from sterilizing the air in the operating room, antisepsis, to sterilizing hands and instruments, asepsis. Steam sterilization was introduced by the Berlin surgeon Ernst von Bergmann in 1886. The debatable "germ theory" gave way to the science of bacteriology, which exploded in Germany during 1874–1886 [1, 2].

This was the golden age of German medicine in the Second Reich and the Austro-Hungarian Dual Monarchy during 1871–1918. Vienna became the medical capital of the Western world, followed by Berlin. After the discovery of cocaine in 1859, its use for topical anesthesia was publicized by Sigmund Freud and an ophthalmologist friend in Vienna in 1884. German surgeons used anesthesia and antisepsis to replace speed and dexterity [2]. During 1850–1890, as many as 40–50 % of American physicians studied in Germany to learn microscopic anatomy, pathology, bacteriology, physiology, and chemistry. Ten thousand Americans were reported to have studied in Vienna alone between 1870 and 1914 [3].

Theodor Billroth (1829–1894) was a Baltic Prussian who, after posts in Berlin and Zurich, took over a chair of surgery in Vienna in 1867 and further revolutionized surgery. He adopted Listerism in 1875 and then developed asepsis in 1880–1890. He began the first formal training of surgical residents and supported the nursing staff. He utilized a bacteriology laboratory, the gram stain, steam sterilization, surgical caps and gowns, and dust filters for the operating room. Led by Billroth, German surgeons entered the abdomen, chest, and brain, areas which in England were proclaimed in 1873 to be "forever shut from the invasion of the wise surgeon." [1] One of the most glaring example of the relative status of surgery in Germany and England came in 1887/1888 [1, 4].

Manuel P. R. Garcia (1805–1906) came from a family of Spanish opera singers. When he realized he did not have the talent for opera, he became a voice teacher. In 1850, he was appointed a professor at the Royal Academy of Music in London. To study the larynx, he used a handheld dental mouth mirror to reflect and focus sunlight on his own vocal cords. He published a paper in 1855 on his visualization of the larynx. No one in London was particularly interested, but they were in Austria.

In 1857, Viennese neurologist Ludwig Türck (1810–1868) tried Garcia's method on patients at the *Allgemeines Krankenhaus* but was unsuccessful. He loaned some mirrors to Johann N. Czermak (1828–1873), a professor of biology/physiology at the University of Pest. Ophthalmologists were using a handheld concave mirror with a hole in the middle to visualize the retina. Czermak attached this concave mirror on his head and used this head mirror to capture and reflect a beam of candlelight onto a dental mirror positioned at the back of the pharynx (throat). This made indirect laryngoscopy relatively easy, and Czermak presented his work at Vienna in 1858 [1, 4]. The use of a head mirror and artificial light instead of capricious sunlight, while freeing the hands of the operator, allowed for the inspection and biopsy

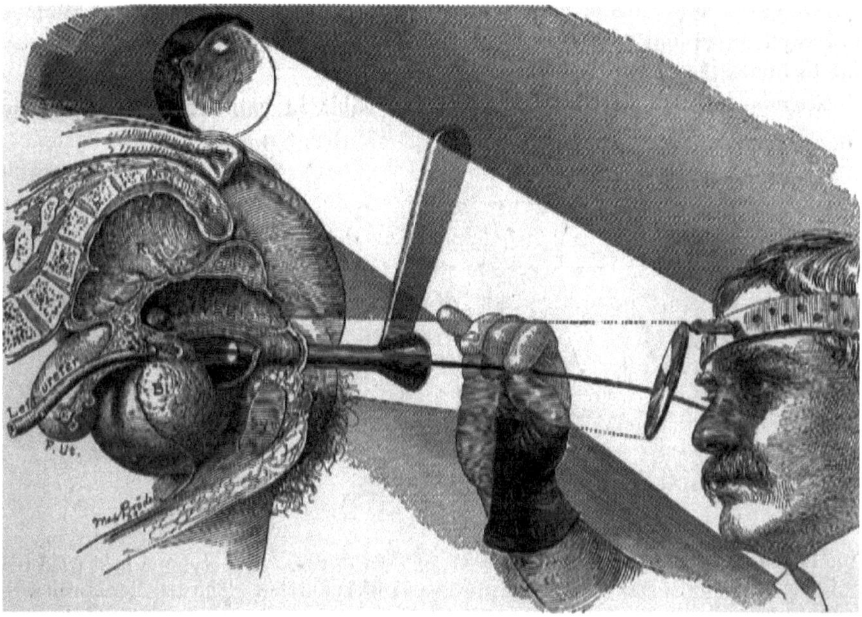

Fig. 7.1 Dr. Kelly using head mirror [5] 1914. (Public domain)

of numerous body orifices. This ushered in the era of endoscopy and physician specialization, starting with (a) eye, ear, nose, and throat and (b) urology. Figure 7.1 shows Dr. H. A. Kelly using the Czermak head mirror to catheterize the ureter [5].

The German kaiser William I (1797–1888) was 90 years old when, during March–May of 1887, his son, the crown prince Frederick (1831–1888), was found to have larynx cancer. A Berlin physician had used indirect laryngoscopy to twice excise a tumor involving one vocal cord. As it had recurred twice, this was considered pathognomonic for malignancy. Billroth had been doing total laryngectomies since 1873, and hemilaryngectomies had recently been done by Ernst von Bergmann, who was consulted in May. He found a larger but still unilateral tumor and recommended surgery. The Crown Princess Victoria, the eldest daughter of Queen Victoria (1831–1901), insisted on a second opinion from the UK, where no one had any surgical experience with the larynx. Morell Mackenzie (1837–1892), from London, had studied in Vienna and Budapest and learned indirect laryngoscopy in 1859. Mackenzie became the London larynx specialist, but his surgeries were confined to tonsillectomies [1, 4].

Mackenzie was summoned to Berlin, where he nixed the surgery and tried a third biopsy which was sent to Rudolph Virchow (1821–1902), the most famous German pathologist. Virchow asked for more tissue as the sample he got did not reveal cancer. This was ignored, and the crown princess moved the prince to the UK, where they all went into denial while the prince got worse. In June of 1887, a German laryngologist and aide to the prince, Dr. Bramann, found the disease was probably now bilateral. In August, he found the entire larynx was involved and only a total laryngectomy would suffice. By September, it was incurable. Mackenzie and the Royal couple continued in denial. By February 1888, the prince had airway obstruction and required a tracheostomy.

In March, the old kaiser died and the dying prince took the throne for 99 days. He died of a tracheoesophageal fistula on 15 June. Instead of the liberal crown prince, the Second Reich was now ruled by his son William/Wilhelm II (1859–1941), who blamed his English mother and her English physicians for the death of his father and for his withered left hand during his breech delivery. However, his mother and all Queen Victoria's children were essentially German. The queen's father was from the House of Hanover and her mother from Saxe-Coburg-Saalfeld, while her husband Prince Albert was from Saxe-Coburg and Gotha. But the errors by English physicians were real.

7.2 New York Cancer Hospital

In New York City, pioneer gynecology surgeon James Marion Sims (1813–1883) became outraged in 1874 when the New York Women's Hospital, which he had organized in 1854, decided to prohibit the admission of cancer patients. At that time, cancer had the same stigma as leprosy, insanity or venereal disease. Public opinion considered cancer to possibly be contagious and/or a venereal disease since it commonly involved the sex organs: uterus, breast, and prostate. In view of recent studies on the sexual transmission of human papillomavirus in cancer of the cervix, vulva, vagina, penis, anus, and oropharynx, the public assumption was not unreasonable. Nevertheless, before he died, Sims was able to interest Mr. and Mrs. John J. Astor and their relatives and friends in creating a New York Cancer Hospital, which was built during 1884–1887. Meanwhile, the deaths of former president U. S. Grant (1885) and crown prince Frederick (1888) from inoperable and untreatable cancer did little to enhance the status of a cancer hospital. The stigma and the lack of success in treatment forced the hospital to alter its mission and its name in order to survive. In 1899, the New York Cancer Hospital became the General Memorial Hospital. The cancer patients, now only 10–12% of the admissions, were partitioned off at the back near the kitchen and refuse collection areas [6, 7].

However, in 1902, the C. P. Huntington family (Southern Pacific Railroad money) contributed funds for cancer research and purchased two machines to begin X-ray treatments. The Huntington money for research was assigned to the Cornell University Medical College for supervision by pathologist James Ewing (1866–1943). Dr. Ewing had graduated from the Columbia Medical School in 1891 and had married in 1900. He was widowed in 1903 and would remain single. The Cornell pathologists averaged six research publications a year, and by *1910,* Ewing was pushing for a clinical cancer research center to complement the pathology research. Ewing's efforts brought him in contact with American industrialist and Canadian citizen Professor James Douglas (1837–1918) of New York City, whose daughter

Naomi had just died of breast cancer. She had been operated on numerous times in New York City before Douglas took her to Guy's Hospital in London for radium treatment. However, Guy's had no radium available in *1909,* so Douglas had to rent 80 mg of radon from the Banc de Radium in Paris. The radium treatment also failed, and she died in the spring of 1910. Douglas endowed a bed at Guy's in her memory and funded a student fellowship in "actinotherapeutics" (radium) [6, 7].

Douglas and Ewing began corresponding in *1910.* Douglas wished to support Ewing in implementing a clinical cancer treatment and research hospital. They approached the General Memorial Hospital in *1912* with that concept. Douglas agreed on an infusion of capital if the hospital (a) returned to the exclusive care of cancer patients, (b) affiliated with Cornell, and (c) appointed Ewing as pathologist and acting head of the hospital. This passed the beleaguered hospital board by one vote [6]. In 1913, the Memorial Hospital became affiliated with the Cornell Medical College and Ewing moved from Cornell to Memorial and became the president of the medical board. By 1916, General Memorial had become the Memorial Hospital for the Treatment of Cancer and Allied Diseases and had returned exclusively to cancer care [6, 7].

Over the next few years, Professor Douglas bequeathed US$ 100,000 in cash (1912), a legacy trust of US$ 390,000 (1914), several grams of radium (1915–1917), a new research laboratory (1917), and new equipment for an expanded X-ray department (1922) [6]. Dr. Ewing was skeptical of radical surgery and supported radium which he termed "X-rays without wires." Ewing and Douglas traveled to Europe in the summer of 1913 to visit radium therapy centers. Douglas was a mining engineer and the president of the Phelps Dodge Company of New York City. He was well equipped to help Dr. Ewing evaluate radium therapy (Fig. 7.2).

Fig. 7.2 Prof. Douglas 1919. (Public domain)

7.3 Professor James Douglas

The lives of Dr. James Douglas Sr. (1800–1886) and his son Professor James Douglas Jr. (1837–1918) seem to have sprung from a novel by Robert Louis Stevenson—from Scotland to Central America, to the Erie Canal, to the Quebec Insane Asylum, to the Arizona Territory and then to New York City and deserved riches. James Sr. was born and raised in Brechin, Scotland, where his father George was a Methodist minister. Seeking a medical career, James Sr. was apprenticed with Dr. Thomas Law of Penrith, England, from age 13 to 18. He then signed on as ship's doctor on a Norwegian whaler which spent the winter of 1818/1819 stuck in the ice between Greenland and the North Pole. As was then the custom, he finished his medical training by attending a few semesters at a university medical school, Edinburgh in his case. James Sr. graduated in 1820 and went to London for more training in 1820/1821 which may have qualified him for the Royal College of Surgeons. He spent 1821/1822 in India and then returned to London, where he fell victim to a giant scam [8–11].

In 1783, the British had obtained logging rights from Spain in the Bay of Honduras. They expanded inland seeking mahogany and, after 1786, began to create the colony of British Honduras, later Belize, to the northeast. By 1822, the Spanish-American colonies were all in revolt, seeking independence. Gregor MacGregor (1786–1845) was a Scot soldier of fortune who had joined the fighting in South America from 1812 to 1820. He then returned to the UK and promoted a mythical colony called Poyais that allegedly existed on the Mosquito Coast off the Bay of Honduras. He recruited colonists and investors. Douglas signed on as chief medical officer. He embarked at Gravesend on 22 November on the *Honduras Packet* with about 80 other colonists. Instead of joining an established colony, they were deposited in the jungle on 23 February *1823* [10, 11].

The local natives furnished them with food until the English rum ran out in March. Dr. Douglas had a house built with sugarcane walls and a palm-leaf roof. He was able to procure food by "bleeding the natives, of which they were very fond and never tired." [11] On 22 March, the desperate settlers were joined by 160 others who arrived on the *Kennersley Castle* and were also abandoned with no provisions. Yellow fever then struck the 240 colonists. Deaths occurred from the fever, starvation, and suicide while one colonist was eaten by an alligator. Several survivors left on foot for British Honduras. On 20 May, a schooner arrived from there but only had room for 57 of the most ill colonists. Dr. Douglas remained on the job but was finally overcome with the fever. A British warship arrived in June and rescued him and the other survivors. More died on the journey back to London, and fewer than 50 of the 240 made it home alive [10, 11].

Dr. Douglas was too ill to travel until September when he found passage to Boston. He embarked on the Erie Canal and headed for Buffalo and Canada to visit some former classmates from Edinburgh. The canal sprang a leak at Utica, New York, and while ashore, he was pressed into emergency surgery practice. The grateful citizens enticed him to stay during 1823–1826, and he married a local girl. In Utica, he practiced dissection on cadavers which were obtained in the usual way, from grave robbers. Unfortunately, a prominent local dignitary had been dug up by mistake and was discovered in the Douglas laboratory during the winter of 1825/1826. Fearing prosecution, Douglas fled with his family due north and crossed the St. Lawrence River into Canada. He kept going along the north side of the river all the way to Quebec City, where he quickly established a flourishing surgical practice. However, his wife died of tuberculosis in 1830. He remarried in 1831 to an Elizabeth Ferguson, and they had two boys who survived childhood. Douglas took charge of the 300-bed Marine and Immigrant Hospital in 1837. He trained medical students there, and it became part of the Quebec Medical School that his friend Dr. Joseph Morrin (1794–1861) had opened in 1848 [12].

The mentally ill people of Quebec were being kept in prison cells or hospital cages throughout the city. In 1845, Douglas signed a city contract to house and care for them, which he did for the next 20 years. He bought a 200-acre estate and manor house on Beauport Road, along the north bank of the St. Lawrence River, some 2.5 miles northeast of the city. It is currently the Centre hospitalier Robert-Giffard. He converted the manor to the Asile de Beauport and hired two physicians to assist him. On 60 of the acres, he built his own estate, Glenalla. Beginning with 81 inmates, he successfully introduced them to fresh air, sunshine, outdoor exercise, domestic chores, and farmwork.

During 1850/1851, Douglas abandoned active medical practice, except for the Beauport Asylum, and he began wintering in the Mediterranean area: Italy, Palestine, and Egypt. During the winters of 1852/1853–1854/1855, he was accompanied by his son James Jr., aged 15–18. Dr. Douglas brought back artifacts from Italy and several mummies from Egypt. One mummy was later sold to the Niagara Falls, Ontario, Museum & Daredevil Hall of Fame. It remained there for another 130 years until it was purchased by a museum in Atlanta and identified as Ramesses I (1292–1200 BC). It was returned to Egypt in 2003 [13].

James Jr. had a German tutor in Quebec, but his education was somewhat desultory until he entered the University of Edinburgh in the fall of 1855. His mother was Presbyterian, the national Church of Scotland, and James was interested in studying for the ministry. He boarded at the home of Commodore Walter Douglas (no relation) of the Cunard Line and a friend of his father. After two semesters, he returned home because of his mother's illness. He finished his BA in 1858 at Queen's College in Kingston, Ontario. Queen's had been founded in 1841/1842 by Presbyterians for the education of ministers. After his mother's death in 1858, James returned to Edinburgh to the Presbyterian theological school. After three more semesters and a doctor of divinity, he changed his mind and declined to be ordained. By 1860, he had married the commodore's daughter Naomi and returned home to help his widowed father [8, 9].

He found that his father had mortgaged all his real estate and invested the family fortune in mining properties to the south of Quebec, an area where many Loyalists had settled after 1783. The Chaudière River runs north from the US border at Maine to the St. Lawrence. Dr. Douglas had bought approximately 55,000 acres along that river after gold had been reported by local farmers. In the same area, Megantic County, the doctor had also taken a 10-year lease on a copper mine.

During 1860/1861, James entered the local Université Laval, a French-language institution since 1663, to study medicine while working with his father at the asylum.

In addition to his medical studies, James Jr. also studied chemistry because of his father's disastrous investments. Fortuitously, James came under the tutelage of the renowned American chemist and geologist Thomas Sterry Hunt (1826–1892) who was then in Quebec City. Hunt was employed during 1847–1868 as a chemist–mineralogist at the Geological Survey of Canada in Montreal. Hunt was also hired by Laval from 1856 to 1868 to teach chemistry, mineralogy, and metallurgy. At Laval, James published his first paper in 1863, about his father's folly, and entitled "The Gold Fields of Canada." Hunt also taught chemistry from 1862 to 1868 at Morrin College. It had been founded by Dr. Joseph Morrin (1794–1861) and was the only English-language college in the city [8, 9, 14].

After his father lost the contract for running the asylum in 1865, James abandoned medical training for the study of chemistry, geology, and metallurgy under Professor Hunt. His father's gold properties were written off, but attempts were made to smelt the copper from the mine that he had leased. A new reduction plant smelter was built, but it burned down in 1867. Douglas and Hunt developed a new process in 1869 for copper extraction, but it proved unsuccessful. Hunt left Canada for Boston, and James was hired at Morrin in 1871–1875 to teach chemistry. In 1871, he and Hunt improved the Hunt–Douglas process, and this was successful. During 1871–1875, the Douglas reputation grew as he published extensively and became a paid consultant. He would become known as Professor Douglas, a mining engineer and chemical geologist, although he lacked a degree in any of these fields.

Dr. Douglas was bankrupt by 1875. James had to auction off everything but a few mummies. With his wife, six children, and an aged father, 38-year-old James immigrated to the USA, where he took a job for US$ 150 a month at a copper smelter in Phoenixville, Pennsylvania, on the Schuylkill River northwest of Philadelphia. Benjamin Silliman Jr. (1816–1885), who had popularized petroleum as a source of illumination, was a chemistry professor at Yale University. C. M. Wheatley owned the local copper mine. Silliman, Wheatley, and Douglas organized the Chemical Copper Company of Phoenixville. The company folded after 8 years when the smelter burned down and the treasurer absconded with the funds. However, during those 8 years, 1875–1883, Douglas and E. Weston invented an electrolytic refining process, the first for copper in the USA. This process made Douglas well known and his consultant work began to furnish a good income [8, 9, 15].

His big break came in 1881 when he was hired by the Phelps Dodge Company of New York City, to evaluate some copper mines in the Arizona Territory [15, 16]. Douglas inspected the copper mines at Morenci and the Atlanta mine at Bisbee and advised Phelps Dodge to procure investment, which they did. Instead of a consulting fee, he took a gamble by accepting 10% of any future copper profits. After 20 years, 1881–1901, he had made a fortune. At Bisbee, Douglas had Phelps Dodge expand by taking over the adjacent Copper Queen during 1884–1886. All the copper mining was consolidated by them in 1885 into a subsidiary, the Copper Queen Co. of New York City, with Douglas as president and with a 10% share. Surrounding claims were acquired as well as mines in Sonora, Mexico. Enlarged and improved smelters were built using European patents and relocated near water, timber, and rail lines. The Bisbee smelter was relocated 26 miles to the newly established town of Douglas in 1901–1905. The El Paso and Southwestern Railroad was built to connect all the mines and smelters to the major rail lines. Eventually, this rail line connected at Tucumcari, NM, with Chicago, IL; at El Paso, TX, with New Orleans, LA; and at Tucson, AZ, with Los Angeles, CA. Southern Pacific leased the line in 1924 and bought it in 1955. Several El Paso and Southwestern Railroad buildings remain on the *National Register of Historic Places* [17–19].

The venerable Phelps Dodge Company had been started in New York City by Anson G. Phelps (1781–1853) as a cotton export–tinware import business in 1812. After his death, the company was run by his sons-in-law, W. Dodge and D. James, and they were succeeded by their sons. Phelps Dodge was directed away from their mercantile origins into a copper mining conglomerate and a railroad owner by Professor James Douglas. By 1906, they announced that their business was now predominantly mining and selling copper [16–19]. In 1908/1909, Professor Douglas took over as president of Phelps Dodge. He then began extensive philanthropic projects which included Queens College and Memorial Hospital of New York City.

In 1915, Dr. Ewing had found Douglas so anemic that he required blood transfusions. It was known that Douglas kept some radium in his Phelps Dodge office in Manhattan, where he maintained a small laboratory. He used radium to make a foot ointment for his wife. He also prepared drinking water laced with radium, which he recommended to visitors. Professor Douglas, still a Canadian citizen, died in New York in 1918 of idiopathic anemia, leaving an estate of US$ 20 million and multiple philanthropic trusts [7–9]. In addition to his academic studies, the professor was awarded many honors [8, 9, 15, 16]:

- Queen's College, Kingston, Ontario; BA degree. 1856–1858
- Edinburgh Presbyterian Theological School; DD. 1858–1860
- Université Laval, Quebec City; medical and geochemistry student. 1861–1866
- Literary and Historical Society of Quebec City; president. 1869, 1874–1875
- Morrin College, Quebec City; professor of chemistry. 1871–1874
- McGill University, Montreal, honorary LLD. 1899
- American Institute of Mining Engineers; president. 1900
- International Mining Congress, Paris; US delegate. 1900
- London Institute of Mining and Metallurgy; Gold Medal. 1908
- Phelps Dodge Company (1881–1918); corporation president. 1908–1916
- American Society of Mechanical Engineers; John Fritz medal. 1915
- Queen's University; chancellor. 1915–1918

Because of his health problems, the professor had moved to chairman of the board in 1916/1917, while his son Walter assumed the presidency of Phelps Dodge. Walter (1870–1946) had received his education in Canada before postgraduate work at Columbia University in New York City. He joined Phelps Dodge in 1894 and became general manager at the Copper Queen. He directed the construction of the El Paso and Southwestern Railroad. In 1940, Walter presided over the dedication ceremonies at the University of Arizona when Phelps Dodge donated a mining and metallurgy building in honor of the professor [9]. Walter's home in Bisbee is on the *National Register of Historic Places* [16, 20]. The professor's other son, Jimmy Stuart Douglas (1867–1949), had a disagreement with Phelps Dodge and left their employ by 1912, when he bought into a copper mine in Jerome, Arizona. During 1914–1938, it paid US$ 55 million in dividends. His estate in Jerome can be visited as an Arizona State Historic Park [20, 21].

7.4 National Radium Institute

After the death of his daughter in 1910, Douglas had directed his research metallurgist, George D. Van Arsdale, in Douglas, Arizona, to work on an economical method to obtain radium from domestic uranium ore. Douglas also consulted with Yale chemist Bertram Borden Boltwood (1870–1927) and one of his recent Ph.D. graduates, Rowland Bosworth, about radium extraction [22]. Van Arsdale selected the low-grade carnotite ore and developed a wet concentration method using hydrochloric acid. In a pilot plant, he was only able to get a 30% recovery, which projected to a cost of US$ 130,000/g, the current market price [23] (Appendix VII).

The federal income tax became law in February of 1913, and President Woodrow Wilson took office in March. The Bureau of Mines published their Bulletin #70 in 1913, and they were concerned about the foreign exploitation of the American radium ore [24]. There was interest in the nationalization of the western uranium mines and in developing an American Radium Institute such as those in Paris, Vienna, and London. This concept was not popular with most western mining men. However, the bureau triumvirate of C. A. Parsons in Washington, DC, and R. B. Moore and K. H. Kithil at the Denver office pushed ahead. In June 1913, they hired radiochemist S. C. Lind. Lind had trained at the Massachusetts Institute of Technology (Boston), Leipzig (Ph.D.), Paris, and Vienna and specialized in research on alpha rays [25].

Lind's job in Denver was to investigate carnotite for the recovery of radium, uranium, and vanadium. He first verified that the Boltwood–Rutherford ratio of radium to the uranium was the same in carnotite as in pechblende: 1:3 million. Some European vendors had claimed the radium content was lower in carnotite. The purchase price of radium-bearing ore was based on uranium content which was easier to measure than radium. The bureau also found that Colorado carnotite with less than 2% uranium was not marketable abroad. Moreover, the vanadium content in carnotite was twice the percentage of uranium and had a ready market. However, the uranium market was questionable. The bureau lacked the funding to do a commercial-scale study of radium production, but Parsons began campaigning in 1913 for private *individuals* to invest in their research [23, 25].

Dr. Howard Atwood Kelly (1858–1943) was also interested in radium production. He is known mostly for being the chief of the Gynecology Service at the Johns Hopkins Hospital and Medical School from 1889 to 1919. However, from 1919 until his retirement in 1940, he practiced surgical and radiation oncology at his own private hospital in Baltimore, the Howard A. Kelly Hospital. Kelly started experimenting with a few milligrams of radium in *1904* in his private practice, but its use was then considered experimental and was reserved for recurrent or inoperable and essentially incurable cancers [26, 27]. In *1907,* Parisian radium therapist Louis Wickham visited Baltimore and convinced Kelly that larger quantities of radium could be effective in deep-seated "surgical cancers." Radium was being used in Europe for cancers of the rectum, bladder, colon, and head and neck as well as the uterus. It was believed that quantities of at least 1 g were needed. Kelly stated: "When Wickham came I at once took him at his word and began investing heavily, embracing the opportunity which drew me out of the narrower limits of my specialty into the newer more extensive fields of ray therapy." [27] With Wickham's help, Kelly began to buy radium, and he hired B. B. Boltwood to assay the radium content [22]. Between 1907 and 1912, Kelly purchased 330 mg of radium bromide.

Kelly and his associate Dr. C. F. Burnam attended summer tutorials by the European experts in 1911–1913 while seeking to obtain more radium. During the summer of 1912, Kelly contracted with the Austrian government to purchase 1 g for delivery in January of 1913. However, due to increased demand, his order was delayed and then aborted by the Austrians in April of 1913. Worldwide demand for radium, costing up to US$ 130,000–180,000 a gram, had increased (Appendix VII). Between April and October of 1913, Standard Chemical radium went on sale, and Kelly became one of their first customers. In May of 1913, he bought 750 mg of radium. He also purchased 100 mg of mesothorium, a radium-228 isotope, in 1913. In December, the USA received its international radium standard from Vienna [23, 27].

This European experience influenced Kelly to begin campaigning for a government-administered *National Radium Institute* to provide radium for American physicians and to encourage the medical application of radium in the USA. This did not win him any friends at Standard, but the Bureau of Mines was listening. Kelly and Douglas met, either through Dr. J. Ewing or through B. B. Boltwood. The Bureau proposed that Dr. Kelly and Professor Douglas furnish the capital to establish a *National Radium Institute* to produce radium for their hospitals in Baltimore and New York City. The bureau would furnish the technical staff and expertise [7, 8, 22, 23]. Dr. Kelly and Mr. Charles L. Parsons, the chief of Mineral Technology in Washington, visited the Paradox Valley to determine if adequate carnotite claims were still available. Many claims had already been taken by Standard and by the radium extraction subsidiary of International Vanadium of Liverpool. However, the Crucible Steel Company of Pittsburgh had been holding some 16 claims in the Paradox since 1910, waiting for the vanadium market to improve. Crucible Steel was willing to lease the claims for 3 years and to purchase at market value *any vanadium or uranium* produced secondary to the radium extraction. Douglas and Dr. Kelly agreed to put up US$ 75,000 each to lease the claims and build an extraction plant in Denver. The *National Radium Institute* was incorporated in October/November of 1913 with H. A. Kelly as president and C. F. Burnam as vice president. James Douglas was a director and his son-in-law Archibald Douglas was a secretary-treasurer [23, 25, 28].

Only ten of the claims were mined as they contained high-grade carnotite ore that averaged 2.5–3 % uranium. The average in the Paradox was less than 2 %. The claims were in Long Park (3.5 miles long by 0.5 miles wide), which was on flat terrain so that the ore could be loaded directly onto wagons without the aid of mules and which had a spring. The Long Park camp became known as the Doctor's Camp because of its many amenities. Douglas and Kelly saw that the miners had heated tents with wood floors and walls, a desirable diet, excellent sanitation, a telephone and phonograph, and a reading room with magazines, books, and all the major city newspapers. Alcohol and gambling were forbidden, but the equipment for baseball, bocce ball, and croquet was provided [25, 28].

The Bureau staff, under the direction of Mr. R. Moore, moved into an office-laboratory at 910 S. 16th Street in downtown Denver during 1912–1916. An old zinc smelter at 500 S. Santa Fe Drive was acquired and converted into the radium extraction plant. It was located along the west side of the Denver & Rio Grande Western Railroad tracks which are now part of the south Denver light-rail line. The plant was 2 miles south of the Bureau's downtown laboratory. The plan was to mine 1000 t of ore and obtain 7.0 g of radium bromide salt. Douglas had found that a hydrochloric acid concentration method was unsatisfactory, so the Denver staff adopted a nitric acid extraction process. This was first tested at the 16th Street laboratory before being used at the plant. For the final separation from barium, they used the Curie method of fractional crystallization, which started at the plant and was completed at 16th Street. The fractionating process at the plant was improved by moving the distillate down through a series of seven bathtubs stacked from floor to ceiling [25].

Improvements and innovations by the Bureau staff doubled the original production figures. Capacity was increased to 3.5 t of ore a day and radium recovery was increased. In January of 1915, Kelly announced receipt of the first 171 mg of radium bromide, equivalent to 91.7 mg of pure radium. Production was so promising that Kelly and Douglas constructed a second plant on site in February of 1915. This required increasing their investment to US$ 250,000 each. By the end of 1916, the Institute had produced 8.5 g of radium bromide. Kelly and Douglas each got 4 g, and 0.5 g went to the Bureau, specifically to S. C. Lind for his research [25]. In the summer of 1916, the Bureau moved from downtown Denver to the Colorado School of Mines in Golden. Dr. Kelly obtained additional radium from the plant in late 1916. The Bureau at Golden mechanically concentrated 100 t of low-grade *pechblende* from Gilpin County (Front Range) mines owned by the Du Ponts. These concentrates and 6 t of other pechblende were processed at the National Radium Institute before it closed, and Kelly bought the 250 mg produced [23]. Kelly's total quantity of radium-226 was now 5.330 g [27] (Table 7.1).

Table 7.1 Radium purchases by H. A. Kelly [27]

Year purchased	Amount (mg)	Vendor
1907	30	Vienna
1907	100	Paris
1912	200	Paris
1913	750	Standard: Pittsburgh
1914–1916	4000	NRI carnotite
1917	250	NRI pechblende from du Ponts

NRI National Radium Institute

During the 3-year run of the *National Radium Institute,* November 1913 to late 1916, the bureau continued to agitate for government involvement in the American radium industry. Congress took up the issue in June 1914 and again in January 1915. The bureau claimed that the *institute* could produce radium for a cost of US$ 39,000 a gram compared to the price charged by Standard of US$ 120,000 a gram. Douglas and Kelly had increased their investment to a total of US$ 500,000 and the *institute* had delivered 8.5 g, which would amount to US$ 59,000 a gram. The bureau claimed that they had reduced this to US$ 39,000 by selling the vanadium by-product to Crucible Steel as previously contracted [25, 29].

The problem with the math was that the *institute* was producing radium bromide which was only 53.6% radium while Standard was charging per gram of pure 100% radium as mandated by the International Standards Committee. Therefore, the *institute* produced only 4.556 g (8.5×0.536) of pure radium at a cost of US$ 109,745.00 a gram (US$ 500,000/4.556 g), which might be reduced to US$ 89,000 with the vanadium sales [25, 29]. Standard acknowledged that the *institute* was producing radium more economically but argued that this was unrealistic for the commercial market. Standard argued (a) the *institute* used higher-grade ore than the average found in the valley, (b) the average carnotite ore has a higher grade of sulfates which also lowers the yield, (c) the nitric acid extraction process is not suitable for lower-grade ores, (d) the cost of research and development was furnished by the government, (e) the *institute* had no marketing expense and no obligation for stockholder profit, and (f) the *institute* freight charges to Denver were less than to Canonsburg. Congress declined to get involved and radium production remained in private hands [23, 29].

However, the *institute* still had some profits coming. Crucible Steel had bought the *institute* vanadium but refused to buy the uranium by-product, stating there was no market for it. Picking a fight with Professor Douglas was probably not wise. He sued them as the *institute* contract specified the sale of vanadium *and* uranium to Crucible Steel. At the courtroom trial, a flamboyant Crucible Company lawyer threw scores of letters onto a table to visibly demonstrate all the refusals the company had received from their entreaties to sell any uranium. The *institute* attorney picked out one on top to scan and found it was from a large steel company anxious to purchase the uranium for steel alloys. After this was read to the court, Crucible was ready to settle. The *National Radium Institute* was given US$ 35,000 in damages and got to keep the 30 t of uranium. It was shipped east to the Douglas family and Lord knows what happened to it [25].

7.5 Dr. H. A. Kelly Hospital

Dr. Kelly was in a financial position to participate in the *National Radium Institute* along with the self-made millionaire Douglas. Kelly was the only boy among five children of a wealthy, privileged, and pious Philadelphia family. He graduated from the University of Pennsylvania, where he majored in natural history and modern European languages [30–32]. He entered the University of Pennsylvania Medical

School at age 19. Kelly was one of 19 college graduates in a class of 136. Before the Flexner Report of 1910, more than half of all medical school freshman had only a high school degree or less [33–35]. Originally, American medical schools were usually a 4-year program with 1–3 years spent in apprenticeship and the rest at a medical school for 1–2 years. The school year varied from November through March or April. Beginning in 1877, Penn required three academic years of 5 months each, with a fourth year being optional [3, 27].

Kelly graduated with the class of 1882 and began a 16-month residency in surgery at the Episcopal Hospital in the seedy Philadelphia mill district of Kensington. After finishing at Episcopal, he opened an office and surgical suite in Kensington, where he began specializing in gynecology because of the endless supply of surgical problems among the prostitutes and poor women. In 1885/1886, he moved his practice to a four-story house on Norris Square, which he purchased and which became the *Kensington Hospital for Women.* Kelly spent the summers of 1886–1888 abroad in Germany and Austria. Kelly utilized surgery professor David Agnew and William Osler, the professor of medicine during 1884–1889 at Penn, as consultants. As Kelly's reputation grew, he became their consultant. He developed into a remarkable surgeon: quick, decisive, and ambidextrous. After only 4 years, 1884–1888, both Osler and Kelly were recognized in Philadelphia for their remarkable abilities [3, 34].

In 1886, Kelly published a paper entitled "Asepsis not anti-sepsis: a plea for principles not paraphernalia in laparotomy." Kelly had been using the Lister system of carbolic acid aerosols and had become aware of its toxicity problems. This paper received widespread attention and accelerated the evolution to asepsis techniques [3]. Dr. Richard A. F. Penrose Sr. (1827–1908), the professor of Obstetrics and Diseases of Women and Children at Penn for 25 years, was getting ready to retire in 1889. Through Osler, Kelly was appointed as an associate professor in the department and heir apparent during 1888–1889. Kelly returned to Danzig, Prussia, in 1889 to marry Laetitia Bredow, the daughter of a Berlin professor [30–32, 34]. She was seldom heard of again and was apparently a traditional hausfrau devoted to *kinder, küche, und kirche,* while raising nine children.

The entire medical education system in the USA was about to change. By the time of the American centennial, attempts were being made to create an independent system that did not rely on Europe for postdoctoral medical training. Baltimore merchant and childless bachelor Johns Hopkins left US$ 7 million to found a university based on the German system of research-focused instruction in 1874. The university opened in 1876 and was to be followed by a hospital and then a medical school along the same principles. When the hospital finally opened in 1889, four outstanding physicians were recruited as department chiefs: bachelors William Welch, William S. Halsted, and William Osler and newlywed H. A. Kelly. All four had obtained postdoctoral training in Germany and had continued to go to Europe every summer for continuing medical education [34, 35].

Pathologist W. H. Welch (1850–1934) and surgeon W. S. Halsted (1852–1919) were old friends from New York City, where they had applied the German training system during 1880–1885. They arrived in Baltimore during 1885–1886. Welch inaugurated the laboratories of chemistry, bacteriology, and microscopic and gross pathology. Halsted lived with Welch and began 4 years of surgical research in

Welch's animal laboratory, while trying to recover from cocaine addiction. Osler was appointed in 1889, and he recommended Kelly to be the professor of obstetrics and gynecology. Kelly got someone else to do the obstetrics, and he then expanded gynecology to include all the diseases of women within the pelvis and the abdomen, while inventing new instruments for their study. Gynecology now included the bladder, ureters, gall bladder, and appendix, which he recommended for elective removal [36]. Figure 7.1 depicts Kelly catheterizing a ureter [5].

After the hospital opened in 1889, there was not enough Hopkins money left for the medical school. Fortuitously, this allowed for the final step in creating the new American medical school model. Four dedicated and wealthy spinsters formed a national committee to raise the necessary US$ 500,000. Mary Garrett alone donated US$ 354,764.50. The women insisted that admission requirements include (a) a 4-year college degree, (b) a reading knowledge of German and/or French, and (c) female students. After some bitter debate and compromise, the terms were accepted. At a time when over half the medical school entrants did not have a high school degree, these entrance requirements were a significant modernization [33–35].

Kelly had broad intellectual interests besides clinical medicine [36]. He was interested in medical history, bibliography, and biography and had an extensive knowledge of geography, mining, minerals, and the natural sciences, especially reptiles, mycology, and botany (Table 7.2). He advanced the use of medical illustration

Table 7.2 Books and papers by H. A. Kelly [36]

Published	Clinical medicine texts
1898	*Operative Gynecology*; 2 volumes; 24 plates
	1906 2nd edition; 703 illustrations; 11 plates
1905	*Vermiform Appendix with E Hurdon*; 399 illustrations; 3 plates
1907	*Gynecology & Abdominal Surgery*; vol. 1 edited with CP Noble
	1908 vol. 2
1908	*Medical Gynecology*; 2 volumes
	1912 2nd edition
1909	*Appendicitis & Other Diseases of the Vermiform Appendix*
1909	*Myomata of the Uterus with TS Cullen*
1908–1913	*Stereo Clinics* vol. I–XX
1914	*Diseases of the Kidneys, Bladder & Ureters with Special Reference to the Diseases in Women with CF Burnam*
1928	*Gynecology with 25 Collaborators*. 20 years after 1908 book
1932	*Electrosurgery* with GE ward
	Non-clinical Texts
1906, 1907, 1912	*Walter Reed & Yellow Fever*
1912–1920–1928–1971	*American Medical Biography*
1914	*American Medical Botanists*
1925	*A Scientific Man & the Bible*
1936	*Snakes of Maryland*
1882–1941	*PAPERS published* = 509 +/−; on radium = 41; nonmedical = 19%

Fig. 7.3 Max Brödel. (Public domain)

by utilizing the famous medical artist *Max Brödel* (1870–1941), who was imported from Leipzig, Germany, in 1894 (Fig. 7.3) [37]. Due to the volume of Kelly's work, two of *Brödel's* classmates from Leipzig were also enlisted: Herman Becker and August Horn [38]. In 1911, a Department of Medical Illustration or Art as Applied to Medicine was created for them at Hopkins. Kelly also produced stereoscopic photographs, which illustrated the operations of major surgeons at work [39]. Twenty volumes of these "Stereo-Clinics" were published during 1908–1913. Many are still available today on eBay.

In 1910, the Flexner Report recommended closing 118 of the 153 orthodox medical schools and all 32 sectarian (unorthodox) schools. The later included 15 homeopathic, 8 osteopathic, and 8 eclectic schools. Abraham Flexner identified Hopkins, where his brother Simon worked, as the American standard that should be emulated [33]. Flexner advocated the closing of all private schools that were staffed only by their physician owners. Medical schools were to be placed in universities only. Flexner also believed that the medical faculty should be engaged in full-time research and that part-time private practice had no place at a medical school. Welch and Halstead, financially independent, disliked patient contact while Osler and Kelly appreciated the income [34]. Kelly wanted the income to provide for his extraordinary lifestyle [26, 27]. This issue caused Kelly to leave Hopkins in 1919 at age 60/61. He then concentrated solely on cancer surgery and radium/radon therapy and became one of the first American radiation oncologists.

Hunter Robb (1863–1940) had joined Kelly in 1886 as an assistant and then accompanied him to Baltimore to serve as his first resident at the Hopkins from 1889 to 1892. Robb (probably with help from Kelly) opened a private hospital, a sanatorium at *1418* Eutaw *Place Boulevard* in the historic neighborhood of Bolton Hill, about 2.5 miles from Hopkins [40]. Kelly moved his family to *1406* Eutaw, a few doors north. When Robb left for Cleveland in 1894, Kelly took over the sanatorium and acquired the adjacent house on the corner, *1420 Eutaw*. Dr. Curtis Burnam, one of Kelly's residents, became his associate there from 1906 until 1940.

The enlarged sanatorium became known as the *Howard A. Kelly Hospital* in *1913*. It was well equipped with operating rooms and various laboratories: bacteriology, pathology, and chemistry. Radium and X-ray therapy were done at the 1418–1420 duplex. It was the first hospital in Baltimore to acquire diagnostic X-ray and deep X-ray therapy equipment. The *hospital* would perform most of the radiotherapy in Maryland for many years [41]. The houses at *1412* and *1414* were acquired to provide in-patient beds, and all four houses were joined at the rear by covered walkways [30]. Patient rooms reportedly looked out on a courtyard that separated 1414 and 1418. The entire complex of the hospital and the clinic from 1412 to 1420 is now a public park, at the corner of Eutaw Boulevard and Morris Street (Fig. 7.4).

In the autumn of 1913, cancer therapists worldwide were entranced by the reports from a Gynecology Congress at Halle, Germany, the *Deutsche Gesellschaft für Gynäkologie*. B. Kronig, C. J. Gauss, E. Bumm, O. Schindler, F. Schauta, and A. Doderlein reported their successful results in the treatment of cervical cancer by using massive doses of gamma rays from radium-228 (mesothorium). One gram or more was placed externally on the skin of the pelvis as an addition to uterine intracavitary radium in the usual milligram doses. In some cases, X-rays were also used externally to boost the dose to the pelvis [43–45]. This use of radium externally became known as tele-curie or tele-gamma therapy. *Mass radium* implied the use of 1 g of more of radium, radon, or mesothorium. For the first time, a surgical cancer, not merely a superficial skin cancer, was successfully treated by radium. The reports were even more astonishing because cancer of the cervix was then one

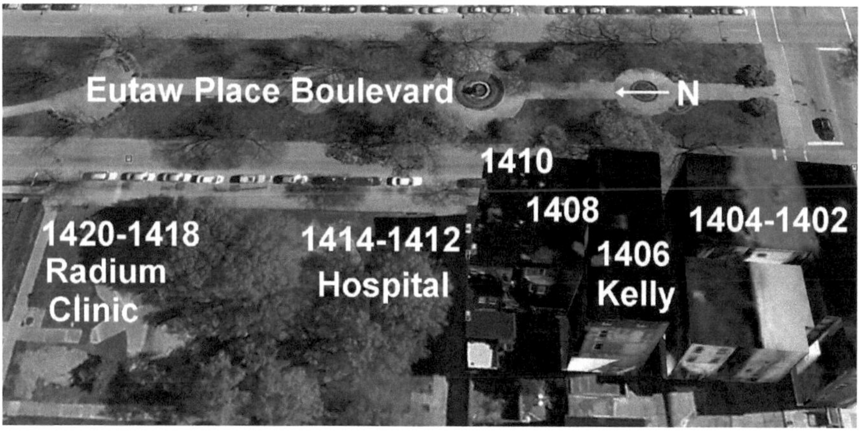

Fig. 7.4 Bolton Hill park (2014); site of Kelly Hospital-Clinic during 1913–51 [42]. (Public domain from US Department of Agriculture Farm Service Agency, edited by author)

of the most common causes of all deaths in women of childbearing age. This situation did not improve in the USA until the early 1940s, and it remains a problem in Latin America, Asia, and Africa. Worldwide, it is currently the second most frequent cause of cancer death in women, second only to tobacco abuse [27].

In the nineteenth century, a surgical hysterectomy for this malignancy was considered an antemortem autopsy. However, anesthesia and antisepsis allowed cancer surgeons to extend the limits of operability. In 1898, Viennese surgeon Ernst Wertheim (1864–1920), and then others, began to develop the radical abdominal hysterectomy for cancer of the cervix by following the Halsted principles of lymphatic drainage. This expanded the operation from the removal of the uterus alone to include all the regional lymphatics as well [1, 2]. Pelvic examinations were infrequent and cytology screening nonexistent, so that 50 % of the patients were inoperable at diagnosis. Among the operable patients, half were cured with the Wertheim, but the operative mortality was 20 % and the morbidity was 30 %. Dr. H. A. Kelly and Hopkins's trainees J. G. Clark, and E. K. Cullen, modified the Wertheim to a German-American procedure and popularized it [46, 47]. The mortality and morbidity improved over time. However, few surgeons attempted it. When Kelly and J. C. Neel reviewed the Hopkins' patient data in 1913, they found that only 57 % were operable, with a mortality of 11 % and an overall cure rate of only 25 %, the same as Dr. Wertheim's [47].

Beginning in 1915, Kelly began to put all his radium into solution to produce radon ($^{222}Rn_{86}$) gas which was then inserted into intracavitary or interstitial applicators or used externally as *mass radium*. The Kelly Hospital had 5 years of experience treating cervical cancer with radium and/or radon when they presented their results to the American Medical Association in October of 1915. Ninety-three percent of the 213 cases had been inoperable and thus incurable. Fifty-five percent of these patients achieved palliation from pain, hemorrhage, and foul discharge; another 6 % had complete local control and 26 % had no evidence of disease at all after the radium/radon therapy. Fourteen other patients had operable disease but were medically inoperable. They all tolerated the less invasive radium treatment with 100 % complete disappearance of disease [47] (Table 7.3).

After an average of 7 years of follow-up, in 1922, the results indicated the same 50 % cure rate with radium as with the Wertheim for operable patients and a 7.5 % cure of the inoperable (incurable) women. Moreover, the radium patients had no mortality and minimal morbidity. In 1926, the Kelly Hospital reported on their second series of patients, who had been treated during 1915–1923 and had an average of 4.5 years of follow-up. The cure rate for incurable disease was now over 10 % and was up to 58 % for operable patients. There was no longer any need for the Wertheim with its morbidity and mortality [41].

By 1917, the Kelly Hospital had 5.43 g of radium bromide (2.91 g pure radium). During 1918–1922, Kelly and his associates started using *mass radium/radon* teletherapy in the treatment of other cancers as well as the uterus [48, 49]. The radon was placed in a lead-shielded cylinder which was suspended from the ceiling and moved about by a pulley to position it over the tumor area (Fig. 7.5). Skin toxic-

Table 7.3 Kelly's results in the treatment of cervical cancer with radium [27, 41, 47]

Patients	First series	First series	Second series
Total # of patients	213	213	429[a] (419)
Treatment period	1/1/1909–1/1/1915	1/1/1909–1/1-1915	1915–1923
Average follow-up	1 year	7 years	4.5 years
Inoperable #	199	199	390
# NED	53	15	40
% NED	26.6	7.5	10.3
Operable #	14	14	29
# NED; radium only	10/10	05/10	10/16
# NED; post op radium	04/04	02/04	07/13
% NED	14/14=100	07/14=50	17/29=58.6

NED no evidence of disease
[a] Ten patients lost to follow up

ity could be reduced by using multiple angled beams over different skin surfaces, known as cross-firing.

This tele-curie or tele-gamma apparatus could also be positioned at varying distances from the patient. By increasing the source (radium) to skin distance (SSD),

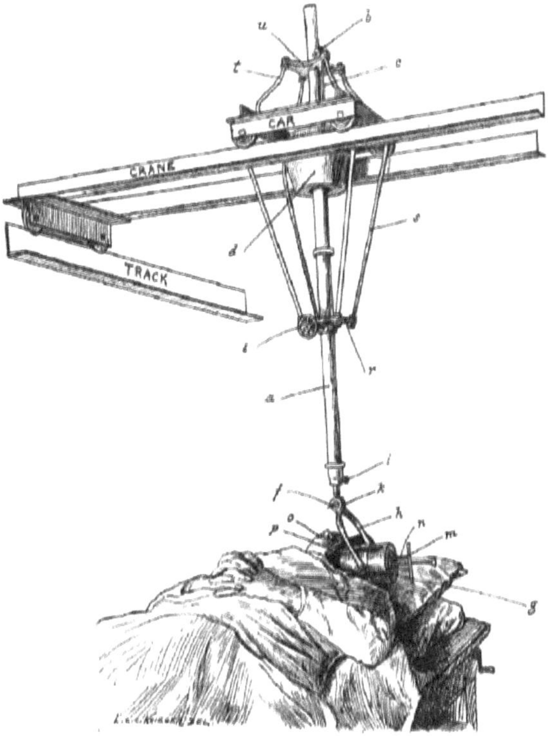

Fig. 7.5 Kelly tele-gamma machine 1923 [48]. (Public domain)

deep-seated cancers within the body get a higher relative percentage of the dose received by the skin. When the SSD is set at 6 cm, a tumor at 10 cm below the skin gets 16.2 % of the skin dose. When the SSD is increased to 15 cm, the same tumor gets 30.8 % of the skin dose [50]. Therefore, the skin dose is reduced and the patient can tolerate higher doses to the tumor. However, increasing the SSD decreased the dose rate, but this could be compensated for by using large quantities of radon. Kelly had enough radium and radon to compensate for this lower dose rate [49].

Fifty papers on radium therapy came from the Kelly Hospital during 1913–1936 and gynecology tumors accounted for only a third of their patients. Kelly and Burnam pioneered radium treatments of the rectum, breast, lung, mediastinum, lymphatics, and abdomen [26, 27, 49]. Kelly retired in 1940 while C. Burnam and W. Neill continued working in the Kelly (Radium) Clinic. They were joined by Kelly's son Edmund (1900–1964), after his army discharge in 1945. Following a fire in 1951, the clinic moved to new quarters in 1952. The clinic buildings were studied for low-grade radioactive contamination by the US Public Health Service before they were demolished [51] and turned into a present-day park (Fig. 7.4).

7.6 Janeway and Tele-Gamma Therapy

Henry H. Janeway MD (1873–1921) graduated from Rutgers in 1889, did postgraduate work at the Yale Sheffield School, and graduated from the Columbia Medical School in 1898 (Fig. 7.6). He had 2 years of surgical training at Roosevelt Hospital with Charles McBurney (1845–1913), the famed appendicitis surgeon. He began a general medical and surgical practice in New Jersey and assisted his father-in-law in dermatology surgery. In 1905, he began a 16-year struggle with recurring ameloblastoma of the mandible that eventually proved fatal [52]. He began working in

Fig. 7.6 Dr. H. H. Janeway [53] 1920. (Public domain)

1906 at the Laboratory for Experimental Research at Columbia. He was an attending surgeon at the New York City Hospital, where he became adept at endoscopic biopsy of the aerodigestive tract and at gastric surgery. He was equally at home in the operating room as in the laboratory, but he was considered eccentric, "a screwball genius," by some of his surgical colleagues [6]. He began using radium in 1910 and caught the attention of Dr. Ewing.

Janeway was hired by Dr. Ewing in 1912 to be the director of Cancer Surgery and Radiation Therapy at Memorial. During his first 2 years at Memorial, 1912–1914, Janeway attended night school for physics training. In 1915, he started radium therapy at Memorial and joined Kelly as one of the first dedicated American radium therapists (radiation oncologists) [45]. By October of 1916, Janeway reported that Memorial had acquired 2.3547 g of radium [54]. This would be equivalent to 4.393 g of radium bromide. Janeway amassed 424 cases by 1917, when he and two associates published their experiences with brachytherapy and with mass radon packs for teletherapy. The book's coauthors were physicist G. Failla (1891–1961) and urologist Dr. B. Barringer (1877–1953) [53].

Janeway was especially interested in head and neck tumors and squamous cell carcinomas of the aerodigestive tract, then known as cancers of the mucous membranes. Between 1910 and 1914, many radium therapists had seen these cancers disappear only to recur. It appeared that larger doses of radium were needed and/or a prolongation of the treatment time. Janeway used dose fractionation to prolong the treatment time. This was accomplished by the use of interstitial (inside the tissues) and intracavitary (within a body cavity) sources. He put radon in glass tubes for the surface irradiation with both beta and gamma rays for superficial tumors of the lip, tongue, tonsil, breast, and skin [52, 54]. He treated cancers of the tongue, tonsil, maxillary sinus, nasal cavity, larynx, and rectum with permanent implants of interstitial gold seeds/tubes which filtered out the painful beta rays. For temporary interstitial implants, he used needles filtered by steel or platinum.

Janeway and Failla were also among the first, along with Kelly, to popularize *mass radium/radon* or the radon pack for deep-seated tumors of the thymus, testis, ovary, lymphatics, bones, and lungs [26, 27, 49]. They kept increasing the SSD to improve the relative depth dose; tele-gamma therapy. One to four grams of radon had to be used because of the drop off in dose rate with the longer SSD. After Janeway's untimely death, Memorial's Failla went on to develop a 4-g tele-gamma unit in 1928 [55–58] (Fig. 7.7).

Roentgen (X-ray) therapy was a hard sell during the tele-gamma era. In 1933, Dr. Ewing summarized the status of roentgen therapy during 1910–1920: "At one period the prescription of dosage was so uncertain and results apparently so capricious that all one could really do was to place the patient under the machine and hope for the best. Patients were often burned from unexpected leaks and on one or more occasions it is said they were actually electrocuted on the treatment table. In cancer the atmosphere of roentgen therapy was quite obnoxious.... The noisy racket of the transformer, the noxious gases exhaled from the tubes in ill ventilated rooms, and the retching of nauseated patients, created a repulsive contrast to the solemn ceremonies of the surgical amphitheater" [52].

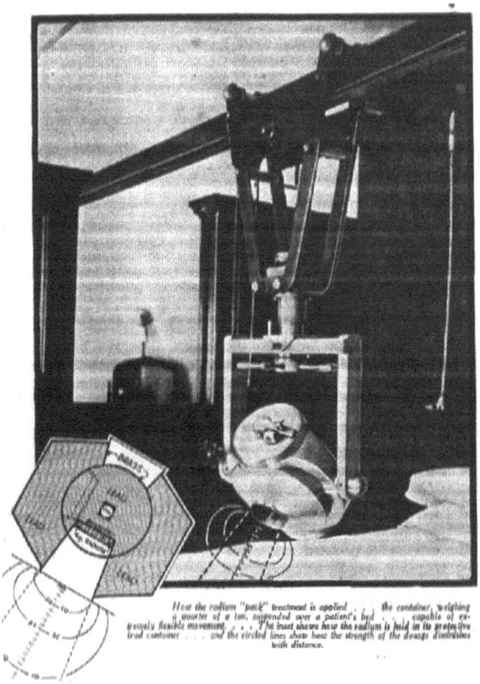

Fig. 7.7 Failla tele-gamma machine [57, 58]. (Public domain)

By 1923, clinicians had realized that the superior results with tele-radium were due to the gamma rays energy of 1.25 MeV. This was far superior to the 140–250 kV X-ray therapy machines of that era. The tele-gamma machines were considered to have more flexibility and maneuverability and did not require a degree in electrical engineering. The radium containers might leak, but this was considered less a risk than electrocution. There had been a radium standard since 1913, whereas X-ray dosage was a mystery and X-ray machines were seldom adequately or uniformly calibrated [52, 55]. Until the physicists and the engineers eventually overcame the X-ray machine difficulties, radium was the therapy of choice for many oncologists.

As radium became cheaper and more plentiful after 1923, tele-gamma radium/radon machines proliferated in the USA and Western Europe [59] (Table 7.4). During World War II, many new artificial unstable isotopes were produced in nuclear reactors and tested as radium substitutes. One of these was cobalt-60, which was cheaper, easier to shield, had a half-life of 5 years, and emitted gamma rays of 1.25 MeV, the same as radium. Tele-cobalt machines came on the market in 1951, just as Failla unveiled an obsolete 50-g tele-radium machine [56]. In the UK, several old tele-radium machines were converted to tele-coblalt, tele-iridium, or tele-cesium as the basic design was the same [59]. Tele-cobalt machines dominated the radiotherapy market for decades until superseded by linear accelerator X-ray machines, with 6–18 MeV energy beams, in North America and Europe. However,

Table 7.4 Tele-gamma machines [59]

In use	Institution: investigators	Grams of radon	SSD (cm)
1917	Memorial, NYC: H Janeway, G Failla	1.0–4.0	1–10
1918	HAKH, Baltimore: H Kelly, C Burnam	1.0–4.0	1–10
1922	Middlesex, London: S Russ	2.5	?
1926	Curie Institute, Paris: C Regaud, R Ferroux, O Monod	4.0	3–10
1928	Villejuief, Paris: S Laborde	1.0–4.0	?
1928	Memorial, NYC: G Failla	4.0	6–20
1930	Westminster, London: R Carling, H Flint, L Grimmett	4.0	10
1931	Brussels: M Cheval, L Mayer	4.0	10
1931	Bellevue, NYC: Ira Kaplan	5.0	6
1935	Philadelphia: W Newcomet	4.0	4–25
1936	Buffalo: B Simpson, M Reinhard	4.5	10–15
1936	Radium Institute, London: L Grimmett	5.0	8.3
1937	Radiumhemmet, Stockholm: R Sievert, S Benner	5.0	?
1938	Westminster, London: H. Flint, C. Wilson	4.0	8
1940	Los Angeles Tumor Institute: A Warner, R Neil	4.0	?
1947–1956	UK: C Wilson	10.0	?
1951	Royal Hospital, London: C Wilson	12.0	?
1954	Roosevelt Hospital, NYC: G Failla, D Quick	50	25

NYC New York City, *SSD* source to skin distance

in many developing countries, tele-cobalt remains the workhorse of the therapy department [59, 60].

References

1. Kennedy M (2004) A brief history of disease, science and medicine. Writer's Collective, Cranston
2. Nuland SB (1988) Doctors: illustrated history of medical pioneers. Black Dog & Levanthal Workman, New York
3. Bordley J, Harvey AM (1976) Two centuries of American medicine: 1776–1976. Saunders, Philadelphia
4. Million RR, Cassisi NJ (1984) Management of head and neck cancers. Lippincott, Philadelphia
5. Kelly HA, Burnam CF (1914) Diseases of the kidneys, ureters, and bladder. Appleton, New York
6. Considine B (1959) That many may live. Memorial Center for Cancer and Allied Diseases, New York

7. del Regato JA (1993) James Ewing. In: Radiation oncologists (Chap. 8). Radiology Centennial Publ., Reston, VA, pp 65–75
8. Langton HH (1940) James Douglas a memoir, privately printed (500 copies). University of Toronto, Toronto
9. Douglas W (1940) University of Arizona dedication of the Douglas memorial building for mines/metallurgy. University of Arizona, Tuscon
10. The Economist (2012) The King of con-men. The Economist 22 Dec 2012. http://www.economist.com/news/christmas-specials/21568583-biggest-fraud-history-warning-professional-and-amateur-investors. Accessed 27 June 2014
11. Douglas J (1868) Account of the attempt to form a settlement on the mosquito shore 1823. Transactions of the literary & historical society of Quebec Series 2 No 6 1868-9 paper I pp 25–40 In Series 2 No. 5–8 1867-71. http://catalog.hathitrust.org/Record/009965458. Accessed 22 Aug 2014
12. Leblond S (1982) James Douglas (1800–1886). In: Dictionary of Canadian biography, vol 11. http://www.biographica/en/bio/douglas_james_1800_86_11E.html. Accessed 18 Feb 2014
13. Mayell H (2003) U.S. museum to return Ramses I mummy to Egypt. http://news.nationalgeographic.com/news/2003/04/0430_030430_royalmummy.html. Accessed 22 Aug 2014
14. Encyclopedia Britannica (1911) Thomas Sterry hunt. http://en.wikipedia.org/wiki/Thomas_Sterry_Hunt. Accessed 19 Feb 2014
15. Arizona Archives Online (2014) Dr. James Douglas collection MS 1031. http://www.azarchivesonline.org/xtf/view?docId=ead/ahssd/ms1031.xml. Accessed 22 Aug 2014
16. Bailey LR (1983) Bisbee; queen of the copper camps. Westernlore, Tuscon
17. Cleland RG (1952) A history of Phelps Dodge 1834-50. Knopf, New York
18. Durham GR (1989) Phelps Dodge corporation. Newcomen Society Publications #1325. Princeton University Press, New York
19. Phelps Dodge Co. (2014) History of Phelps Dodge corporation-company profile. http://www.referenceforbusiness.com/history2/63/Phelps-Dodge-Corporation.html. Accessed 28 Feb 2014
20. Arizona Daily Star (1913) Walter Douglas. http://www.onlinebiographies.info/az/douglas_w.htm. Accessed 18 Feb 2014
21. Young HV (1964) Ghosts of Cleopatra hill. Jerome Historical Society, Jerome
22. Badash L (1979) Radioactivity in America. Johns Hopkins, Baltimore
23. Landa E (1987) From buried treasure to buried waste. Colo Sch Mine Q 82(2):77
24. Moore RB, Kithil KL (1913) A preliminary report on uranium, radium, and vanadium, Bulletin 70. Department of Interior, Bureau of Mines, Washington Government Printing Office, Washington
25. Lind SC (1972) The memoirs of Samuel Colville Lind. J TN Acad Sci 47:1–40
26. Aronowitz JA, Robison RF (2010) Howard Kelly establishes gynecologic brachytherapy in the United States. Brachythrapy 9(2):178–184
27. Robison RF (2010) Howard Atwood Kelly (1858–1943). Nowotw J Oncol 60(1):21e–35e
28. Bruyn K (1955) Uranium country. University of Colorado Press, Boulder
29. Viol CH (1915) The radium situation in America. Radium (Pittsburgh) 4:105–120
30. Davis A (1959) Dr. Kelly of Hopkins: surgeon, scientist, Christian. Johns Hopkins, Baltimore
31. TeLinde RW (1954) Howard Atwood Kelly. Am J Obstet Gynecol 68:1203–1211
32. Bent AF (1996) Howard Atwood Kelly. Int Urogynecol J 7:48–61
33. Flexner A (1972) Medical education in the United States and Canada. Arno Press & The New York Times, New York
34. Bliss M (1999) William Osler. Oxford University Press, New York
35. Harvey AM, Brieger GH, Abrams SL, McKusick VA (1989) A model of its kind. A centennial history of medicine at Johns Hopkins, vol 1. Johns Hopkins, Baltimore
36. Blogg MW (1919) Bibliography of H.A. Kelly. Bull Johns Hopkins Hosp 30:293–344
37. Johns Hopkins University (2010) Department of art as applied to medicine. http://www.hopkinsmedicine.org/medart/HistoryArchives.htm. Accessed 22 Aug 2014
38. Cullen TS (1945) Max Brödel 1870–1941. Bull Med Libr Assoc 33(1):1–29

39. Burns SB (2004) Oncology tumors and treatment: a photographic history. Burns Archive Press, New York
40. Bolton Hill Online (2014) Bolton Hill. http://www.boltonhill.org/. Accessed 22 Aug 2014
41. Burnam CF (1936) Early experiences with radium (Janeway lecture). Am J Roentgenol 36:437–452
42. Baltimore MD Google Maps (2014) Bolton Hill Baltimore. USDA farm service maps. http://Adwww.wow.com/USDA+Farm+Service+Maps. Accessed 15 Feb 2014
43. Gauss CJ (1913) Gynaekologische tiefentherapy (deep therapy). Strahlentherapie 2:623–641
44. Kronig B (1914) The Roentgen rays, radium, and mesothorium in the treatment of uterine fibroids and malignant tumors. Am J Obstet 69:205–214
45. Janeway HH (1919) The treatment of uterine cancer by radium. Surg Gyn Obst 29:242–265
46. Clark JG (1895) A more radical method of performing hysterectomy for cancer of the uterus. John Hopkins Bull 6:120–124
47. Kelly HA, Burnam CF (1915) Radium in the treatment of carcinoma of the cervix, uteri, and vagina. J Am Med Assoc 65:1874–1878
48. Burnam CF, Ward GE (1923) Recent developments in protective methods and appliances in radium therapy. Am J Roentgenol 10:625–632
49. Robison RF (2000) American radium engenders telecurie therapy during World War I. Med Phys 27(6):1212–1216
50. Hendee WR (1970) Medical radiation physics. Year Book Medical, Chicago
51. Gallager RG, Mitchell RZ, Doyle HN (1955) Radioactive contamination in a radium therapy clinic. U.S. Public Health Rep 70(7):617–624
52. Ewing J (1934) Early experiences in radiation therapy. Am J Roentgenol 31:153–163
53. Viol CH, Cameron WH (1921) Dr. Henry H. Janeway (obituary). Radium (Pittsburgh) 16:81–82
54. Janeway HH, Barringer BS, Failla G (1917) Radium therapy in cancer at memorial hospital 1915–1916. Paul Hoeber, New York
55. Failla G (1928) Design of a well-protected radium "pack". Am J Roentgenol Rad Ther 20(2):128–141
56. del Regato JA (1990). Gioaccino Failla. Int J Radiat Oncol Biol Phys 19(6):1609–1620
57. Watson DL (1931) Handling precious radium in safety. Every Week Magazine, NEA Services, Cleveland, p 3
58. Radium Belge (1931). Radium. Union Miniere du Haut Katanga, Radium Department, Brussels
59. Robison RF (1995) The race for megavoltage. Acta Oncol 34(8):1055–1074
60. Page BR et al (2014) Cobalt, linac or other. Int J Radiat Oncol Biol Phys 89(3):476–480

Chapter 8
Congo and Arctic Uranium Sales

8.1 Belgium Radium Monopoly

Between 1880 and 1914, almost all of Africa was partitioned and occupied by the European colonial powers. King Leopold II of Belgium claimed the Congo Free State, which was 75–80 times the size of Belgium, as his personal possession in 1885. The American newspaper reporter Henry M. Stanley had helped secure the area for Leopold, who had financed Stanley's African explorations and search for Dr. Livingston. Stanley spent at least 5 years exploring the Congo basin and tributaries. The Congo occupied 906,000 mi.^{2}; equal to the USA east of the Mississippi River, and equal to the combined areas of Spain, France, Germany, Norway, and Sweden. The Congo straddles the equator from latitude 6°N to 14°S and extends east from meridian 12°E to 32°E (Fig. 8.1). Originally, it was exploited only for rubber and ivory. Leopold was forced to release it to the Belgian government in 1908. Between 1885 and independence in 1960, the Belgian Congo was considered a textbook example of colonial brutality and abuse of human rights [1–6].

The *Union Miniere du Haut Katanga,* the Mineral Union of the High Katanga, was created in 1906 by a merger of King Leopold's mineral exploitation company and a British mining company, Tanganyika Concessions Ltd. The latter was granted mining concessions in the Katanga Province of the southern Congo in 1900. They had British mining interests in Rhodesia, South Africa, and the UK and had followed mineral veins from Rhodesia into Katanga. They supplied 50% of the capital and received 40% of the profits while controlling the mining operations in the Katanga Province for the next 30 years. The Belgium half of the *Union Miniere* was eventually taken over by the *Societe Generale de Belgique,* Belgium's largest holding company. It was one of the largest investment consortiums in Europe and one of its oldest. They had international investors as well as the financial support of the Belgian government. Copper, tin, bismuth, cobalt, and zinc were mined initially because of their shallow location. But iron, tungsten, gold, silver, and coal were also mineral assets in Katanga. For decades, the mineral wealth of Katanga Province was exploited by European investors, who hired mercenaries to control forced labor by the natives [3–6].

R. F. Robison, *Mining and Selling Radium and Uranium,*
DOI 10.1007/978-3-319-11830-7_8

Fig. 8.1 The Democratic Republic of the Congo. (Map courtesy of the US Central Intelligence Agency World Factbook; modified by author [41]; public domain)

Pechblende was first discovered in Katanga province by the *Union Miniere* in 1913 but it was temporarily overlooked in favor of the more valuable copper, rubber, and ivory. It was discovered again in 1915 by Major Robert Rich Sharp, who was employed by Tanganyika Concessions. Allegedly, he was fresh out of Oxford and had completed only one semester of geology [6]. This time, pechblende was fully recognized for its radium potential but copper was the number one priority during World War I. At the conclusion of the war, the *Union Miniere* was able to exploit the ore which was at least 45–50 % U_3O_8. The Congo ore was a mixture of pechblende, carnotite, tobernite, and other variants of uranium phosphates and silicates [6, 7]. Arrangements were made with the *Societe Generale Metallurgique de Hobeken* to develop a radium extraction factory in Olen, Belgium. The *Societe Generale de Belgique* had a controlling financial interest in both the *Generale Metallurgique de Hobeken* and the *Union Miniere*. However, the postwar economic depression of 1920–1921 slowed the development of the radium business [3–5].

The first 12 t of Congo uranium ore arrived in Antwerp for transport to Olen in December of 1921. Only the richest ore went to Belgium, while the lower grade ore was left in dumps at Katanga. The *Union Miniere* could get a gram of radium from 10 t of this ore compared to the 125–500 t of American carnotite necessary to produce 1 g. American radium was selling at US$ 120,000 a gram. The Belgians threatened to lower the price of radium, so low as to make the American product valueless. To prove it, they started selling at US$ 70,000 a gram. The Belgians convinced the two largest American radium producers, Standard Chemical and the Radium Company of Colorado that they could not compete against the vast and rich reserves of Congo ore mined by cheap labor and/or slave labor [7–9].

The *Union Miniere* created a radium sales and marketing division, *Radium Belge,* in 1922 [10]. By 1925, the Belgian sales probably accounted for 80 % of world market. However, production exceeded demand. This led to intermittent production at Katanga and no uranium was exported during 1925–1927, 1933, and 1936. The *Radium Belge* started an aggressive marketing campaign based upon the American experience. Radium therapy had originally been only for wealthy patients treated by wealthy private physicians with their own personal radium stock. The *Radium Belge* sales approach was to create a more socialistic approach to radium therapy whereby patients could be treated at large medical centers, university hospitals, or cancer institutes funded by the government. Poor patients would be treated for free. Hospitals or cancer centers could buy radium on a payment plan or receive loans of radium from *Radium Belge* agents. They offered delivery within 24 h for radium or radon in glass tubes, while other therapy devices might require 3–4 weeks for delivery. They guaranteed the radium at 99.5 % purity as tested by the University of Brussels [5, 10].

The Americans were offered a 5-year contract, 1922–1927, to painlessly exit the radium business if they became technical advisors and sales representatives for the Belgians. Standard Chemical and the Radium Company of Colorado accepted. With American technical and factory expertise, the Belgians produced their first commercial radium in December 1922 and started sales in 1923 [5, 7–10]. Coincident with the demise of the American companies was an increasing interest by American physicians in radium therapy. This led to the necessity of importing radium from Belgium (Table 8.1). The Belgian cost of production with the rich ore and cheap labor was estimated at US$ 20,000/g [5, 8, 9]. After 1933, the price dropped to US$ 50,000/g (Appendix VII).

Table 8.1 Belgium and the USA radium production: 1923–1938 [5, 8]

	Belgium production: in grams [5]		USA production: in grams [8]		USA imports: in grams [8]
Years	Annual	Total	Annual	Total	Annual
1923	20	20	122	195.0	08.8
1924	22	42	03.4	198.4	08.1
1925	20	62	03.0	201.4	09.3
1926	20	82	01.7	203.1	11.0
1927	25	107	0		07.3
1928	42	149	0		11.0
1929	58	207	0		10.7
1930	60	267	0		16.9
1931	60	327	0		15.0
1932	52	379	0		?
1933	48	427	0		?
1934	57	484	0		?
1935–1937	180	664	0	203.1	?

8.2 Eldorado Gold Mining Company

Canada, The New Dominion of 1867, was sparsely inhabited to the west and north. Its citizens hoped that there might be a wealth of minerals out there in that magnificent wilderness in addition to hostile Indians, dangerous animals, and a terrible climate. They were willing to invest in speculative mining ventures for fun and for profit; especially with shares selling for under US$ 2.00. And such a venture was the Eldorado Gold Mining Company, created in 1926 by the LaBine brothers. The LaBines were not Sorbonne graduates. Coming from a farming family in Ontario, they had learned geology and mining from "on the job training." Gilbert had "some high school" and Charlie was an army veteran. Their treasury had US$ 300,000 in 1929 due to their optimistic stockholders but they had yet to find gold [11].

In 1900/1901, the region of Great Bear Lake, located in Canada's Northwest Territories, had been explored by geologists of the Camsell Expedition who noted colored rocks suggesting iron, copper, cobalt, and uranium; red, green, pink, and yellow, respectively. The lake had been discovered by Europeans before 1780 and it was home to lots of bears. It is 200 mi. across from east to west and 110 mi., north to south, with five bays (Fig. 8.2) [13]. It is the eighth largest of the world's major

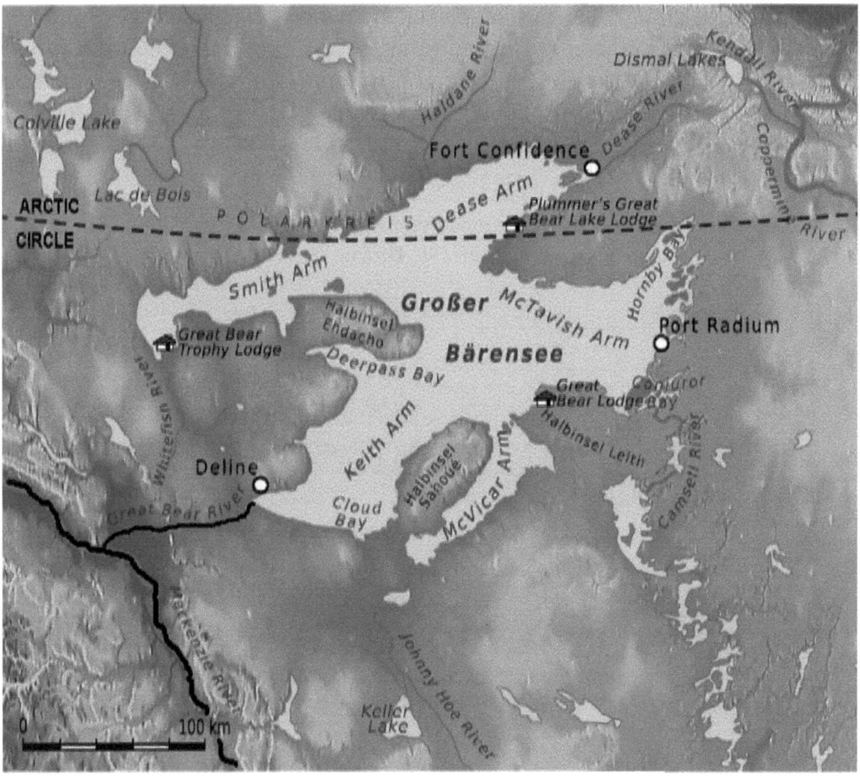

Fig. 8.2 Great Bear Lake. (Map courtesy of Tom Gonzales [13])

lakes, about the size of the Netherlands, with summer temperatures averaging a high of 66 °F and winter highs of − 16 [12]. Two different fur companies had a presence on the lake. At the northeast corner, the Hudson Bay Company had Fort Confidence. The rival Northwest Company had a post at the southwest corner, Deline, where Bear River exited on its way west to the Mackenzie River. The latter is 1000 mi. long. It originates in Great Slave Lake and empties into the Arctic Ocean. Tributaries all over Northwest Canada feed into the Mackenzie. The temperature of the lake water seldom exceeds 39 °F (4 °C) and the lake is open for navigation from July to October, maybe.

Gilbert hired an airplane to take him up to Great Bear Lake on the Arctic Circle, at the extreme eastern edge of the famed Canadian Shield mineral belt, to look for signs of copper and silver. In 1929, LaBine saw some colored rocks of interest from the airplane. After spending the winter of 1929/1930 at the Ottawa Archives reviewing the old Camsell geology report, LaBine returned in 1930 with a geologist. At the Lake, he staked several claims after finding silver and some pechblende ore that was 57–64 % uranium. The uranium ore got the most attention, because in 1930 "the great attraction of radium was its high price" [11].

By February of 1931, the LaBine brothers and the Eldorado Company had staked out 54 50-acre claims in the Echo Bay area, later to be known as *Port Radium*. During 1931, they got assistance and encouragement from the Department of Mines in Ottawa. Ore samples were measuring 28–84 % uranium and the government mining agency thought they could come up with a refining process that was more economical and efficient than what the Belgians had at Olen. Eldorado then became a victim of wrong information on three issues. (a) The uranium ore from the lake would average only 1–2 %, not 60 %. (b) The Belgium Congo mines were not exhausted as some Canadians had hoped. (c) The current price of radium at US$ 70,000/g could and would be dropped by Radium Belge in the face of any competition. With this misinformation, Eldorado began mining at Great Bear Lake in 1932.

Transportation of men and equipment coming into the area and of ore going out was arranged by air or by boat during the short summers. Eldorado established LaBine Point on the eastern side of the lake to facilitate airplane and boat connections. The Canadian government helped by setting up a Royal Canadian Mounted Police Post at adjacent Cameron Bay, which became a settlement for the miners. The Mounties set up a wireless station and a weather station. Entrepreneurs opened a grocery that sold liquor and a restaurant, which had dance, music, and women. Cameron Bay and the Eldorado properties were combined into the settlement known as *Port Radium*.

The La Bines contacted Marcel Pochon to start a refinery. He had become unemployed at the Cornwall refinery of a British–French radium company in Paris (Chap. 4) and was convinced to move his family to *Port Hope*, Ontario. His contract stipulated that his salary was to continue even if he became disabled from radium poisoning! He started work on the refinery on 1 June 1932, and used an extraction process devised by the Department of Mines in Ottawa. This process would also allow for recovery of the silver contaminant in the pechblende ore. The Arctic ore was sent south to the nearest railhead and then by train to *Port Hope*.

Fig. 8.3 Gilbert LaBine (*GL*), Marcel Pochon (*MP*), and Charles LaBine (*CL*); from *left* to *right*, in 1936 [11]. Eldorado nuclear. (Public domain)

Port Hope was a great location, with two railroads serving the harbors on the north shore of Lake Ontario. Across the Lake were Rochester and Niagara, New York. Three large buildings with lakefront and rail connections were for sale by the bankrupt Morrow Seed Company, which to sweeten the deal, threw in 75 bags of Golden Eye Bean seeds and a weight scale! The little town was found to be full of eccentrics and had been nicknamed "the Open Air Asylum." Many elderly Canadian gentry had retired there and rich Americans came in every summer. Additionally, there were professors from Port Hope's biggest treasure, the Trinity College School, which had been founded in 1865. Sir William Osler, during 1865–67, was one of its many illustrious students. The refinery became operational in 1933 and Pochon ran it until 1945 (Fig. 8.3) [11].

In 1933, Eldorado produced their first 3 g of radium and lost US$ 62,000 for the year. Their radium production was only about 33 g during the first 4 years from 1933 to 1936. They were supplied with ore that had only 1–2% uranium. Additionally, the price of radium was falling from US$ 70,000/g in 1930 to about US$ 33,000/g in 1936. However, things were looking up in late 1936 when Port Radium got a new bunkhouse with running water, central heat, and flush toilets. To celebrate the great year, they were having the mine manager and his wife entertained all the miners on Christmas Eve. The centerpiece of the manager's home was the huge fireplace and mantle of native stone. As the roaring Christmas fire blazed away to the good cheer of all, suddenly the fireplace and the entire floor dropped two feet into the muck and mire. The heat had melted the permafrost foundation [11].

In 1937, their production system was perfected and Eldorado had 24 g for sale at US$ 21,000/g. In the spring of 1938, Eldorado was underbid on a radium sale to Finland when *Radium Belge* lowered the price to $ 7000/g. Eldorado then went to Brussels to seek an agreement. Unlike the American radium factories, the Canadians got a deal in April of 1938. First, the Belgians showed them the 180 g of radium they had on the shelf, ready for purchase. Then, they toured the refinery with its 2500 t of Congo ore with 80 % uranium waiting to be processed. Then, instead of forcing Eldorado out of the radium business, the Belgians offered them a cartel arrangement. Each company retained its home market. Profits elsewhere would go 60 % to *Union Miniere*. The price would be fixed at US$ 20,000–35,000/g, depending on demand, and Brussels would serve as a clearinghouse for orders. This purchasing process would also include the 2 g or so a year produced at Jachymov, Czechoslovakia. This remarkable arrangement could have been due to the fact that 30 % of the shares in the *Societe Generale de Belgique,* the parent company of *Union Miniere*, were held in the UK. This may have influenced the *Societe Generale de Belgique* toward accommodation with a Commonwealth member. Or it is possible that the agreement was the brainchild of the chief sales agent of *Radium Belge,* Boris Pregel. In October of 1938, German radium companies (Chaps. 9, 10) took control of the Czechoslovakia mine, removing them from the cartel agreement.

Whatever the reason for this good fortune, Eldorado had a record years profit for 1938 (Table 8.2). However, international radium sales ceased with the start of World War II in September of 1939. Canada declared war on Germany in October. Eldorado profits fell 50 % in 1939. In January of 1940, the *Union Miniere* shut down the Olen refinery and in May, the Germans occupied Belgium. The cartel was dead. North and South America were the only markets left open for radium. Eldorado shut down the mines at *Port Radium* in 1940 but kept refining uranium ore at *Port Hope* for possible radium sales in North America. And the Great Bear Lake ore had some silver and copper, which was sent to Tacoma, Washington, to be extracted.

In 1940, Eldorado had some requests for its uranium *residue* left after the radium and silver extraction. It was considered a waste product and piles of it were stacked

Table 8.2 Eldorado and Belgium radium production: 1933–1949. [5, 11, 12]

Years	Eldorado production in grams [11]	Eldorado profit in US$ [11]	Belgium production total in grams [5]
1930	–	–	267
1933	3.02	Loss 62,000	427
1934	?	–	484
1935	13.0	–	544
1936	15.0	–	604
1937	24.0	–	664
1938	33.6	339,000	Reserves = 180
1939	34.5	15,400	Reserves = 120
1940	–	50,000	Reserves confiscated
1941	–	167,000	
1942–1945	–	–	
1946	19.86	574,083	
1947	–	1,600,000	
1948	–	858,000	
1949	–	400,000	
1950–1954	Sales cease		

around the Port Hope refinery; in the yard, in silos, and at the harbor. They sold small amounts to (a) a local art school needing red and orange pigment, (b) the National Research Council in Ottawa, and (c) the physics department at Columbia University in New York City. They shipped 10–15 t to the UK for physics research by the Military Application of Uranium Detonation (MAUD) Committee, which was investigating nuclear fission. However, the Eldorado profits for 1940 dropped precipitously. The Company was in dire circumstances when a savior arrived from Europe.

In November of 1940, Eldorado acquired a Boris Pregel (1893–1976) as their exclusive worldwide sales agent. Pregel had been the chief sales agent for *Radium Belge* for about 17 years; 1923–1940. With the Nazi invasion of Belgium and France, both *Radium Belge* and Pregel were out of work. Pregel moved from Paris to New York City. His contract with the distressed Eldorado gave him the privilege to buy from Eldorado and sell through his own company. Eldorado was desperate and did not know they were sitting on a "gold mine" of uranium. By December of 1940, Pregel had formed his own sales company, Canadian Radium & Uranium Company, with offices in the Rockefeller Center [11].

During 1941, Pregel sold 12–13 t of Eldorado uranium to the US Uranium Committee investigating nuclear fission. Eldorado profits tripled for 1941. On 9 December 1941, Pregel had tried to interest the US Office of Production Management in buying uranium but they were not interested. Then the US Uranium Committee tried to bypass Pregel and order directly from Eldorado. Pregel put a stop to this infringement on his contract. He then sold 60 t to Vannevar Bush (1890–1974) and the Uranium Committee in March of 1942. This order allowed Eldorado to reopen the *Port Radium* mine. All during 1941 and 1942, Pregel kept abreast of the progress in uranium research. He had been involved with the business and the science of radioactive minerals all his adult life. However, he was a Russian-Jewish foreigner and there was a war going on. He was suspect and not trusted by three different governments: the USA, Canada, and the UK [14–16].

8.3 Uranium Supplies

In 1938, the map of Europe was changing. Since 1933, Germany had become a dictatorship under Adolph Hitler and his fascist Nationalsozialistische Deutsche Arbeiterpartei (National Socialist German Workers' Party) or Nazi political party. Hitler was anxious to reverse the outcome of the 1914–1918 World War and regain the German territory of the Second Reich, 1870–1918. He may even have had visions of a new Heilige Romische Reich. In 1935, Germany reoccupied the Saar-Rhineland area taken from them in 1918. In March 1938, they annexed what was left of Austria in an Anschluss (Union). In September of 1938, Hitler wanted the Sudetenland of Czechoslovakia, the Sudetes Mountain area that stretches from the Carpathians to the Erzgebirge and which had a German speaking population of 2.3 million. They represented a third of population of Bohemia. France and the UK acquiesced on 29 September to a Czech evacuation within 10 days. Hitler claimed that his only territorial ambitions were in the German-speaking areas of Europe. In 1939, this was to include (a) all of Czechoslovakia and (b) the western half of Poland, both former Austrian possessions [14–16].

In January of 1939, German physicist Otto Hahn (1879–1968) published the first paper on the fission of uranium by neutrons. His longtime collaborator at the Kaiser Wilhelm Institute in the Berlin suburbs, Jewess Lise Meitner (1878–1968) had immigrated to Copenhagen to escape Nazi imprisonment or worse. When Hahn notified her over the 1938 Christmas vacation period of his puzzling results, she and her nephew Otto Frisch correctly interpreted the results to show that the neutrons had split uranium into two large fragments rather than just knocking off a section of the huge uranium atom. As the fission process produced even more neutrons, the possibility of a chain reaction releasing tremendous energy was perceived [17]. Uranium was now in demand for physics study and as a possible commercial energy source. And it might be the basis of an explosive device.

In March of 1939, Hitler demanded the German annexation of the entire state of Czechoslovakia, formally Austrian until 1918. With Nazi encouragement, Slovakia and Ruthenia seceded from the Czech state, leaving Bohemia and Moravia as a German Protectorate (satellite). The Czechs would not forget the Slovak actions. Germany and fascist Hungary would annex all the pieces of Czechoslovakia. The Bohemian radium mines at St. Joachimsthal had been taken over by Germany in October of 1938 but this did not seem significant until the discovery of fission in January of 1939.

Physics was not yet a strong suit in the USA. Many of the best physicists in Europe and Germany were Jewish and they were emigrating as fast as they could. Einstein had been at Princeton since 1932. Hungarian Jewish physicists Eugene Wigner (in 1930), Edward Teller (in 1935), and Leo Szilard (in 1939) all came to the USA where they were now working on fission physics and the possibility of a nuclear bomb [14–16]. In July of 1939, Teller and Wigner called on Einstein at Princeton to discuss uranium and the possibility of the bomb. In August of 1939, *Teller and Szilard* drove out to Long Island to get Einstein to sign a letter they had written. The letter was to alert President F. D. Roosevelt to the possibility that German physicists, probably the best in the world, could utilize Joachimsthal uranium to make a nuclear bomb [18]. The Einstein letter of 2 August was accompanied by a Szilard memorandum of 15 August. Szilard's memorandum noted the need to secure supplies of uranium. He estimated that Canada had about 3000 t of good uranium ore, German Bohemia about 1500 t, Russia an unknown amount, and the Belgian Congo the most significant quantities. Szilard suggested that since the Congo pechblende was used to produce radium, the American acquisition of Congo ore could be camouflaged as a plan for the "extraction of radium for medical use" [18]. The letter, written by the Hungarians, said:

> ...leads me to expect that the element uranium may be turned into a new and important source of energy.... The United States has only very poor ores of uranium in moderate quantities. There is some good ore in Canada and the former Czechoslovakia, while the most important source of uranium is Belgian Congo. And...giving particular attention to

the problem of securing a supply of uranium ore for the United States. I understand that Germany has actually stopped the sale of uranium from the Czechoslovakia mines, which she has taken over.

The letter was to be personally delivered to Roosevelt by Alexander Sachs, a Jewish scholar, economist, and confidant of the president. On 1 September 1939, Hitler invaded western Poland while the Soviet Union (1922–1991) invaded eastern Poland, the Baltic Republics, and Finland. Sachs could not get to the president until *11 October* 1939. Sachs had prepared an 800-word summation of the science involved in the letter and memorandum. Sachs emphasized (a) energy production, (b) the medical use of radioisotopes, and (c) bombs of unprecedented potency and scope. He recommended a committee to serve as a liaison between the Administration and the scientists pursuing this work. Roosevelt immediately organized an "Advisory Committee on Uranium" to meet on *21 October* 1939. Nine people were selected including Teller, Wigner, Szilard, Sachs, the Army, the Navy, a representative from the Department of Terrestrial Magnetism (physics) at the Carnegie Institute in Washington. L.J. Briggs from the Bureau of Standards (physics) was to act as chairman. Sachs also recommended that arrangements be made with Belgium for uranium ore [15].

Edgar Sengier (1879–1963) was born in Kortrijk, Belgium, near the French border. He graduated in 1903 with a degree in mining engineering from the Catholic University of Leuven, a French language institution. Sengier had gone to work for the *Union Miniere* in 1911. By 1939, he was one of the directors of both the *Union Miniere* and its parent, the *Societe Generale de Belgique*. They operated the Shinkolobwe (the fruit that scalds) Mine in the Congo as a private preserve or "state within a state." The *Union Miniere* produced 7 % of the world's copper and 99 % of its cobalt. Sengier shut down the uranium mining at the Shinkolobwe mine in the Congo in late 1938 because of a large unsold inventory of 180 g of radium at the Belgium refinery. By 1939, the unsold radium reserves were down to 120 g. Also, at the Olen refinery there were 3500 t of partially processed uranium ore. With 120 g of unsold radium, there was no hurry to process the uranium. At Shinkolobwe, there were about 6000 t of raw ore stockpiled above ground. *Union Miniere* had let the mine flood, which happens when mines are not being actively pumped dry. Essential equipment had been sent to *Union Miniere* cobalt and copper mines [6, 14–16].

After the Hahn publication, British, French, Italian, and German physicists were all in competition to crack open the atomic nucleus [17]. As early as May of 1939, Sir Henry Tizard was interested in procuring uranium ore for research efforts in the UK. Tizard had occasion to talk to Edgar Sengier, who was visiting London for a meeting of the *Union Miniere*. Sengier turned down a purchase offer from Tizard, perhaps because of the low price. The *Union Miniere* was well aware of the potential value of their Congo uranium ore. Shortly after turning down the British offer, Sengier was approached by a French nuclear physicist team led by Frederick Joliot-Curie, the 1935 Nobel winner in physics. Sengier tentatively agreed to a joint venture with the French to develop a nuclear bomb to be tested in the Sahara. *Union Miniere* was to get half the patent royalties in exchange for furnishing the uranium [14–16]. The start of World War II on 1 September 1939 put a stop to that idea, and Sengier moved to New York City in a hurry in October. Sengier and his wife moved into the Ambassador Hotel in New York City where he opened an office, The African Metals Company, in the Cunard Building at 25 Broadway. He was to direct the business of the *Societe Generale de Belgique* and the *Union Miniere* if and when Belgium was occupied by the Germans. In January of 1940 the production of radium was halted and Sengier ordered the 120 g of radium at Olen to be sent to *Radium Belge* agents in England and the USA. The uranium ore at Olen was also to be shipped to the USA, but the Germans got there first, in May 1940. The Belgium and Paris radium refineries were confiscated in May and June of 1940, respectively [6, 14–16].

By September of 1940, Sengier had begun to worry about a German invasion of Africa to confiscate Belgium mineral wealth. In September and October he started sending 1250 t of uranium ore by rail to the port of Lobito in Portuguese, Angola. From there, the 2000 steel drums were shipped on two freighters to Port Richmond on Staten Island, New York City. They were stored in a three-story warehouse belonging to the agri-business concern, Archer Daniels Midland, the only place available on short notice. The drums were conspicuously stamped "Uranium ore—product of the Belgium Congo." There they sat for 2 years: from October of 1940 until September of 1942 [6, 15].

The American interest in the uranium bomb had started with the Uranium Committee in October of 1939 but gained added interest in April of 1940 when it was found that the Kaiser Wilhelm Institute in Berlin was starting uranium research. In June of 1940, the National Defense Research Council, under Chairman J. B. Conant, took charge of the Uranium Committee investigations. In November of 1940, Boris Pregel became the exclusive sales agent for Canadian (Eldorado) radium and uranium. He also bought into their *Port Hope* refinery, managed by his old friend from Paris, Marcel Ponchon. In May of 1941, he sold the Uranium Committee 7 t of uranium ore and in December, another 5 t. The ore came from *Port Radium* and contained only 1–5 % uranium oxide when sent to the refinery at *Port Hope* [11, 14–16].

While Britain was under siege during 1940/1941, the UK moved their nuclear research team to Montreal, Canada, and they tried to get the Americans interested in the uranium bomb. By *November* of 1941, the American interest was growing and the National Defense Research Council was superseded by the Office of Scientific Research and Development (OSRD), led by Vannevar Bush. They were given the authority to pursue engineering development of a bomb. However, the OSRD scientists were convinced that a chain reaction would be impossible without an enormous quantity of uranium becoming available. After Pearl Harbor, 7 *December* of 1941, the scientists recommended that the Army get involved in the construction of pilot plants for uranium production by isotope separation and plutonium production. It had taken 3 years since the discovery of fission in December of 1938 for the USA to become concerned about uranium sources; Table 8.3.

The Canadian uranium mines at *Port Radium* had been shut down in 1940 due to the loss of radium sales. However, their refinery at *Port Hope*, Ontario, was still functioning while processing the uranium ordered by the Americans. In order to keep their uranium supply lines open, the Americans ordered another 60 t of

Table 8.3 American interest in Uranium: 1938–1941 [14–18]

Year	Month	Uranium research developments
1938	October	Germany annexes Jachymov UR mines
	December	Fission of uranium achieved in Berlin. Published January 1939
1939	January–February	Fission explained by Jewish refugees
	June	Union Miniere and French discuss fission bomb tests in Sahara
	July	Jewish émigré physicists in USA concerned
	August	Einstein letter written by Szilard and teller
	September 1	Germany and the Soviet Union start WW II by invading Poland, the Baltics, and Finland
	October	Sengier moves Union Miniere to New York City
	October 11	Sachs delivers Jewish letter to President Roosevelt
	October 21	President appoints Advisory Committee on Uranium
1939–1940	Winter	"Phony War"
1940	April	Berlin Kaiser Wilhelm Institute starts uranium research
	May	Germans seize Belgium, radium refinery and uranium ore
	June	Germans occupy Paris and radium refineries
	June	NDRC supersedes Uranium committee
	September–October	Congo uranium shipped to Staten Island
1941	March–December	Uranium Committee buys uranium from Canada
	November	OSRD supersedes NDRC to pursue bomb development
	December 7	Pearl Harbor; OSRD wants uranium

NDRC National Defense Research Committee, *OSRD* Office of Scientific Research and Development

Canadian ore from Pregel in *March of 1942*. This allowed the mines on Great Bear Lake to be reopened. Pregel sold the US Army 350 t of uranium ore in *July* and another 500 t in *December* of 1942. The refinery at *Port Hope* could only process 120–145 t a month. It would take at least a half-year to process these 850 t. So the mine and the refinery were preempted and not available to furnish uranium for the British.

In June of 1942, OSRD thought the bomb was indeed possible and might be ready toward the war's end. By 11 August of 1942, the bomb was no longer considered a theoretical physics problem but a production problem for the Army. Between June and August, the Manhattan Engineering District (Manhattan Project) was created to produce the bomb. The Manhattan Project was put under the direction of Major General Leslie Groves and the US Army Corps of Engineers. Edgar Sengier had been busy selling copper and cobalt and other Congo metals to the Allies, while trying to get some interest in his uranium. In March of 1942, he met with US State Department officials about their desire to purchase Congo cobalt. Sengier informed the State Department that his uranium was more important than the cobalt they were buying. No response, as the State Department had no knowledge of the bomb program. In April of 1942, Sengier again suggested that his uranium should be transferred to Fort Knox for safe keeping. No response. On 21 April, he wrote the State Department to say *"As I told you previously these ores containing radium and uranium are very valuable"* [14–16].

Five months later, on 7 September of 1942, the State Department contacted the Manhattan Project about a request by Sengier to ship several tons of Congo ore from Staten Island to *Port Hope* for refining. Pregel, an enterprising entrepreneur was attempting to buy the Belgian uranium stored on Staten Island and ship it to Canada for processing. He could then sell the processed uranium to the American, British, and Canadian governments for their nuclear bomb research. If the various governments were too inept to start securing uranium, Pregel was willing to profit as the middle man. When the State Department notified the Army and General Groves about the request, they were shocked to learn about all the uranium sitting on Staten Island [14–16].

Brigadier General Leslie R. Groves started work on 17 September 1942. On that date, he was informed of the Congo ore on Staten Island, which was 65–80% uranium oxide, the richest in the world. The next day, he arranged to buy the 1250 t on Staten Island and order 400 t a month from the Congo, where there were at least another 1000 t of high-grade ore above ground. Additionally, there were another 2000 lbs. of waste tailings that contained *only* 20% uranium oxide. These had been rejected by the *Union Miniere* in favor of the 65% rich ore! The American government used fast freighters to outrun the German U-boats from the Congo to New York and only lost one freighter during the war. The price was US$ 2080/t [6].

This rich Congo ore was sent to *Port Hope* and the *Port Radium* ore from Great Bear Lake was put on hold. Now the *Port Hope* refinery would be preempted for the next *5 years*. In September of 1942, the British had contracted to have 350 t of uranium refined at *Port Hope* for the combined UK–Canadian research joint venture in Montreal, which would be started in October of 1942. However, their contract was suspended because the USA was sending the Congo ore to *Port Hope* and had first priority. The British were mad and the Canadians were embarrassed. The UK now pressed their Commonwealth partner and erstwhile colony to take action.

In 1943, the USA was finally able to secure a long-term contract for the Congo ore. The Army Corps of Engineers spent US$ 13 million reopening the abandoned and flooded mines at Shinkolobwe [6]. During 1942–1944, another 30,000 t of ore were mined. Airports were opened at Elizabethville and Leopoldville while a port was established on the Congo River at Matadi [6] (Fig. 8.1). The Congo would furnish most of the uranium for the Manhattan Project. However, Canada would furnish about a third of the uranium. In 1942, the Vitro Chemical Company in Canonsburg, Pennsylvania, was also engaged by the Manhattan Project to produce uranium from Congo and Canadian pechblende and from American carnotite with 1–2% uranium oxide. The Colorado uranium mines furnished 200 t of uranium residues, discarded after vanadium extraction [15].

In August of 1943, the Americans agreed to furnish uranium to the British–Canadian research effort for their project in Montreal: the Combined Policy Committee Agreement. Meanwhile, the Canadians had decided to nationalize Eldorado so as to have more complete control over their uranium ore. Private entrepreneurs like the LaBines and the Pregels were now to be removed from the war effort. All three countries engaged in nuclear research did not want Boris Pregel to sell uranium. It was alright for the Belgium and Canadian governments to make a profit on their ore but no individual was to profit. There was to be no "war profiteering" from uranium and there was *not* going to be a uranium business similar to the radium business. Indeed, there had been many radium merchants but there were few businessmen that could trade in uranium. Boris Pregel was unique. He had the money and the knowledge to do so and that was his livelihood. It was inevitable that he would incur government animosity.

8.4 Boris Pregel

Odessa was first occupied by the Russians, under Empress Catherine, in 1789 and it had one of the largest Jewish populations in the Russian Pale of Settlement [19–21]. The Pale, Latin for enclosure, extended roughly from Odessa to Warsaw to Riga to Rostov-on-Don. It was the idea of Catherine the Great (1729–1796). She and Tsar Nicholas I (1796–1855) greatly expanded Russian territory by taking the Crimea, Ukraine, Moldova, Belarus, Lithuania, and half of Poland [22]. These areas contained about 5 million Jews, about 40–75 % of all the Jews in the world [23]. She decided they should stay there. This abusive restriction remained until the collapse of the Russian Empire in 1917/1918. The Germans forced the Treaty of Brest-Litovsk on the Bolsheviks in March of 1918 and this ended the Pale and temporarily freed Finland, the Baltics, Poland, Belarus, and the Ukraine from Russia [19–23] (Fig. 8.4).

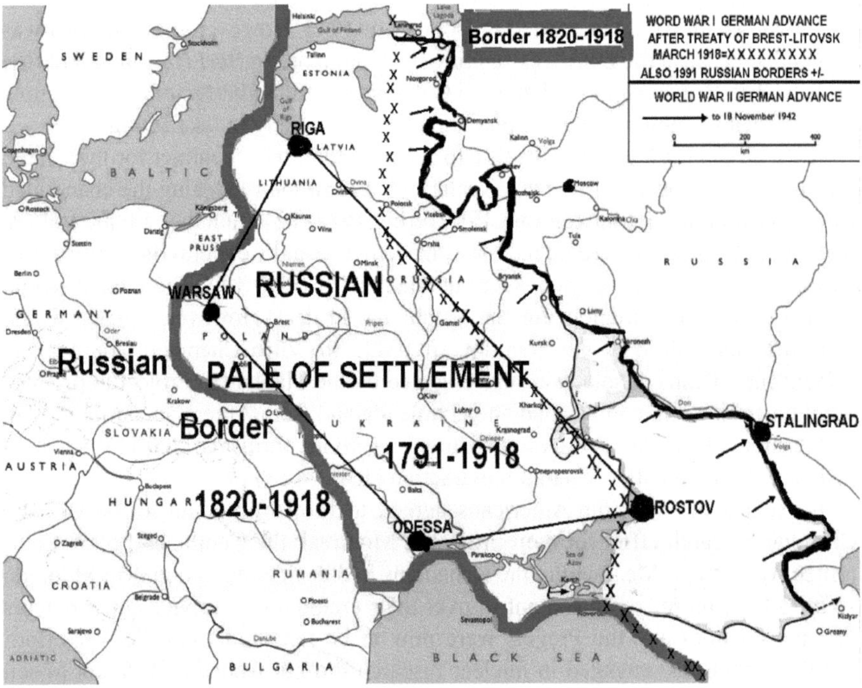

Fig. 8.4 Russian border (1820–1918) and Pale of Settlement (1772–1918). (Public domain; modified by author)

By 1900, the local Jews in Odessa were dominating the local grain export business and much of the medical (70%) and legal (56%) professions [24–26]. The Pregel (pronounced Prez-hell) business was the export of edible vegetable oils. Boris (1893–1976) and his younger siblings, Sophia and Alexander, were born in Odessa to a well-to-do Jewish merchant family. As the universities were closed to Jews, Boris went to Belgium for an engineering degree (1916) at the Universities of Brussels and Liege; French language schools. During World War I, he advanced in rank to become a colonel of Engineering at the one Russian aircraft factory. During the ensuing civil war, October of 1917 to 1922, the White Russian Pregels fled to Constantinople and then Romania. Boris and his sister Sophia (1894–1972) were in Paris by 1922 [27, 28]. The Odessa Jews were considered relatively apathetic toward Judaism and the upper socioeconomic class embraced the Russian culture. After the Bolshevik Revolution in 1917–1921, many of these Jews were leaders in the Russian émigré groups in Paris and the USA [29].

Boris, having studied in Belgium, found work with *Radium Belge,* who employed him as a sales agent. His brother Alexander (1907–1998) stayed in Constantinople long enough to graduate from the historic English language Robert's College there. He eventually went to the University of Minnesota where he earned a Masters degree in engineering in 1929. Alex then joined his brother in business in Paris [30]. Boris lived in Paris but functioned as an international salesman for *Radium Belge*. He traveled on a Romanian passport. He studied radioactivity and its minerals. He developed contacts and friendly relationships with the physicists working in the field, including Marcel Pochon. Pregel was credited with furnishing Madam Curie with the long-term loan of a 5 g radium source, used for her studies. He also marketed and then helped design and build radium centers worldwide, such as the Sophiahemmet in Stockholm. He established numerous ancillary companies to compliment the radioactive mineral business: calibration and measuring devices, source standards, etc. He also was involved in industrial uses of radioactivity to detect welding and structural imperfections [31–34].

With the start of World War II, the Pregels moved from Paris to New York City. Boris took over as the exclusive Eldorado sales agent in November of 1940 [11]. Three weeks later, he and his brother Alex established their own company the Canadian Radium and Uranium Company, with business offices at 630 Fifth Ave, New York City. The brothers opened their own refinery at Mount Kisko, New York, and had a laboratory on E. 16th Street in Manhattan for the manufacture of radium applicators and other commercial radioactive devices [35–37]. Dr. B. A. Veebrink was director of the laboratory (Fig. 8.5) [35]. The Pregel brothers had transferred their business interests from Europe to North America and in spite of the war; they were able to continue dealing in radioactivity and rare minerals.

The Manhattan Project purchased its uranium from the Congo and from Canada but they contracted only for the uranium content of these ores. All other minerals in the ore remained the property of the original owner. *Union Miniere* was the owner of the African ore and Pregel's Canadian Radium and Uranium Company was the owner of the Canadian ore. The refinery at Port Hope stored the residues remaining after the uranium was extracted. The Canadian residues were sent to Mt. Kisko for extraction of the radium and other radioactive minerals, such as polonium. Radium

Fig. 8.5 B. A. Veebrink, Boris Pregel, and Alexander Pregel, (*left* to *right*) at the New York City laboratory of their Canadian Uranium and Radium Co. in 1947 [35]. (Public domain)

and polonium were sold by Pregel to (a) the Manhattan Project, (b) Great Britain and (c) commercially. Pregel also bought into the Port Hope refinery so that it could be upgraded to process all the Eldorado uranium being sold to the USA. This investment earned Pregel 83 % of the refinery profits [27–34].

At the Fall meeting in 1942 of the Radiological Society of North America in Chicago, the Pregels had a technical exhibit demonstrating ore from Canada, radium containers, accessory equipment, testing equipment, and the use of radium for making industrial radiographs [38]. The Canadian Radium and Uranium Company claimed they were the only remaining active producer and vendor for *radium* in the free world, as Belgium and Czechoslovakia were occupied by Nazi Germany. The Pregels also offered for lease a radium standard and the only available polonium standard. Later, they manufactured signs that used radioactive or ultraviolet light. They continued in business up into the 1960s, when they were primarily doing radium recovery from discarded medical or military instruments and watch dials [34–37].

The Canadian government secretly and slowly began to buy up the Eldorado shares in late 1942 [11]. It would take them a year to take complete control of the LaBine business. They were pushed to do this by the British, who did not want the Americans or Pregel to have such a dominant control of the uranium. The British figured they would get a better deal when Canada controlled Eldorado. As part of their campaign against the Americans, the British started complaining, from September of 1942 through July of 1943, about what a rascal Boris Pregel was. LaBine shared the written accusations of the British with the Canadian government, which stated that the British document was "sloppily prepared, excessively dependent on inaccurate hearsay and easy to refute in detail." LaBine stated: "So many of its details were so inaccurate that doubt was cast on the validity of the whole document" [11, 14, 27].

In June of 1943, the USA set about trying to find the location of the entire world's supply of uranium: the Murray Hill Area study [16], see Chap. 10. They started investigating the Canadian Northwest Territories. Pregel, through his Company, had already purchased some uranium claims in the Northwest Territories, including the Great Bear Lake area. The Canadians were incensed. On 15 September of 1943 they announced that "*all radioactive materials found in the Northwest Territories were reserved to the Crown.*" On 28 January, 1944, the Eldorado Mining & Refining Company, as it was known since 1943, became a Crown Company. All the outstanding shares had been purchased at US$ 1.35, whereas they once had been 60–80 cents [11]. Private entrepreneur Pregel believed this was illegal and unconstitutional and he thought General Groves might agree, since the Americans were also being kicked out of Canada. Groves ignored Pregel and disregarded his advice that in the postwar period, the USA would want private prospectors out hunting for uranium. Pregel was later found to be quite correct about the future government-sponsored uranium frenzy in the American Southwest during the Cold War [11, 14, 27].

In January of 1944, Eldorado was taken over by the Crown and the LaBines were out. Disposing of Pregel was more difficult. He was excluded from all future contracts but he maintained those already in force. His contract ran until 1947 but the Canadian government sought to renegotiate it while they conducted an exhaustive survey of Pregel's business dealings with the LaBines. LaBine had been instructed to run Eldorado on a nonprofit but no loss basis during the war. Whereas the LaBines did the best accounting they could, following those instructions required an expertise the small company did not have and they could not match the unlimited number of government clerks now going over the books. The government was shocked to find that Eldorado had lost money every year and that Pregel received 83 % of the profit from Port Hope. Under pressure, the Eldorado secretary-treasurer, Mr. French, quit in January of 1945. Extensive investigations of Pregel were conducted twice by different government agencies and there was no conclusive evidence of wrongdoing. Nevertheless, the government wanted Pregel to relinquish his contract with Eldorado. He refused unless they paid him a monthly salary for the time left on his 1940–1947 contract.

The Canadian government waited until the war ended, but then brought criminal charges in December of 1945 against Mr. French, the Eldorado secretary-treasurer, accusing him of fraud and secret commissions. In February of 1946, they also charged Pregel, Mr. French, and M. Pochon with criminal conspiracy to defraud Eldorado and issued warrants for them. In the state of New York, they filed a civil case against the three and against the Canadian Radium and Uranium Company and its successor, International Rare Metals Refinery Inc., charging fraud. Canada wanted US$ 2,663,362.00 in damages. Pregel counter-sued for US$ 1,171,078.00. The two sides were in court during all of 1946 [11]. In March 1947, they settled out of court. All the criminal charges were dropped. In the civil suit, Pregel paid Eldorado US$ 576,000 in cash and US$ 537,044.48 "in kind," which may have been in uranium and/or radium stocks. Pochon returned to Port Hope, where there was much mutual admiration [11].

After the war, the Pregels continued to prosper in the radioactive mineral business. In 1948, the Canadian Radium and Uranium Company purchased the old Caribou silver mine in Nederland, Colorado, which had some uranium. The Pregels also purchased other claims in the American Southwest. Their company became one of the largest concerns in the mining and refining of radioactive minerals. They were also involved in making neutron and polonium sources and making Standards available for calibration and measurements. Boris was the founder and president of Conrad-Hanovia Inc. of Newark, which made ultraviolet equipment. In 1949, Pregel was on hand when General D. D. Eisenhower was awarded the Grand Medal of the Association Des Ingenieurs, Docteurs de France for his engineering work in World War II. The only other Americans to receive the Medal were Pregel and Herbert Hoover [30, 31, 34].

In 1950, the House Un-American Activities Committee investigated the sale of some 0.5 t (0.23 kg) (1000 lbs) of uranium to the Soviet Union by the Canadian Radium and Uranium Company in 1943. This was obviously necessary to keep the Russians from suspecting anything about its value. The sale had been approved by the Manhattan Project and paid for by Lend Lease, according to Secretary of State Dean Acheson. The Pregels were never called to testify [27, 31, 39]. In 1957, New York state targeted the Canadian Radium and Uranium Company in Mt. Kisko, for exposing employees to excess radiation while working with radioactive materials for medical, research and industrial use. This was the first prosecution by the State and the Pregels pleaded guilty. The judge dismissed the case as there was no evidence of any harm done and nothing further could be gained by a fine [35–37, 39].

Boris acquired a personal fortune, according to the Federal Bureau of Investigation, and was known for his lifelong philanthropy. Before the war, he was made an officer of the French Legion of Honor for his services as head of the International Organization to Fight Cancer. He was vice president of the French Pro-Palestinian Committee and the vice president of the Jewish Health Society. He was awarded gold medals by the City of Paris and the French Professional Engineers. He received the Grand Cross of Order of Merit for Research and Invention [32–34]. In New York, he was president of the French University (Ecole Libre) in New York City. He served as vice president of the American Geographical Society. He was a Trustee of the New School for Social Research [28–31, 34]. From 1966 to 1976 he was president of the American Division of the World Academy of Arts and Science. He was president of the New York *Academy of Sciences* in 1958 and chairman of the Board from 1959 to 1965. After that, he was honorary chairman of the Board of Governors for life. He is still remembered today for the annual *Boris Pregel Award* for applied science and technology [34]. Boris died in 1976 at 83 and his brother Alex in 1998 at 91. The eulogy for the brothers by Alex's widow seemed apt: "*Considering they were 'on the run' all their lives, they accomplished a lot in between*" [30]. Leo Szilard, J.B. Conant, A. Sachs, G. B. Pegram and the Smythe Report all credited Pregel for his crucial role in making uranium available to the American war effort [15, 16].

8.5 Congo Crisis

Edgar Sengier (1879–1963) returned to Belgium in 1945 and was awarded significant decorations and medals for his wartime service. He continued as a director of the Société Générale and the Union Minière until 1949. Sengier remained on the company's administrative board until 1960 before retiring in Cannes, where he died [6, 27, 40]. In 1959, the Belgian profit from *Union Miniere* was 3.5 billion Belgian francs. The taxes paid to the Congo on this income furnished 50–66 % of the countries revenue. *Union Miniere* had annual sales of US$ 200 million in 1960, and they produced 60 % of the uranium, 73 % of the cobalt, and 10 % of the copper for the Western democracies. In the Congo, *Union Miniere* had 24 affiliates: chemical factories, railroads, and hydroelectric plants [4, 5, 40, 41].

The primitive Congolese sought independence in the 1950s and the United Nations pressured Belgium to reluctantly acquiesce on 30 June 1960. At the independence ceremonies, the Belgian king and Prime Minister Patrice Lumumba traded insults. The king extolled the virtues of his great uncle Leopold, one of the worst tyrants in history, while Lumumba spoke of the Belgium oppression and exploitation. *Lumumba* created an all-black *Armie Nationale Congolese*, devoid of any experienced officers. Joseph Mobutu was made colonel and chief of staff. The 100,000 white Belgians in the country began to get nervous. From 1960 to 1966 the Congo was engaged in a war of independence that became known as the Congo Crisis [41–43]. It resulted in the death of the Congo prime minister, the United Nations secretary-general Dag Hammarskjold and possibly 100,000 others [44, 45]. The Crises ended on 25 November 1965, when Colonel Joseph Mobutu seized power via the US Central Intelligence Agency and began a 30-year dictatorship from 1966 to 1996 [41–43].

In 1966, Mobutu nationalized all the major industries including *Union Miniere,* from which he seized $ 800 million of their funds. It became the state owned *Gecamines* (*Societe generale des Carriers et des Mines*). With the nationalization and plunder by Mobutu the *Gecamines* approached bankruptcy. *Gecamines* has recovered somewhat since Mobutu and is still one of the leading industries in the impoverished Democratic Republic of the Congo. The *Societe Generale de Belgique* was able to salvage some of the *Union Miniere* assets and resurrect the company. The *Societe Generale de Belgique* then merged Union Miniere with three of their other companies into UMICORE in 1989. This multinational materials technology company is based in Brussels [41–43].

Mobutu changed the name of the country to Zaire and replaced all the European names with African names in 1971. Goodbye Leopoldville, Stanleyville, and Elizabethville. Hello Kinshasa, Kisangani, and Lubumbashi. Mobutu amassed a personal fortune thought to be $ 4–5 billion while the country became even more destitute. When the Cold War ended in 1991, the West quit supporting him. He was overthrown in 1997, and died shortly afterward of prostate cancer [6].

Unfortunately, things only got worse for the second largest country in Africa whose 75 million people are among the poorest in the world in spite of their mineral riches. There were motley Congo wars from August of 1998 to July of 2003. Since 2003 there has been a modicum of stability under leader Joseph Kabila, who was actually duly elected in 2006 and 2011 [41–43].

However, the country has porous borders and weak defenses especially in the east, where neighbors Uganda, Zimbabwe, and Rwanda have routinely and systematically exploited the resources of diamonds, cobalt, gold and other minerals. Uganda and Rwanda have been invading the Congo for the last 18 years while pursuing their Tutsi and Hutu rebels. Over five million Congolese have died in the continuous warfare, over half of them children under 15. Rape is rampant and the country is a human rights disaster. Efforts by the United Nations, the European Union, and the USA have been of no help. As of 2012, the Congo was ranked as the world's least developed country. The Congo governments since 1877 have been [4]:

- 1877–1908 African Congo free state
- 1908–1960 Belgian Congo (King Leopold)
- 1960–1965 Republic of Congo at Leopoldville (P.M. Lumumba)
- 1965–1971 Democratic Republic of Congo (Colonel Mobuto)
- 1971–1996 Republic of Zaire (Col. Mobuto)
- 2003–Democratic Republic Congo (President J. Kaliba)

In 1959, in a misguided attempt at anticolonialism, the Belgium government was pressured into placing a nuclear reactor at the University of Leopoldville. During the Crisis of 1960–1965 an American Central Intelligence Agency agent was instructed to remove the uranium fuel rods from the Leopoldville reactor in order to turn it off; a task for which he had neither training nor desire to attempt. Not to worry, the rods were stolen [6]. The Union Miniere began sealing their mine shafts with concrete. The abandoned mining pits are now being scavenged daily by anywhere from 50,000 to 70,000 people digging for copper and cobalt, which are then smuggled out of the country. Uranium is also being smuggled out in spite of frequent confiscations at the border and United Nation's investigations [6, 46]. Between 2001 and 2009, over 50 instances of uranium smuggling were identified at the borders of the Democratic Republic of the Congo [6]. In July of 2006, the United Nations reported a huge shipment of smuggled uranium from Lubumbashi and was discovered at the Tanzania border bound for Iran [47].

References

1. Bruce M (1958) The shaping of the modern world, 1870–1914. Random House, New York
2. Bauer L (1935) Leopold the unloved, king of the Belgians and of wealth. Little Brown, Boston
3. Emerson B (1979) Leopold II of the Belgians, king of colonialism. St. Martin's, New York
4. Depelchin J (1992) From the Congo free state to Zaire 1885–1974. Codesria, Dakar
5. Adams A (1993) The origin and the early development of the Belgium radium industry. Environ Int 19:491–501
6. Zoellner T (2009) Uranium. Penguin, New York
7. Landa ER (1987) Buried treasure to buried waste-the rise and fall of the radium industry. Colo Sch Mines Q 82:1–77

8. Sayers RR (1933) Radium in medical use in the United States. Radiology 20:305–309
9. Lounsbury JE (1938) Famous Pittsburgh industries, the Standard Chemical Co. Crucible 22:86–89, 109–113, 134–137
10. Radium Belge (~1931) Radium. Union Miniere du Haut Katanga. Radium Dept., Brussels
11. Bothwell R (1984) Eldorado-Canada's national uranium company. University of Toronto, Toronto
12. Johnson L (1975) The Great Bear Lake: its place in history. Arctic 28(4):231–244. http://www.pubs.aina.ucalgary.ca/arctic/Arctic28-4-230.pdf. Accessed 22 Aug 2014
13. Gonzales T (2011) The Great Bear Lake: its place in history NWT-explorers. http://www.en.wikipedia.org/wiki/File:Great_Bear_Lake_(de).png. Accessed 15 June 2014
14. Groves LR (1962) Now it can be told. Harper & Row, New York
15. Rhodes R (1986) The making of the atomic bomb. Simon & Schuster, New York
16. Brown AC, MacDonald CB (eds) (1977) The secret history of the atomic bomb. Dial/James Wade, New York
17. Segre E (1980) From x-rays to quarks. Freeman, San Francisco
18. Feld BT, Szilard GW (eds) (1972) The collected works of Leo Szilard [1898–1964]. MIT Press, Cambridge
19. Jewish Virtual Library (2008) The pale of settlement. http://www.jewishvirtuallibrary.org/jsource/History/pale.html. Accessed 21 June 2012
20. JewishEncyclopedia.com (1906) Pale of settlement. http://www.jewishencyclopedia.com/articles/11862-pale-of-settlement. Accessed 21 June 2012
21. Klier J (2010) Pale of settlement. YIVO institute for Jewish research http://www.yivoencyclopedia.org/article.aspx/Pale_of_Settlement. Accessed 21 June 2014
22. Pogonowski IC (1988) Poland. A historical atlas revised edition. Dorset, New York
23. Pogonowski IC (1993) Jews in Poland, a documentary history. Hippocrene, New York
24. JewishEncyclopedia.com (1906) Odessa. http://www.jewishencyclopedia.com/articles/11660-odessa. Accessed 23 July 2012
25. Jewish Virtual Library (2008) Odessa. http://www.jewishvirtuallibrary.org/jsource/judaica/ejud_0002_0015_0_15016.html. Accessed 23 July 2012
26. Zipperstein SJ (2010) Odessa. YIVO institute for Jewish research http://www.yivoencyclopedia.org/article.aspx/Odessa. Accessed 21 June 2012
27. Hunt VR (1995) The Manhattan-uranium connection. http://www3.amherst.edu/~mrhunt/uranium/. Accessed 21 June 2012
28. Weart S (1975) Oral history transcript of an interview of Boris Pregel, the Neils Bohr library & archives. http://www.aip.org/history/ohilist/4833.html. Accessed 21 June 2012
29. Gauchman J (2005) Alexandra Pregel. http://www.bnphoto.org/pregel/Bio.htm. Accessed 21 June 2012
30. Moorhead S (1998) Alexander Pregel. Sarasota Herald-Tribune, 11 June. http://www.news.google.com/newspapers?nid=1755&dat…id. Accessed 21 June 2012
31. Edson P (1949) Pregel colorful figure behind uranium sales. The Owosso (MI) Argus-Press, p 9, 13 Dec. http://news.google.com/newspapers?id=SF4iAAAAIBAJ&sjid=ZqoFAAAAIBAJ&pg=864,14725 74. Accessed 21 June 2012
32. Jewish Telegraphic Agency (1939) Boris Pregel French philanthropist named to Legion of Honor. 24 Aug 1939. http://www.cdn.jta.org/archive_pdfs/1939/1939-08-24_020.pdf. Accessed 21 June 2012
33. Korf SA (1977) Boris Pregel. Phys Today 30(5):98–99
34. New York Times (1976) Boris Pregel obituary. 9 Dec 1976. http://www.query.nytimes.com/gst/abstract.html?res=9C0CE3DB123EE334BC4153DFB467838D669EDE. Accessed 21 June 2012
35. Walker CL (1947) You can rent a little atomic energy. Nation's Bus 35(10):60–65. Nations Business 1912 to 1999, October 1947, Hagley Digital Archives & Hagley Museum and Library. http://digital.hagley.org/cdm/compoundobject/collection/p16038coll3/id/104751/rec/3. Accessed 21 June 2014

36. DOE Office of Legacy Management (2009) International Rare Metals Refinery, Inc. (Mt. Kisko NY). http://www.lm.doe.gov/Considered_Sites/International_Rare_Metals_Refinery_Inc_-_NY_38.aspx (a) http://www.lm.doe.gov › LM Sites › Considered Sites. Accessed 21 June 2014
37. Wall St J (2013) International Rare Metals Refinery, Inc. http://www.projects.wsj.com/wastelands/site/212-international-rare-metals-refinery-inc/. Accessed 21 June 2014
38. Scott WR (1943) The technical exhibits. Radiology 40(Jan):94–95
39. New York Times (1998) Alexander Pregel, uranium dealer, dies at 91. http://www.nytimes.com/1998/06/28/us/alexander-pregel-uranium-dealer-dies-at-91.html. Accessed 21 June 2012
40. American Heritage Foundation (2014) Edgar Sengier. http://www.atomicheritage.org/profile/edgar-sengier. Accessed 22 Aug 2014
41. United States Central Intelligence Agency (2011) World Fact Book Democratic Republic of the Congo. http://www.cia.gov/library/publications/the-world-factbook/geos/cg.html. Accessed 15 June 2014
42. Trueman C (2012) The United Nations and the Congo. History Learning Site. http://www.historylearningsite.co.uk/united_nations_congo.htm. Accessed 26 June 2014
43. U.S. Dept. State (2013) Milestones 1961–1968 The Congo, decolonization & Cold War. http://www.history.state.gov/milestones/1961-1968/pl-480. Accessed 26 June 2014
44. Björkdahl G (2012) I have no doubt Dag Hammarskjöld's plane was brought down. The Guardian: 2011-08-17. http://www.theguardian.com/world/2011/aug/17/dag-hammarskjold-crash-goran-bjorkdahl. Accessed 17 Aug 2012
45. Borger J (2014) Dag Hammarskjöld's plane may have been shot down: theguardian.com, Friday 4 April 2014 09.27 EDT. http://www.theguardian.com›News›World news. Accessed 26 June 2014
46. Slavin B (2007) Funds told to cut ties to Iran. USA Today, 7 March
47. Swain J, Deppard D et al (2006) Iran's plot to mine uranium in Africa. Sunday London Times, 6 Aug

Chapter 9
Miner's Plague

9.1 Agricola

The printing press was invented around 1440 in Mainz, Germany, and quickly spread throughout the Holy Roman Empire including northern Italy. This new tool was quickly utilized by the Protestant Reformation, the mining industry, and the medical profession. During the first half of the sixteenth century, the Heiliges Römisches Reich was blessed with three luminaries and outstanding authors, whose printed works found a favorable audience. This triumvirate included Martin Luther (1483–1546) in religion, Paracelsus (1493–1541) in medicine, and Agricola (1494–1556) in mining.

Agricola became one of the most venerated mining scholars of his age. His publications became the standard reference texts for over 200 years. He was a Saxon physician turned metallurgist and mineralogist who took residence in Sankt Joachimsthal (SJT) Bohemia during 1527–1530. He was George Bauer, who later became known by his Latin pen name, Georgius Agricola. His family name, Bauer, meant peasant or farmer, but he became highly educated and was anything but a peasant. By 1530, he had adopted the name Agricola (Latin: farmer). He was born in Glauchau, Saxony, some 7 miles from Zwickau, where he is buried, and 17 miles from Chemnitz, where he would live most of his life. He would remain a Roman Catholic all his life, even though Saxony and Bohemia were then fast becoming the cradle of the Protestant Reformation [1–6].

He was educated in the Classics at the University of Leipzig (founded 1409), receiving his BA in 1518. He then taught Latin and Greek at the Zwickau Municipal School during 1518–1522 and became a Rector (Principle) there. He returned to Leipzig in 1522/1523 as a lecturer and to study the natural sciences as well. After the untimely death of a colleague at the University and amidst increasing religious controversy, Bauer left Saxony for Catholic Italy. He studied the sciences as well as Hebrew and Arabic at Padua, Venice, and at the University of Bologna, where he obtained a medical degree in 1526. In Italy, he met the famous *Erasmus of Rotterdam,* who was now the editor of Froben's Press in Basel [2].

After returning to Saxony, Bauer received an appointment as the city physician for the Hapsburg's newly acquired Imperial mining town of *Sankt Joachimsthal.*

R. F. Robison, *Mining and Selling Radium and Uranium,*
DOI 10.1007/978-3-319-11830-7_9

Table 9.1 Agricola bibliography [2]

A.	Published works in mining–mineralogy–geology: #10
1530	Bermannus sive de re mettallica dialogues (dialog on metallurgy)
1533	De Mensuis et Pondeibur (Greek/Roman weights and measures)
1550	De Precio Metallorum et Monetis (second edition of the above)
1546	One volume of 5 works: revised editions were reprinted 1558 1-De Ortu et Causes Subterraneorum (physical geology/erosion); 5 books 2-De Natura Eorum quae Effluent ex Terra (subterranean water and gases); 4 books, 83 pages 3-De Natura Fossiliium (classification of 80 minerals = fossils); 10 books, 217 pages 4-De Rerum Metallicarum Interpretatio (German glossary for 500 Latin terms) 5-De Veteribus et Novis Metallis (history of mining), 2 books, 31 pages
1549	De Animantibus subterraneis (subterranean animals) 23 pages
1556	De Re Metallica (nature of metals/minerals); 12 books
B.	*Published on other subjects: #5; including medical text De Peste in 1554*

He served from 1527 to 1530. He spent all of his free time visiting mines and smelters and reading the Greek and Latin authorities. *Sankt Joachimsthal* was then the most prolific metal mining district in Central Europe and it was within 50 miles of the other mining centers of Freiberg, Schneeberg, Geyer, Annaberg, and Altenburg. In 1530, he published the first of the ten known texts by him on mining and metallurgy. It was published by Froben in Basel with a forward by Erasmus. It was in the form of a dialogue with a "most learned miner" on mineralogy, mining terms, and history. His mining text of 1546 consisted of five works collected in one book [2] (Table 9.1).

Presumably, Bauer spent 1530–1533 in travel and study. After his second book in 1533, he obtained a position in Chemnitz, Saxony, as their city physician, a post he held until his death. At this time, he may have been subsidized by the Roman Catholic Elector of Saxony. In 1541 the Duke Maurice, a Protestant, became the Saxon Elector. He and his brother Augustus, who would succeed him as Elector in 1553 both became patrons of Bauer. From 1533 through 1546, Bauer was also called to various districts as a mining consultant. Bauer was probably married in SJT and then married again in 1543 to a Chemnitz widow; he had several children [1–6].

His magnum opus was De Re Metallica. He probably began work on De Re Metallica during 1530–1533, finished it in 1550, and sent it to press in 1553. It was published in 1556, just after his death and was composed of 12 books (chapters). The 624 pages included almost 300 wonderful woodcuts [1]. The delay in publishing was probably due to the preparation of the entertaining, illuminating, and detailed woodcuts, which are used to illustrate the text. Fourteen dogs, two goats powering a tread mill, two birds, one peacock, six women, one wife, a mother and baby, and a couple kissing are depicted among the 292 otherwise businesslike woodcuts. All types and manner of tools, instruments, and machines are illustrated. It was written in Latin for publication in Basel but was soon reprinted in German and Italian for international publication.

Table 9.2 Table of contents for D.R.M [2]

Book	Subject: # of pages	Wood-cuts #	Book	Subject: # of pages	Wood-cuts#
I	Art of mining/metallurgy: 24	00	VII	Assaying: 48	14
II	Miners and prospecting: 18	01	VIII	Extraction: 86	65
III	Recognizing ore veins: 34	34	IX	Smelting: 86	34
IV	Claims, officials and laws: 24	09	X	Precious metals separation: 52	17
V	Underground mining; surveying: 48	23	XI	Separating Ag, Cu, Fe: 54	19
VI	Tools and machines: 70	50	XII	Making salt, glass: 48	26

In 1912, an English translation of De Re Metallica was published by the *Mining Magazine* in London. The translation was accomplished, over 5 years, by Stanford University graduate and mining engineer Herbert C. Hoover and his wife Lou Henry Hoover, a geologist and Latinist. Hoover later served as the 31st US president during 1929–1932, amidst a lengthy career as a public servant. The 1912 Hoover translation was reprinted in 1950 by Dover Publications in New York [2]. This edition contains 638 pages with 292 woodcuts and a 13-page biography of Bauer.

The Table of Contents for De Re Metallica is given in Table 9.2. Book VI on iron tools and machines has two woodcuts which may be the first depictions of the use of special wheeled cars on timbered railroad tracks; mine carts. Book VI also discusses the various machines then utilized in the mines for pumping water out and fresh air in. Until the eighteenth century and the invention of the steam engine, the accumulation of water in deep mines would eventually force their abandonment. Agricola devoted 15 pages to the construction of three different types of ventilating machines which could be used to bring fresh air into the mines using: (a) wind, (b) fans, and (c) bellows (Fig. 9.1).

Agricola devoted only the last 4.5 pages of Book VI to the ailments and accidents of miners. Half of that text is used to describe the poisonous dust that affects the lungs. Agricola recognized the dangers to the lungs from dust, stagnant air, and toxic vapors and fumes. He cautioned that: "…the dust which is stirred and beaten up by digging penetrates into the windpipe and lungs and produces difficulty in breathing…." And "If the dust has corrosive qualities, it eats away the lungs and implants consumption within the body; hence in the mines of the Carpathian Mountains women are found who have married seven husbands all of whom this terrible consumption carried off to a premature death" [2]. The "widows of the Carpathians" is an often repeated story but seemingly unrelated to radiation which takes 15–20 years of exposure to produce lung cancer. The Carpathian miners were more likely to have died from acute toxins or tuberculosis than exposure to radioactive ore, which is not known to be present in the Carpathians.

Fig. 9.1 Bellow machines for mine ventilation, 1565, [1, 2]. (Public domain)

For minerals like arsenic and zinc that are toxic to the skin, Agricola recommended boots and gloves. For the dust, he advised loose veils (masks) over the face and eyes. Ever the Classicist, he cited Pliny (23–79 AD) for recommending veils of bladder-skin worn across the face to protect the trachea and lungs. Lace veils were also made locally but despised by the miners. For the stagnant air, he advised the aforementioned ventilating machines. For really noxious fumes, he suggested the digging of an adjacent air shaft to draw off the bad air. In 1956, 400 years after this sage advice, uranium miners in Saxony and the USA were still working in unventilated mines.

Bauer, at age 52 had begun to serve his patron Maurice as a diplomat and a public servant. In *1546,* he was elected Burgher (Mayor) of Protestant Chemnitz and the Elector appointed him Burgomaster of the district for four terms. He was made a member of the Diet in Freiberg. He was sent on diplomatic missions by the Protestant Duke to the Catholic Holy Roman Emperor, Charles V, and to his brother, King Ferdinand of Austria-Bohemia. Bauer remained a staunch Roman Catholic while serving his Protestant Sovereign and patron in a Saxony that was militantly Protestant. However, his friend Fabricius reported that Bauer "would not tolerate with patience that anyone should discuss ecclesiastical matters with him."

Agricola died in 1556 at age 62 and may have suffered a stroke after a theological discourse with a Protestant cleric. The Hoovers reported that he died after a 4-day fever. His burial proved a problem for his Protestant patron. Bauer's family evidently wanted him buried at the Cathedral Church of St. Jacob (Sankt Jacobkirke) in Chemnitz. There were few if any Catholics left in Chemnitz and the Cathedral was now Protestant. In spite of his prominence in the Protestant city and his position as Burgomaster, Agricola was denied burial at St. Jacobs. Instead, he was interred 50 km away; just outside the Catholic Cathedral at Zeitz [2–5].

During 1908–1912, the Hoovers found little in Saxony to commemorate Agricola. A memorial brass plate in the Zeitz Cathedral had disappeared by 1686. At Chemnitz in 1822, a commemorative marble tablet had been placed behind the altar at St. Jacob; of all places! This was later removed to the nearby Historical Museum. In the twenty-first century, another team of science-historians did find Agricola memorials in Zeitz, Chemnitz, and Glauchau [3, 4] (Fig. 9.2). A replica of the stolen brass plate was returned to the Zeitz Cathedral in 1935. Chemnitz now features plaques and a bust of Agricola in the Agricola Cloister, which is located across the street from the old city hall, adjacent to St. Jacob. In Glauchau, the Castle now contains a historical museum with an entire room devoted to Agricola and a large statue of him now stands near the train station (Fig. 9.3).

Fig. 9.2 Jim and Virginia Marshall from the University of North Texas. (Courtesy J. L. and V. R. Marshall, "Rediscovery of the Elements," 2010, ISBN 978-0-615-30795)

Fig. 9.3 Statue of Agricola in Glauchau [3]. (Courtesy J. L. and V. R. Marshall, "Rediscovery of the Elements," 2010, ISBN 978-0-615-30795)

9.2 Miner's Medical Maladies

In 1524, a brief leaflet (*flugblatt*) on an occupational hazard was published in Augsburg, Bavaria. The author was Ulrich Ellenbog (1459–1499) the town physician. He compiled the pamphlet in 1473 as a warning to the Augsburg goldsmiths about the dangers from the vapors and fumes of the metals, with which they were working. A few simple precautions and remedies were offered [7]. Like many items written in the late fifteenth century, it was not printed until the number of printing presses multiplied throughout Western Europe. The pamphlet appears to be the beginning of a medical interest in occupational diseases [7].

A contemporary of Agricola was Paracelsus (1493–1541), another physician with a background in minerals and metals, but one who applied that knowledge to pharmacology and patient therapy rather than to writing texts on mining technology. Philippus Aureolus Theophrastus Bombastus von Hohenheim (1493–1541) was born in Switzerland to a father, who was allegedly a physician. As was then the custom, Phillip later adopted a Latin pen name: Paracelsus. After his medical (MD) degree at the *University of Ferrara* in 1515/1516, he became an itinerant, truculent iconoclast, who said much but could afford to publish little. By attacking the status quo of medieval medicine, he earned the sobriquet by Osler as the "Luther of medicine, for when authority was paramount he stood out for independent study." After his death, at age 47, his reputation soared during the Renaissance. It was an era which sought to question and cast off the authoritative teachings of the past and replace them with newly acquired information [8–10].

During 1589–1591, Basel publisher Johann Huser edited and printed ten volumes of his work. Since then, multiple other volumes of his disorganized writings have been collected by various scholars of the arcane sixteenth century [11, 12]. In the USA, there are collections at Hahnemann University in Philadelphia, at Washington University of St. Louis, and at the National Library of Medicine. The University of Zurich has established a Paracelsus Repository, which has 30 volumes of his collected works [13]. The five most well-known tracts of Paracelsus were about (a) the sensible treatment of wounds, (b) chemical and mineral therapy, (c) the treatment of syphilis (lues), (d) mental illness, and (e) the diseases of miners. His most famous aphorism was "the dose makes the poison." He may not have coined the anonymous "one night with Venus, a lifetime with Mercury" which refers to the treatment of lues, but it certainly fits his persona. [8–10, 14, 15].

In 1537, Paracelsus is credited with writing the first *monograph* on the diseases of a specific occupational group: miners. He had worked at a smelting plant in the Tyrol and at the Fugger mines in Vallach. His monograph, "*Von der Bergsucht oder Bergkrankheiten drey Bucher* " or "About miner's illness or miner's diseases, three books," was published posthumously in *1567*. The first two books deal with the diseases of miners and smelters while the third discusses the toxicity of mercury. He wrote in the German vernacular and this was translated into English in 1941 by George Rosen [15]. He identified the predominant problem for miners as lung disease from airborne toxins, known as well as unknown. He recognized the toxic nature of arsenic, sulfur, mercury, and antimony. He was silent regarding prevention. He had no helpful suggestions for avoiding the dust problems and never mentions masks or ventilation. For inhaled dust, he was keen on the healing power of plain salt water used externally or within the nose, nasopharynx and oropharynx. He was more concerned with pharmacological therapy than with avoiding exposure [15, 16].

Table 9.3 Ramzzini tradesmen disorders [7, 16, 17]

1710 edition	1710	1710	1713 edition
Gilders	Midwifes	Runners	Printers
Inunction healers	Wet-nurses	Grooms	Scribes-notaries
Chemists	Vinters-brewers	Porters	Candy M.
Potters	Bakers-millers	Athletes	Weavers
Tinsmiths	Starch M.	Eye strain	Coppersmiths
Glass W.	Grain W.	Voice abuse	Wood W.
Mirror M.	Stone cutters	Farmers	Sharps grinders
Painters	Gypsum-Lyme W.	Fishermen	Brick M.
Sulfur W.	Laundresses	Camp diseases	Well diggers
Blacksmiths	Bathmen	Metal Miners	Sailors
Apothecaries	Salt M.	Corpse bearers	Hunters
Fullers	Standing W.	Jews	Soap makers
Tobacco workers	Sedentary W.	Scholars	1714 edition
Privy—Cesspit cleaners	Dirty: oil presser, tanner, cheese M.	Flax, Hemp and Silk W.	Princes

W. workers, M. makers

Bernardino Ramazzini, 1633–1714, was truly the Father of Occupational Medicine as no one before him had written a *comprehensive text* on occupational diseases and their prevention. Unlike Paracelsus, who traveled the globe, Ramazzini never left northern Italy and rarely left home [7]. Born in Capri, with a medical degree from Parma, he was appointed at age 49 to an academic chair at the University of Modena, about 33 miles from Parma. At age 67, he was selected to the Practical Medicine Chair at the celebrated University of Padua, some 70 miles away [7, 16]. That same year, 1700, he published his magnum opus, "*De Morbis Artificum Diatriba,*" or "*Diseases of Workers*" which addressed the acquired ills of all manner of tradesmen [17]. In the first edition, he identified the health hazards of 42 different occupations, including those of scholars whom he believed to suffer melancholia. He added another 12 occupations to the list for a second edition in 1713 (Table 9.3). Demonstrating his practical nature, Ramazzini published both editions in a *pocket size* so that practitioners would have it at hand. "What is your occupation?" became an essential part of a patient's medical history and physical exam [7, 17].

The very first chapter of this book is devoted to miners. In addition, more than half the chapters identify the airborne particles from materials handled and taken in "*per os et nares*" (by mouth and nose) as hazardous. He included dust from mineral, vegetable, and animal origin. He described lung fibrosis from dust inhalation as a disease of miners and stone masons. In addition to the injurious effects of noxious vapors, Ramazzini also investigated the effect of body position on a worker's musculoskeletal system. He studied the problems of violent and irregular motions and the unnatural postures required of workers: what is now called *ergonomics.* Ramazzini was more concerned with prevention than treatment. He advocated personal cleanliness, hand and face washing, clean clothes on weekends, ventilation, temperature control, rest, exercise, and moderation in diet and drink. And, if all else failed, "quit the job" [7, 17].

He was the personal physician to the Lords of Este, who subsidized his studies. By 1714, it had gradually occurred to Ramazzini that his benefactors were all dying at ages 34–48. For his wealthy but unhealthy patrons, he wrote about avoiding the "Diseases of Princes": stone, gout, and arthritis. As there seemed to be a limited market for this text, his publisher declined this one. Ramazzini published it himself and it was a best seller. However, no Prince was likely to have read it because it advocated sobriety, exercise, and moderation [7, 16, 17].

9.3 Radon

Radon was discovered between 1899 and 1908 as a gaseous emanation (decay product) from first radium, then from thorium, and finally from actinium. Friedrich Ernst Dorn (1848–1916) and the Curies found it in 1899 as a decay product of radium. During 1900–1903, Ernest Rutherford and R. B. Owens, in Montreal, found it as a product from the decay of thorium and radium. André-Louis Debierne (1874–1949) found it as a decay product from actinium in 1903; see Appendix IV; Radioactive Decay Tables. The element was described as an emanation (gas) decay product of radium, thorium, and actinium until 1923 when "radon" was accepted as less confusing [18, 19]. Like all the elements beyond # 83, there are no stable isotopes of element # 86, radon (Rn). Instead, there are 36 radioactive (unstable) isotopes of radon that have an atomic mass between 193 and 228. Each isotope has its own characteristic half-life ($T^1/_2$). Radon-222 has the longest half-life and therefore it is used to represent radon in the Periodic Table [19–22].

Radon-222 creates *four* very short-lived daughters during its decay into Pb-210, which has a relatively long half-life of 22.3 years. Two of the daughters decay by *alpha* particle emission and two decay by *beta* emission. The release of some energy through gamma rays usually accompanies both types of decay. The two radon daughters that decay by *alpha* particle emission are Po-218 and Po-214, both of which emit high-energy alpha particles of 6.0–7.7 million electron volts (MeV) [20–23]. See Appendix V Uranium Decay Emissions.

Inhalation of radon gas on a continual basis can lead to lung cancer. Radon itself, a gas, does not adhere to the respiratory tree and its alpha particles are not a problem. However, Po-218 and Po-214 have become known as the "deadly daughters of radon" because of their association with lung cancer in underground miners. The polonium daughters of radon are solid particles, which can adhere to both dust particles and respiratory epithelium. The alpha particle is essentially a helium nucleus with two protons and two neutrons; $^4He_2$. When discharged from the unstable polonium isotope, the alpha particles travel only a few centimeters and are stopped by lodging in the superficial layers of the respiratory tree. The barrage of alphas on the bronchial epithelium leads to bronchogenic carcinoma after 15–20 years; sooner if one smokes tobacco as well. Polonium has also been detected as an agricultural contaminant on tobacco leaves and subsequently in cigarettes. Moreover, polonium has been used as a poison by some espionage agents attempting a covert murder [22–24]

Uranium is now known to be ubiquitous in the earth's crust. It is more common than tin and 500 times more common than gold. Wherever uranium or thorium is found, the decay products of radium and radon occur. In areas where the soil has granite or shale, there are higher concentrations of uranium. A square mile of surface soil, to a depth of 6 in., can contain as much as 1 g of radium, which is releasing radon. Radon gas migrates freely through faults and fragmented soils but may accumulate in caves or in water. Water will release the radon into the atmosphere by volatilization [22–24].

E. Rutherford, in 1900, suggested that radon might be found in the famous mineral baths and springs of continental Europe. Mineral baths have had a long and colorful history in Europe since the Middle Ages. By 1904, Vienna physicist Heinrich Mache (1876–1954) had found radon at Bad Gastein in Austria. Mache performed the first measurements of radon concentrations and introduced the Mache-Einheit (ME) or Mache unit. The Mache unit (ME) has now been replaced in Europe by the Becquerel unit and in the USA by the Curie unit. The Curie unit (Ci) is currently defined as 2.2 trillion decay events per minute, the activity of 1 g of radium [19, 20]. That is such an awkward number that Curies are usually expressed in picoCi (10^{-12}), nanoCi (10^{-9}), microCuries (10^{-6}) or milliCuries (10^{-3}):

- One Curie unit is 2.2 trillion disintegrations per second. One picoCi is 10^{-12} disintegrations/sec.
- One Becquerel (Bq) unit is one disintegration per second or 60 a minute.
- One Bq = 27 picoCi/L.
- One picoCi = one pCi = 0.037 Bq.
- The old ME unit is equal to 364 picoCi/L or 13.45 Bq/L.

The first radon treatment centers were at the traditional mineral spas. The Germans have always been keen on these spas as a treatment for arthritis and musculoskeletal discomfort. By 1906, the radon content of over 30 springs (*Quelle*) in 11 spas in Austria, Bohemia, Germany, and Italy had been published [22, 25, 26] (Table 9.4). At *Oberschlema,* Saxony, they sank 12 wells and built the Radiumbad Spa Hotel, which was only 4 h by train from Berlin. By 1918, they were seeing 3000 patients annually and even in 1943, saw 17,000 [25, 26]. Patients could bathe in radon-rich water without danger as it is quite volatile in open air. Inhaling radon was limited

Table 9.4 Radon content of spa springs: 1906 [22, 25, 26]

Quelle (springs):	pCi/L	Bq/L
Oberschlema, Bad Schlema: Hindenburg Quelle	4.91 million	182,000
Bad Gastein, Austria	2.16 million	80,000
Baden-Baden, Germany	1.89 million	70,000
Bad Bramberg, Germany: Wettiner Quelle	700,000	26,000

to a "therapeutic dose" of 50 ME (18,200 pCi/L). For a spa to be classified as having therapeutic levels of radon in their mineral water, the concentration of radon had to be over 2000 pCi/L. The average indoor/household concentration of radon is 1–2 pCi/L.

Radon spas were also opened at mining centers where the radon-rich water was siphoned off from the mines. Heinrich Mache and Stefan Meyer detected radon in the mines of *Sankt Joachimsthal,* Bohemia, in 1905. Their investigation was probably more concerned with the exploitation of the radioactivity for a spa than with any consideration of a health hazard. The Svornost mine there dated back to 1530 when it was mined for silver. In 1864, it had become flooded when a strong current of spring water was encountered at a depth of 500 m. After the identification of radon, a local entrepreneur began to collect the spring water in wooden buckets and offer healing baths in his bakery. Or one could bathe in a water barrel in which uranium ore residues were wrapped in a cloth bag [22, 25, 26]. An official spa/bath house was established in 1906 with two bath tubs and 43 patients were treated. Horse power was used to haul buckets of water out of the mine until 1908 when a pipeline was installed to carry the water to bath cabins. The Radium Palace/Spa Hotel opened in 1912, and 2476 patients were treated in 1913. Today, there are four springs and multiple new hotels servicing the world famous Jachmov Health Spa [4].

9.4 Schneeberg Lung Disease

Since the sixteenth century, Erzgebirge Saxon miners were reported to be at risk of dying prematurely from a chronic disease of mysterious origin. It was characterized by a productive cough and then progressive dyspnea and weight loss. Local physicians in the area of Schneeberg, Saxony, began to refer to it as Schneeberg lung disease, in order to differentiate it from the vague and nonspecific terms then in use: *Begrsucht and Bergkrankheit.* As the steam engine was being developed during the eighteenth and nineteenth centuries, water could be removed from deep in the mines. Mine shafts began to be dug deeper and deeper as mining ventures increased in the Schneeberg District.

Simultaneously, the incidence of Schneeberg lung disease and the death rate for the miners increased [6, 27–30].

Around 1860, a new institute for pathology opened at the University of Leipzig. Local physicians in the Schneeberg district began to do autopsies on miners with Schneeberg lung disease. They sent the lungs to Leipzig, where the cause of death in most cases was found to be bronchogenic carcinoma, then a rare disease. In 1879, the local mine physician, Dr. Fred Hartung, and the district public health official, Dr. Walter Hesse, published the autopsy reports; lung cancer was responsible for 75% of the deaths in the local miners. It might have been higher if the miners had not been dying prematurely of other problems, such as accidents and other lung diseases. Miners' life expectancy was only 20 years after starting work underground [27–30].

Hartung and Hesse reported that the autopsied lungs were black and that the dust inhaled by the miners measured 6 g per shift of 7 h. The physicians thought that the cancers were due to both: (a) inhaled arsenic, a known carcinogen, and (b) co-existent pneumoconiosis. Friedrich Albert Zenker had coined the term "pneumoconiosis" in 1867 as a general term for the destruction of lung tissue from inhaled dust containing various metals or fibers [12]. Arsenic, nickel, and silica dust were also suspected etiological agents for Schneeberg lung disease. Harting and Hesse sent an inquiry to Joachimsthal about their incidence of lung cancer and got a "negative reply" [29, 30].

Radon was identified in the Schneeberg mines in 1909. In 1913, Carl Schiffner from the Freiberg Mining College (founded 1765) reported on the high *radon* levels in the air and water of the Schneeberg mines that were measured during 1908–1912. Also, in 1913 Alfred Arnstein reported statistical data concerning the cause of death in Schneeberg miners from 1875 to 1912. Lung cancer accounted for 42 % (276/665) of all the deaths and another 18 % (119/665) died of lung disease of unknown cause including suspected lung cancer; giving a possible total of 60 %. This study from Freiberg began 30 years of classical studies, mostly by German scientists, of the relationship between radon and lung cancer [29–31]; see *Supplemental Material.* The studies on Schneeberg lung disease (SLD) may be summarized as follows:

1879: Hartung and Hesse	Found SLD is lung cancer and kills 75 % of miners
1909: C. Schiffner et al.	Found radon elevated at Schneeberg mines
1913: A. Arnstein	Found radon elevations and lung cancer continues
1913: H. E. Muller	Found lung cancer related to radioactivity in local mines
1924: P.Ludwig and E. Lorenser	Found SLD could be caused by radioactivity
1926: Rostoski et al.	Found high radon levels persisted; cancer due to several factors
1929: J. Lowy	Found lung cancer present at Jachymov
1930: H. Siki	Found Jachymov lung cancer present and due to radon
1935: K. Neitzel	Found radon responsible for SLD
1935: E. Lange	Found entire Erzgebirge radioactive and SLD due to radon
1939: B. Rajewsky	Radon accepted as the etiology

SLD Schneeberg lung disease

In 1913, a local official, H. E. Muller, of the Miners Social Service in Schneeberg noted that the lung cancers seemed to have increased since the advent of *radium* mining; he suspected a connection. Muller wrote: "Hence I consider Schneeberg lung disease to be a particular kind of occupational disease, acquired in mines having rock strata containing radium and hence showing high emanation (radon) levels" [30]. In 1921, physician Margaret Uhlig, who was born and raised in the Schneeberg area and practiced there, again suggested that radon was the etiological agent responsible for Schneeberg lung disease. She also complained to the State Health authorities that the mine owners had done nothing to eradicate the radon hazard first noted in 1913 by Schiffner [29, 30].

Table 9.5 Radon activity in Erzgebirge 1908–1935 [29–35]

Years	Area mines (study group)	pCi/L	ME
1908–1924	Schneeberg range (Freiberg)	410–18,200	1.1–*50*
1929–1930	Jachymov range (Prague)	410–18,200	1.1–*50*
1930	Czech suggested standard	364	1.0
1934	Schneeberg average: Neitzel	14,560	*40*
1935	Neitzel: Schneeberg avg. during week On Monday morning	12,740–16,000 40,400	35–44 111
1935	Erzgebirge miner's yearly dose: Lange	30,576,000	84,000
1940	Sudetengau (Brandt and Rajewsky)		3–5

One ME = 364 pCi/L

In 1924, physicists at the Radium Institute of the *Freiberg* Mining College found that the radon levels in the Schneeberg mines were at the same levels as when measured by Schiffner during 1908–1912; from 410 to 18,200 pCi/L or from 1 to 50 ME (Table 9.5). The worst mine was nicknamed the *Todesschacht* or Death Shaft because of high worker mortality there. It had 148 ME of radon. The Institute physicists, Paul Ludwig and Eduard Lorenser, came to the conclusion that Schneeberg lung disease could be caused by radioactivity; uranium decay. They stated: "It seemed natural to suspect…lung cancer arises from the continuous inhalation of air, in the mines, containing radiation." However, the medical physicians at the Institute disagreed, as they also found pneumoconiosis (lung fibrosis) on all the X-rays of all the lung cancer victims. They believed that the cancers were the result of the pneumoconiosis [29–31].

In 1926, a team of experts from Dresden (Rostoski, Saupe, and Schmorl) reported on five detailed surveys of Schneeberg lung disease during 1922–1926. It appeared that 71 % (15/21) of the deaths were due to lung cancer, a percentage similar to the 1879 report of 60 %. Pneumoconiosis was seen on X-ray in 20 % of the miners (34/154) including the 15 with cancer. The Dresden experts concluded that pneumoconiosis was "preparing the way for the lung cancer." They also studied 167 dye factory and spa workers in nearby Oberschlema as well as 186 other non-miners in the area, during 1922–1926. In this group of above-ground workers, they found only two lung cancers in the 353 workers [29–30].

Workman's Compensation was a social insurance system that had been pioneered in Germany in 1883 as part of their national health care system. This was known as *Krankenkassen* or Sickness Funds. The German government, in 1925, compiled a list of the occupational diseases that could be considered for Workman's Compensation. In 1926, Schneeberg lung disease was now included for compensation; but only if there was concomitant pneumoconiosis. Radon was *not* recognized as an etiology [27–31].

After World War I, 1914–1918, the Austrian-Hungarian Empire was dismembered and Czechoslovakia was created. In spite of the principles of self-determination being championed by American President Woodrow Wilson, the Sudetenland with its German majority was incorporated into the new Czechoslovakia. However, the Bohemian miner's lung cancer problems now came under increased study by the Czechs. Measurements of the radon levels in the mines of Sankt Joachimsthal, now Czech Jachymov, were initiated in 1924 and continued until the 1938 Nazi occupation. From 1924 to 1939, Jachymov had about 400 miners and 60 workers in their radium factory [31]. In 1929, Dr. Julius Löwy of Prague reported two cases of lung cancer among the Jachymov miners, the first such evidence of Schneeberg lung disease in Bohemia [32]. He stated that radon exposure at Jachymov and Schneeberg were similar; Jachymov with *40 ME,* Schneeberg with *50 ME.*

Subsequent surveys at Jachymov confirmed an incidence of lung cancer equal to that at Schneeberg, 30 km. away. During *1929–1930,* it was shown that the cause of death for at least 47% of the miners dying at Jachymov was lung cancer. Dr. August Pirchan of the Sate Radium Institute in Prague and Dr. Herman Sikl, the pathologist at Prague's Czech University reported their autopsy results in 1932 [32]. It was found that the average induction time for miners to develop lung cancer was 17 years, with a minimum of 13 years. Further autopsies on 63 miners from 1928 to 1938 revealed lung cancer as the cause of death in 45.4% [33, 34]

The Jachymov mines had the same high radon levels as at Schneeberg; up to 50 Mache units or 18,200 pCi/L (Table 9.5). However, unlike the findings at Schneeberg, they found no evidence of pneumoconiosis such as silicosis or anthracosis and none of arsenic poisoning. Since there was no significant pneumoconiosis at Jachymov, strong support was given to a radon etiology in the Czech mines [32–34]. Beginning in 1930, the Czech government started an extensive ventilation project which attempted to reduce the airborne radon to 364 pCi/L or one ME, at most of the working mines. *In June 1932,* the Czech government designated lung cancer in Erzgebirge miners as a compensable occupational disease. This was the first official acknowledgement of the radon etiology of Schneeberg lung disease or the medieval Miner's Plague [33, 34].

9.5 Third Reich Occupational Medicine

The takeover of Germany by the Nazi Party, the NSDAP or Nationalsozialistische Deutsche Arbeiterpartei, in 1933 solved the unemployment woes of the now extinct Weimar Republic, just as Adolph Hitler had promised. The 60-h-work week was the answer [27, 28]. Because of a poor economy, most of the German mines had closed by 1928. The Saxon mines were reopened in 1933 for tungsten and radium and there was renewed interest in worker safety and compensation. Erich Neitzel of Berlin worked for the Reich Labor Ministry. He had measured radon levels in the Schneeberg mines from *1923 to 1934* and found an average radon concentration of 40 ME or 14,560 pCi/L. In 1935, Neitzel reported the radon concentrations in different locations within the mine; 2.6–11.7 ME at the bottom of the mine shafts and 36–46.5 ME at the drill sites. The drilling workers had 21–29 ME in their urine. In 1937, he reported on a Schneeberg mine with a radon concentration of 35–44 ME during the work week. On Monday morning, after the ventilating fans were off over the weekend, the concentration was *111 ME* (40,404 pCi/L). Neitzel believed that Schneeberg lung disease was due to the daughters of radon emitting alpha particles into the lungs [29, 30].

Dr. Kurt Lange was the chief of surgery at the State Hospital in Aue, Saxony. He presented a study on the relationship between the Erzgebirge geology and Schneeberg lung disease to the Reich Committee for Cancer Therapy in Berlin. This was published in 1935 and appears to be the definitive statement on the etiology of Schneeberg lung disease. Lange reported on the apparent relationship between the enormous radioactivity of the Ore Mountains and the lung cancer rates in Schneeberg and Jachymov. He noted that, more or less, *the entire Erzgebirge was radioactive and on a scale unmatched anywhere else but in the Katanga Province in the Congo* [35]! Lange's work was known to both Dr. Hueper and physicist E. Lorenz [29, 31] at the National Cancer Institute in the USA but it remained unknown to the American atomic bomb physicists and geologists during World War II; see Chap 10.

Lange prepared a geological map of the Erzgebirge, unfortunately not illustrated, which indicated the relative radioactivity of the mountain areas. Upon this, he superimposed the location of the 266 lung cancer victims that he had attended over the 3.5 years he had been in Aue. He found the incidence of lung cancer to be three times higher in the radium-rich areas than in the radium-poor areas of shale. The radioactivity came from the Erzgebirge granite (tourmaline), which contained 4.57×10^{-12} g of radium per gram of granite. The surrounding phyllite shale and glimmer shale had somewhat lesser activity [35].

The mines at Schneeberg, Wittigsthal, and Jachymov as well as the springs at Oberschlema (now Bad Schlema) were all in the phyllite shale district. These shale areas contained the old mines that formerly had been worked for silver, cobalt, arsenic, nickel, and tungsten. Now, at greater depths, they were being mined for uranium (*uranpecherze*). Lange reported the radium emanation (radon) was strongest in the region of Jachymov; 905 pCi/L which was equivalent to 9.05×10^{-10} g of radium [35]. The springs of the Erzgebirge were also radioactive; as they were advertised to be. Readings of 70 ME (25,480 pCi/L) were not unusual. In the highly radioactive spring waters at *Oberschlema,* the emanation (radon) activity could reach 12,000 ME (4,368,000 pCi/L); equivalent to the gas from 4.37×10^{-6} g of radium. At the Oberschlema spa baths, the radon activity was 300–5550 ME/L [35]. However, the springs were above ground and well ventilated, which rendered them essentially harmless. Apparently, only the geologists in the Soviet Union were intrigued by the high radon levels at Oberschlema.

Lange believed that the high mortality of the Erzgebirge miners from lung cancer was due to the radon radioactivity. Accepting the average concentration as 40 ME/L of air, and with a work load of 7 h a day for 300 days a year, Lange calculated the yearly dose of inhaled radon was 84,000 ME (30,576,000 pCi/L), the equivalent of 0.03 mg of radium (Table 9.5) [29–35]. Lange considered this enormous when compared to the recommended therapeutic dose of inhaled radon at the spas, which was a single exposure of 50 ME.

Fig. 9.4 Sudetenland cession to Germany 1938 [36]. (Public domain)

In 1938, the German population was estimated at 23 % of what was then Czechoslovakia and most of them were in the Bohemian Sudetenland. German dictator Adolf Hitler redefined "Sudetenland" to include the entire mountain periphery of Czechoslovakia [36]. In October of 1938, these areas were occupied by Nazi Germany and renamed to *Sudetengau* (*Gau: District*, Fig. 9.4). In the Sudetengau, German miners joined the pro-Nazi, anti-Czech Sudetendeutsche Partei and claimed that they were not being protected from the known mining hazards. In reality, the Jachymov mines had improved after *1932* while the Saxon mines had not. In November of 1938, the Nazi Deutsche Arbeiterfront promised improvement in the Joachimsthal mines and in the miner's standard of living and pensions [34]. Four German companies took over the Jachymov radium mines in 1938/1939 [34]. H. L. Himmler (1900–1945), Reichsfuhrer of the Nazi Schutzstaffel (SS) and Chief of the German Police, was urged in November of 1938 to send concentration camp prisoners to work in the mines in Jachymov in order to free the local Germans from the hazards. In March of 1939, Germany took over the rest of Bohemia and Moravia as a Nazi-administered "Protectorate" and part of the Reich. Slovakia was made a separate puppet state. During 1938–1945, renewed investigations into the etiology of lung cancer at Jachymov and Schneeberg were then undertaken by several renowned German scientists, including Dr. Artur Brandt and Boris Rajewsky [28–30, 34, 37, 38].

Dr. Brandt was a German government Industrial Health physician and the head of a research institute in *Dresden*. In 1938, he published the first experimental evidence that 25 % of the mice living for 1 year in the mine shafts would develop lung

Table 9.6 The radon concentration safety standards of 1930–1945 [37–38]

Year	Investigator	pCi/L	ME
1930	Czech suggested standard	364	one
1939	Rajewsky tolerance	10,920	30
1940	Sudetengau (Jachymov) tolerance	<1820	<5
1940	Evans/Goodman (USA) tolerance	10–1000	0.03–2.75
1940	Oberschlema Monitor Station	1092–5460	3–15
World war II	German mines	<1820	<5

One ME = 364 pCi/L

cancer [28]. He set up an Occupational Health Service in the Bohemian Sudetengau in July of 1939. He found that the Jachymov miners with over 20 years of exposure suffered lung cancer, blood dyscrasias and, in over 80 % of the miners, silicosis. He found that the mine improvements made by the Czech government after 1930 had reduced the radioactivity at Jachymov to less than at Schneeberg. Brandt wanted to amend the 1936 Reich Regulations on Occupational Diseases to include lung cancer, especially now that Jachymov was part of the Reich. Additional measures to limit exposure and to increase miner's compensation were shelved as the War effort began to overwhelm Germany [30].

Boris Rajewsky was the director of the Kaiser-Wilhelm Institute of Biophysics at *Frankfurt.* During the summers of 1937 and 1938, Rajewsky and coworkers measured the radon content of underground air and water at 137 sites *in both* Schneeberg and Jachymov and reported them in 1939. He reported that the extremely high radon concentrations were almost unchanged since Schiffner's 1908–1912 studies. Rajewsky accepted that radon, in all probability might be at least one of the causes of lung cancer at Schneeberg [28, 30, 37, 38]. On the basis of the radon measurements and the animal experiments, Rajewsky recommended, in 1939, a threshold or tolerance dose for German miners of 30 ME (Table 9.6) [37–39].

In March 1940, Dr. A. Brandt and B. Rajewsky reported at a Nazi Deutsche Arbeiterfront Conference on the Joachimsthal Sickness and the radon measurements there. The life expectancy for the Sudetengau underground miners was 39 years. Two thousand measurements gave readings that averaged *3–5 ME,* with an occasional reading of 18 ME, which were all lower than at Schneeberg, Germany. A level of less than 3 ME (1092 pCi/L) was considered harmless. Brandt and Rajewsky confirmed that the lung cancers, silicosis, and blood dyscrasias were linked to the mine environment [30, 37, 38]. Also in March of 1940, Americans R. D. Evans and Clark Goodman at the Massachusetts Institute of Technology physics department published their studies on the lung cancer hazard from radon. They suggested that *2.7 ME* (1000 pCi/L) was a dangerous level and that 0.027 ME (10 pCi/L) was safe [39].

In November of 1940, the Karlsbad, Bohemia, Mining Office, under Nazi German control, issued regulations outlining preventative measures to protect the German miners in the *Sudetengau.* Exposure was to be limited to less than *5 ME* (1820 pCi/L) and anything above 22 ME (8000 pCi/L) was to be reported.

Ventilation was increased and dust decreased. The miners were to have more physical exams, better hygiene conditions, decreased working hours, and more vacations [34, 37, 38].

To monitor the German mines at *Schneeberg* for compliance to the less the *5 ME* standard, Rajewsky set up a research station at the nearby Oberschlema Radium Spa. The chief researcher there was an Adolf Krebs. During World War II, he was charged with controlling the radon levels in all the German mines. Improved ventilation was instituted and radon exposure limits were redefined *at 3–15 ME,* which became mandatory. Krebs reported the average radon concentration in the mines was less than *5 ME* during World War II. The German standards were the second attempt, after the 1930–1932 Czech standards, to protect underground workers [37, 38].

This accomplishment is marred by the fact that by 1942 there were reported to be five million foreign workers conscripted in Germany as well as concentration camp and prisoner-of-war slave laborers [27, 28]. There were no safety standards for these victimized workers. French prisoners of war were used in the mines in 1940 and then prisoners from the Soviet Union during 1942–1944 [27, 28, 34]. In 1943, Rajewsky suggested that the German miners be rotated between underground and surface work. That same year the German Ministry of Economy offered retirement after 15 years of work underground with a pension of 80 Marks per month [37, 38]. Few benefitted from this largesse as retired miners were immediately drafted into the 1943 German Army and welcomed to the Eastern Front.

In March and April of 1945, the Erzgebirge would be invaded by two enemy armies that were both oblivious of the enormous radioactivity on the Saxon side. Jachymov was a primary target for the Red Army because Premier Stalin knew about the American uranium bomb and the Soviet Union had no known uranium. However, the Soviet geologists knew Jachymov and knew about the radon at the world famous spa at Oberschlema. Therefore, as soon as the American army withdrew from Saxony in June of 1945, an army of Soviet geologists descended on the Saxon Erzgebirge; see Chap. 10.

9.6 Supplemental Material: Classic German Studies on the Radon Hazard: 1879–1939

Year	Author(s); study: journal
1879	Hartung FH and Hesse W "Der Lungenkrebs, die Bergkrankheit in den Schneeberger Gruben." Vierteljahresschrift fur gerichtliche Medizin u offentliche Sanitatswcsen 30: 296–309; 31: 102–132, 313–337. (1879)
1913	Schiffner Carl, Weidig M, Friedrich R "Radioaktive Wasser in Sachsen, I-IV." Freiberg i Sa: Graz und Gerlach (1913)

Year	Author(s); study; journal
1913	Arnstein Alfred "Uber de sogenannten "Schneeberger Lungenkrebs." Verhandlungen der deutschen pathologischen Gesellschaft, 16: 332–342, (1913)
1913	Muller HE In Statsarchiv Dresden, Aussenstelle Freiberg (Bergarchiv), Fasc, LBA-AB 298 (1913–1923), folio 7 verso
1924	Ludewig P and Lorenser E "Untersuchungen der Grubenluft in den Schneeberger Gruben auf den Gehalt an Radium-emanation." Ein Betrag zur Frage nach der Entschung des Schneeberger Lungenkrebs." Strahlentherapy 17: 428–435 (1924)
1926	Rostoski Saupe el al. "Der Bergkrankheit der Erzbergleute in Schneeberg in Sachsen." Zeitschrift fur Krebsforschung 23: 360–384. (1926)
1929	Lowry Julius "Uber die Joachimstaler Bergkrankheit; vorlaufige Mittellung." Medizinische Klinik 25; pp 141–142. (1929)
1930	Sikl H "Uber den Lungerkrebs der Bergleute in Joachinsthal." Zeitschrift fur Krebsforschung 32; 609–613 (1930)
1935	Neitzel K "Berufsschadigungen durch radioaktive Substanzen." Arbeits-medizin: Abhandlungen uber Berufskrankenheiten und der deren Verhtung. pp. 6–46.(1935)
1935	Lange E "Krebser-krankungen und geologishe Verhaltnisse im Erzgebirge." Zeitschrift fur Krebsforshung. 42: 306–310 (1935)
1939	Rajewsky Boris "Bericht Uber die Schneeberger Untersuchungen." Zeitschrift fur Krebsforschung 49; 315–340. (1939)

References

1. Hoover HC, Hoover LH (1912) De Re Metallica by Georgius Agricola (translated from the first Latin edition of 1556). Mining Magazine, London (Agricola G (MDLVI) De Re Metallica. Froben, Basel. English edition)
2. Hoover HC, Hoover LH (1950) De re metallica by Georgius Agricola. Dover, New York
3. Marshall JL, Marshall VR (2005) Agricola. Hexag Alpha Chi Sigma 96(3):58–61
4. Marshall JL, Marshall VR (2008) Jachymov (Joachimsthal), Czech Republic. Hexag Alpha Chi Sigma 99(4):68–71, 75
5. Marshall JL (2002) Discovery of the elements, 2nd edn. Pearson Custom Publ., Boston
6. Rosen G (1943) The history of miner's diseases. Schuman's, New York
7. Wright WC (1940) Diseases of workers. University of Chicago Press. (Latin edition: Ramazzini B (1713) De morbis artificum diatribe. Padua)
8. Stillman JM (1920) Paracelsus. Open Court, Chicago
9. Cockren A (2014) Paracelsus: alchemical genius of the middle ages. http://www.alchemylab.com/paracelsus.htm. Accessed 25 June 2014

10. Debus AG (1998) Paracelsus, five hundred years: three American exhibits. http://www.nlm.nih.gov/exhibition/paracelsus/collections.html
11. Garrison FH (1929) An introduction to the history of medicine, 4th edn. Saunders, London
12. Norman JM (ed) (1991) Morton's medical bibliography, 5th edn. Scholar Press, Hants
13. Gantenbein UL (2012) Zurich Paracelsus project. http://www.paracelsus.uzh.ch/. Accessed 26 June 2014
14. Pacher HM (1951) Magic into science, the story of Paracelsus. H Schuman, New York
15. Rosen G (1941) Von der bergsucht oder bergkrqankheiten, English trans. In: Sigerist HE (ed) Four treatises of Theophrastus von Hohenheim, called Paracelsus. Johns Hopkins, Baltimore
16. Gochfeld M (2005) Chronological history of occupational medicine. J Occ Environ Med 47(2):96–114
17. Ramazzini B (1700) De Morbis artificium diatribe. Capponi, Modena
18. Segre E (1980) From x-rays to quarks. WH Freeman, San Francisco
19. Asimov I (1988) Understanding physics, vol I, II, III. Dorset Press, New York
20. Johns HE, Cunningham JR (1983) The physics of radiology, 4th edn. Thomas, Springfield
21. Johns HE (1961) The physics of radiology, 2nd edn. Thomas, Springfield
22. McPherson MJ (undated) Chapter 13: Radiation and Radon Gas. http://www.mvsengineering.com/files/Subsurface…/MVS-SVE_Chapter13.pdf. Accessed 26 June 2014
23. Harley N, Harley JH (1990) Potential lung cancer risk from indoor radon exposure. CA-Cancer J Clin 40(5):265–276
24. Adloff JP, Kauffman GB (2007) The troubled story of polonium. Chem Educ 12:94–101
25. Becker K (2004) One century of radon therapy. Int J Low Radiat 1(3):333–357
26. Becker K (2004) Subsurface Ventilation Engineering. http://www.mvsengineering.com/files/Subsurface…/MVS-SVE_Chapter00.pdf. Accessed 19 Nov 2012
27. Proctor RN (1995) Cancer wars. Basic Books, New York
28. Proctor RN (1999) The nazi war on cancer. Princeton Univ. Press, Princeton
29. Hueper WC (1942) Occupational tumors and allied diseases. Thomas, Springfield
30. Schuttmann W (1993) Schneeberg lung disease and uranium mining in the saxon ore mountains. Am J Ind Med 2:355–68
31. Lorenz E (1944) Radioactivity and lung cancer. J NCI 5(1):1–15
32. Pirchan A, Sikl H (1932) Cancer of the lung in miners of Jachymov. Am J Cancer XVI(4):681–722
33. Bohounek F (1970) History of the exposure of miners to radon. Health Phys 19(7):56–57
34. Zeman Z, Karlsch R (2008) Uranium matters: central European uranium in international politics. Central European University Press, New York, pp 1900–1960
35. Lange E (1935) Krebser-krankungen und geologishe Verhaltnisse im Erzgebirge. Z Krebsforsch 42:306–310
36. Caldwell E, Bourke-White M (1939) North of the Danube. Viking Press, New York
37. Rajewsky B (1939) Bericht uber die schneeberger untersuchungen. Z Krebsforsch 49:315–40
38. Danzer H (1954) Boris Rajewsky zum 60 Geburtstag. Strahlentherapie 94:3–11
39. Evans RD, Goodman C (1940) Determinations of the thoron content of air and its bearing on lung cancer. J Indus Hyg Toxic 22(3):89–99

Chapter 10
Cold War Uranium Blunders

On 5 March 1946, Winston Churchill, speaking at Westminster College in Fulton, Missouri, noted that in Europe: "From Stettin in the Baltic to Trieste in the Adriatic an iron curtain has descended across the Continent." During the ensuing 45 years, an economic and military competition along with mutual antagonism and distrust persisted between the North Atlantic democracies and the Communist totalitarian regimes of the Warsaw Pact: the Soviet Union, Poland, Czechoslovakia, Hungary, Romania, Bulgaria, Albania, and the German Democratic Republic. By 1990–1991, the Communist regimes were no longer economically sustainable and they collapsed. As World War III was avoided, this era was deemed the "Cold War."

10.1 Dr. W. C. Hueper

The USA lagged behind Europe in almost all the scientific disciplines, including physics and occupational medicine, until the upheavals of the two twentieth-century World Wars. The destruction of large areas of Europe enforced emigration, economic depression, and unemployment as well as anti-Semitism, all led to a net gain of scientists in the USA [1, 2]. However, occupational medicine was not to develop in the USA until after World War II. The first American to make significant contributions in the field was German immigrant Dr. W. C. Hueper [3]. Wilhelm Carl Hueper (1894–1979) was born on 4 November 1894, in Schwerin, Mecklenburg, Germany. His father was a railroad official. He started medical school in 1913 at the University of Marburg. After 1.5 years of study, he was called to service in the First World War. He served as a medic and spent time as a prisoner of war. After the 1918 armistice he enrolled at the University of Rostock Medical School for 1918/1919. He finished medical school at the University of Kiel, where he graduated in 1920. Hueper was a surgical intern at Witten during 1920/1921. He then spent 2 years as a pathology resident in Berlin at the *Auguste Viktoria Krankenhaus* from 1921 to 1923. He immigrated to Chicago and finished his pathology training at the Evangelical Deaconess Hospital there. From 1924 to 1930, he worked as a pathologist at the

R. F. Robison, *Mining and Selling Radium and Uranium*,
DOI 10.1007/978-3-319-11830-7_10

Loyola School of Medicine, Mercy Hospital in Chicago. His research interest was in the etiology of lung cancer, which was rapidly increasing after World War I [4].

In 1930, he moved to Philadelphia to work at a new Cancer Research Laboratory at the University of Pennsylvania, which was being financed by the DuPont Co. from nearby Wilmington, Delaware. He subsequently informed DuPont about the probability of their chemicals causing bladder cancer, as was being reported in Germany. However, there had been no bladder cancers observed in Wilmington. In 1933, Hueper was laid off. He went back to Germany to look for scientific research work but quickly returned in 1934. Meanwhile, DuPont had found 23 workers with bladder cancer. They rehired Hueper at their new Haskell Laboratory of Industrial Toxicology in Wilmington. He began testing the toxicity of the DuPont chemicals in the dog lab. When he tried to evaluate the actual working conditions on the factory floor, he was ejected. He was forbidden to return or to study or publish anything on human toxicity. His employer threatened him with legal action and then fired him in 1938. His research on human toxicity was continued by others at the more discrete University of Cincinnati, where it remained secret and unpublished for years [3, 5, 6].

Hueper then found work from 1938 to 1948 at the Warner Institute for Therapeutic Research in New York City as the assistant director and principle pathologist. He was in charge of pathology, pharmacology, roentgenology, bacteriology, and research. His thorough and consistent review of the German literature was utilized during 1944–1950 by the American Medical Association. He was engaged to translate and write abstracts in English for their pathology journal. In 1948, he became the first director of the section of environmental cancer at the fledgling National Cancer Institute which had been founded in 1937. He spent the next 20 years there.

In 1942, while employed at the Warner Institute, Hueper published the first comprehensive American textbook on occupational toxicology: "Occupational Tumors and Allied Diseases" [7]. From the German studies he learned of the presence of radon in the rocks, water, and air of the Schneeberg mines and its unavoidable inhalation. Mine ventilation was poor, if present at all, and the miners disregarded the hated masks. Hueper also noted that beyond the mines, large areas of the Saxon Erzgebirge had radioactive rocks (tumulate granite) as well as springs and brooks. The fact that this indicated significant uranium throughout the northern slopes of the Erzgebirge failed to capture the attention of others interested in uranium in 1942.

In his book, Hueper devoted over 30 pages to the effect of radioactive substances on the respiratory system. He implicated radon as the etiological agent in miner's lung cancer. He said it three times. (1) "This actinic factor (radioactivity), in recent years has gained increasingly in importance and the evidence available at present supports the conception that it (radioactivity) is the chief cause of the Schneeberg cancers." (2) "Considering that all conditions as to site, history, geological formation of the mines, and the type and duration of exposure of the workers are either very similar to or even identical in Joachimsthal with those existing in the Schneeberg district and realizing that the most important and unique common factor in both mining activities is the presence of radioactive matter in the form of radon—it becomes increasingly certain that the causative agent which elicits the neoplastic response in the lungs of the miners of both regions is the radiating energy of radon inhaled." (3)

"...the evidence of the statistical and clinical observations on hand does not leave any doubt that these agents (radioactive substances) represent the chief cause of the pulmonary malignancies observed in workers (miners, artisans, chemists, technicians, etc.) exposed to radioactive matter for occupational reasons" [7].

Hueper reported that in 1940, Robley Evans and others had addressed the health hazards of radium dial painters and they had stated that a prolonged exposure to 1000 pCi/L of radon might cause neoplastic pulmonary lesions in man. Hueper suggested that a concentration of 100 pCi/L or less was necessary in order to remain well within a safe range [7]. He noted that polonium had been verified as the pulmonary carcinogen at Schneeberg and Jachymov in 1939 by Nazi Germany occupational health scientists. After radon was inhaled by the miners, it deposited polonium in the pulmonary tree. The daughters of polonium were the source of the deadly alpha particles, which initiated lung cancer, Chap. 9. Two years after Hueper's book was published, another research worker at the National Cancer Institute came to an exactly opposite conclusion [8]. Hueper's opinion was based on his knowledge of the German literature and his career-long interest in lung cancer etiology. Egon Lorenz, PhD, was the senior biophysicist at the National Cancer Institute. He published his review of the German literature on the lung cancer deaths at Schneeberg and Jachymov in the *Journal of the National Cancer Institute* in 1944. He reviewed 39 German papers including all the classic radon studies; see Chap. 9. In addition, he consulted 18 English-language authors including Robley R. Evans and H. S. Martland.

Somehow, Lorenz was not able to appreciate the German data, either through a language problem or lack of epidemiology training. Although a physicist, Lorenz seemed unaware of the different toxicities of (a) X-rays, (b) gamma rays, and (c) alpha particles. After citing all the evidence incriminating radon, Lorenz surmised that radon could *not* be the sole cause of Schneeberg lung disease. He favored pneumoconiosis, arsenic, and/or hereditary factors due to local Sudeten inbreeding. His failure to understand the alpha particle toxicity and to support Hueper put American uranium miners in deadly peril [5, 9].

10.2 Murray Hill Survey

When the US Army took over the Manhattan Project to produce a nuclear/atomic bomb/device, they quickly realized that procuring the necessary quantity of uranium was their number one priority. At the time, uranium was considered a rare mineral, found in significant quantities in only seven countries: the Belgian Congo, Czechoslovakia, the southwestern USA, the Canadian Northwest Territories, Cornwall, Portugal, and Madagascar. These were the countries where uranium had been mined for its radium content (Appendix III). General Leslie Groves, the head of the Manhattan Project appointed Major Paul L. Guarin, to head a covert operation to locate and evaluate the quality and quantity of the world's resources of uranium ore [10–13]. The operation was code named the "Murray Hill Area" after an obscure neighborhood in Manhattan, New York City. Guarin was a former geologist with Shell Oil Company. He added geologists George W. Bain from Amherst and George Selfridge from the University of Utah as assistants. The operation was active from June of 1943 until October of 1944. The staff and facilities of the Union Carbide and Carbon Corporation were hired by the US Army for the job of geographic exploration and subsequent development of raw material. Union Carbide and Carbon Corporation formed a subsidiary, United Mines Development Corporation, with offices in New York City, to do the work, which was organized into four divisions:

1. Bibliographic search: 67,000 volumes of geology reports from over 50 countries; 50 % non-English
2. Field exploration in 37 countries: collecting soil samples from 20 friendly or neutral countries
3. Exploration research: assaying the foreign soil samples
4. Metallurgical research: new methods to identify uranium deposits

Twenty friendly or neutral foreign countries were then graded according to their potential as a source of uranium ore. The Belgian Congo was identified as an excellent source of uranium. The USA, Canada, and *Sweden* were considered good sources. *Czechoslovakia,* Portugal, and South Africa were considered as fair sources. Other sites were poor, unknown, or unviable. Only 13 countries were evaluated for uranium while eight were studied for thorium. In the autumn of 1944, three British experts joined in the survey and they helped shape the October 1944 report [13]. The USA, Canada, and the UK were hoping that they could monopolize all of the world's uranium so as to prevent the Soviet Union and others from making atomic bombs.

The Murray Hill survey confirmed what the *radium* production companies knew in 1923 about the Belgium Congo, Canada, and the USA. The 1944 report recognized the Congo ore as dominant, with Canada the next best source. Canadian properties in the Great Bear Lake area of the Northwest Territories had been purchased by the Americans in 1943. However, in September of 1943 the Canadian government indicated it preferred to conduct its own uranium explorations and the properties were returned to Canadian possession; see Chap. 8.

American deposits were described as low grade, costly to mine, and not worth developing unless the Congo source was lost. The study identified the carnotite and roscoelite ores of the Colorado Plateau, the Paradox Basin, and Four Corners states as having potential. These areas had originally been mined primarily for *radium only*. However, since 1923, the carnotite ore was mined primarily for *vanadium*. The Murray Hill survey identified a geographical belt in which the uranium to vanadium ratio was high. This belt extended north from the Slick Rock area of Colorado through the Paradox Valley to Uravan, Gateway, and the Utah border; the Carnotite-Uravan Mineral Belt, Chap. 5. Subsequently, properties therein were acquired by the American government for uranium prospecting.

The Murray Hill survey estimated that the historic mines at Jachymov, Czechoslovakia, and all of Eastern Europe could be *no more than 5 %* of the world supply [11]. Thus, the Union Carbide geology consultants seriously underestimated Czechoslovakia and they over-estimated Sweden. The USA had been kicked out of Canada, so the consultants were unable to predict the major deposits that were found there; see Chap. 8. Uranium would turn out to be more common than tin and almost 500 times more common than gold. At least 100 billion t are now known to exist [12].

The biggest blunder of the survey was the total lack of consideration of the radioactive Erzgebirge of the Sudetenland. Because of the multiple studies on Schneeberg lung disease, Saxony was known to be highly radioactive since at least 1935 [14] and Hueper had even published it in English [7]. The US Army and the Manhattan Project were thus unaware that Schneeberg lung cancer was due to radon and that radon came from uranium; and therefore Schneeberg, Saxony, was located in a uranium belt. Nor were they seemingly aware that uranium had first been discovered in ore from the Saxon Erzgebirge in 1789 and that radioactive spas were thriving in Saxony even during the war. As a consequence, the US Army was unable to advise President Roosevelt in his discussions for the eventual post-war partition of Nazi Germany. East Germany including the entire Erzgebirge was allotted to the Soviet Union with no knowledge that Saxony contained the richest pechblende outside of the Congo. The Soviets were preoccupied with Berlin and Saxony might have been negotiable if the Murray Hill geologists had done their homework.

10.3 Jachymov Ignored

The occupation and partition of Germany and Berlin had been discussed by the Allies during meetings at Casablanca in January of 1943 and at Tehran in November. Germany and Berlin were to be divided into three zones of occupation. All three major Allies sent representatives to a European Advisory Commission, which was set up in Moscow in late 1943 and continued in London during 1944, to plan the partitions of Germany. By November of 1944 agreement had been reached based on educated guesses of where in Germany the invading armies would meet. The Normandy invasion had placed British forces in north France and the American Army to the south. Therefore, the partition was basically to have the British occupy the northwest, the Americans the south, and the Soviet Union the east. This gave the Soviets 40 % of the land and 36 % of the population [15, 16].

Thus, there was no real discussion at the Yalta meeting in February of 1945, only the conformation and signing of the 1944 agreements. The Soviets considered the French as essentially noncombatants. However, the USA and the British would allow France to occupy a small western section of both Germany and Berlin. None of the diplomats had a clue about uranium in the Saxon Erzgebirge. Plans were also made for the partition of Austria and Vienna, 1946–1955 [17]. This was the setting for the Graham Greene novel "The Third Man." Czechoslovakia was neither partitioned nor planned for prolonged occupation; but someone forgot to tell the Soviets. The Red Army would remain there until 1989/1990 [17]. None of the three western Allies were interested in the Czech uranium at Jachymov, but the Soviets were [16].

Table 10.1 Atomic bombs (1945–1949) [2, 10, 11]

Year	Month	Site	Fission material	Nickname
1945	16 July	New Mexico	Plutonium-239	Trinity
1945	6 Aug	Hiroshima	Uranium-235	Little boy
1945	9 Aug	Nagasaki	Plutonium-239	Fat man
1949	23 Aug	USSR	Plutonium-239	Joe 1

The sites for the Allied meetings were all at Stalin's convenience, as he did not trust airplanes [16]. American President F. D. Roosevelt was dying of heart failure in early 1943 but made a miraculous recovery after seeing a cardiologist [18]. This allowed him to run for an ill-advised fourth term as the American president. At Yalta on 4–11 February 1945, he was only 8 weeks away from death. He seemed little interested in the future of Eastern Europe and was perhaps more concerned about the Pacific theater. How much Roosevelt was influenced by Soviet spies in his New Deal government, such as Alger Hiss and a personal friend named Harry Dexter White (1892–1948), is unknown [19, 20]. April of 1945 brought an end to World War II. Roosevelt died on 12 April. German dictator A. Hitler committed suicide on 30 April during the Soviet Battle for Berlin; 16 April to 1 May. On 25 April, the American and Soviet Armies made contact on the Elbe River at Torgau. The unconditional surrender was signed on 7/8 May [21–23].

The next and last meeting of the Allies was in Potsdam (Berlin suburb) from 17 July to 2 August of 1945 and involved new leadership from the UK and the USA. At Potsdam, President Truman informed Stalin about the atomic bomb (Table 10.1). It is suspected that Stalin knew more about the bomb than did Truman because of the myriad of ideological Communist spies at Los Alamos, in the New Deal administration, and in the British government [2, 19, 20]. The Soviets had begun a nuclear research project in *March of 1943,* after information provided by their various spies, especially those at the University of California in Berkeley. The Soviets actually had two separate and complete sets of instructions for bomb production; both directly from Los Alamos. British physicist and German émigré Klaus Fuchs (1911–1988) provided one set of plans through his contacts Sgt. David Greenglass, and lab chemist Harry Gold (1910–1972) to the Rosenbergs, Julius and Ethel. She was Greenglasses' sister. The other set was sent by Harvard prodigy Theodore Alvin Hall (nee Holtzberg) (1925–1999) through Lona and Morris Cohen [19, 20].

Even though the USA would not perfect the atomic bomb until 16 July of 1945, the Soviet spies had warned Stalin what was coming. The Soviets were apprehensive that the Western democracies might turn against them because of their military aggression in 1939/1940 against Poland, the Baltics, and Finland. Getting the bomb became their number-one priority for self-defense. The big problem, as one Soviet spy noted, was the apparent total lack of any uranium in the Soviet Union. Therefore, the savvy Soviets were looking for uranium long before the July bomb explosion in New Mexico. The Soviets would have their own bomb within 4 years.

As the Red Army began preparing for the Battle for Berlin, 16 April to 1 May, another Red Army occupied Vienna on 13 April. The venerable Welsbach Company

Fig. 10.1 US Army advance line and occupation zones [24]. (US Army, public domain, modified by author)

in nearby Auer had been engaged in uranium research. The Soviets were able to confiscate at least 100 t of enriched uranium. The entire Auer Company was dismantled and shipped to Elektrostal, just east of Moscow, along with its scientific and technical staffs. The Soviet scientists later reported that this acquisition shortened the work on their atomic bomb by a *year* [15, [16].

The American First Army was at the Elbe River in April, where they were ordered to stop and wait for the Soviets so as not to alarm the Red Army [24] (Fig. 10.1). In response to that order, the American Third Army under General Patton began to move south into Czechoslovakia. By May 4/5, the Third Army had crossed over the eastern mountain border of Bavaria and entered the Erzgebirge and western Czechoslovakia, where the Sudeten German civilians remained hostile. Patton requested to advance to Prague. For various political reasons, Supreme Commander Eisenhower believed he had to hold Patton back from liberating Prague. Patton was ordered to advance no farther than a line from Karlsbad to Pilsen to Ceske Budejovice. This line was 81 miles west of Prague [21–23]. The American advance line was only 10 miles from John-George, Saxony and 18 miles from Jachymov. The Czech and Saxon Erzgebirge were not occupied by either the American or Soviet Army. The US Army was then ordered to pull back on 12 May and leave Czechoslovakia and Saxony for the Soviets to occupy. The Red Army set up their headquarters in Karlsbad, Czechoslovakia, on 11 June [21].

Jachymov was a primary Soviet target because of its famous uranium mines, which had been operational since 1850–1855. On 26/27 August, 3 weeks after Hiroshima, the Soviets were in Jachymov inspecting the Svornost uranium mine in the town center. Soviet geologists were sent immediately to investigate the mines that had been under German control since 1938/1939. The Red Army took over the mines and by 13 September they had 60 troops exploring several different shafts [15, 16]. By November of 1945, the Soviets had pressured the Czech government into a secret agreement giving the Soviets complete control of all three of the uranium mines in Czechoslovakia: Horni Slavkov, Pribram, and Jachymov. They immediately began shipping pechblende to Russia [15, 16, 25]. During 1946–1949, Czech miners were supplemented by 10,000 German prisoners of war. By 1949 they had run out of prisoners. However, after the Communist takeover in 1948, the Czech government operated 17 camps to provide labor for the three Soviet uranium mining centers. The Czechs conscripted political and criminal prisoners to supplement the civilian workers. Opponents of the Communist regime and petty criminals were promptly sentenced to underground slave labor as uranium miners. The average life expectancy for Jáchymov miners during this period was estimated at 42 years [16, 25].

The Soviets were in a hurry until August of 1949, when they detonated their first atomic bomb. By the time Stalin died in March of 1953, they had 120 bombs. The Czech mining operations peaked in 1953, when they utilized 13,821 political prisoners. About 50,000 political prisoners were used from 1949 to 1963. There were only 2923 political prisoners left by 1960 and all were pardoned in 1963. The Czech Communists then hired paid workers from 1963 to 1990. There were 65,000 civilian miners at Jachmov and Horni Slavkow and another 65,000 at Pribram, the largest center. The Czechs furnished 96,660 t of uranium for the Soviets between 1945 and 1990 [16, 25].

When the Soviet Union collapsed during 1989/1990, the Czechs threw the Communists out and the mines closed in 1990. Jachymov was then returned to the Czechs. Most of the spa buildings and half the town of Jachymov had been destroyed by the mining operations and then buried under tons of radioactive waste. The Czechs reclaimed the radon-rich waters of the famous Svornost mine, the world's oldest uranium mine. The waters were again pumped above ground to newly rebuilt spas. The old Schlick Mint building, Chap. 1, was converted from a mint museum to a political expose of the Soviet abuses during 1949–1961 [26, 27]

10.4 Partition of Germany

During 1941–1945, the Russians suffered the most from Nazi aggression and many believed that was appropriate [22] (Table 10.2). However, there had been no way to dislodge the Nazis from Europe without the Red Army, which was welcomed into joining the Allies in 1941. The Baltic states, Finland and eastern Poland had been part of the Czarist Russian Empire before World War I and had been invaded by the Soviets in 1939. After Germany's defeat, the Soviets wanted their old 1914 borders established plus some new buffer states. They retained the three Baltic states and a slice of Finland that they had confiscated in 1939. The eastern border of Germany with Poland was moved west to the Oder–Neisse River line. Russia then helped itself to a section of eastern Poland and secured from eastern Germany another Baltic port, Konigsberg, now Kallingrad. The eastern 40 % of Germany was assigned to be occupied by the Soviet military while the USA occupied the Southwest and the UK, the Northwest of Germany [17, 28, 29]; Fig. 10.1.

Of all the belligerents, the Soviet Union was the most chronically impoverished and backward, even before the carnage started. Why an advanced nation like Germany would invade the Russian backwater mystified even the semi-illiterate Soviet-occupying force. Now, all four Allies wanted some payback or reparations. They all agreed upon dismantling the German military-industrial powers, such as Krupp, Siemens, Daimler-Benz, and I. G. Farben. The Russian demands for reparations exceeded those of the other Allies, and perhaps with good reason. The Soviets wanted (1) an atomic bomb, (2) a rocket program, (3) military modernization, and (4) civilian modernization [15, 16]. After the costly victory over Germany, Stalin seized the opportunity to confiscate and transfer back to the Soviet Union all the German industry, technology, and scientists he could locate. Soviet troops removed 1.28 million t of material and 3.6 million t of equipment. During 1945 and 1946, over 17,000 entire factories were shipped to Mother Russia. Since Germany was

Table 10.2 World War II casualties during 1939–1945, combined military and civilian deaths, in millions of people [22]

North Atlantic allies	# Dead	% Loss of population	Central Europe	# Dead	% loss of population	Eastern Europe	# Dead	% Loss of population
France	0.567	1.4 % of 42	Germany	5.6–7.0	8.1 % of 70	USSR	23.4	14
UK	0.451	<1 % of 48	Romania	0.800	4.0 % of 20	Poland	5.72	16.5
USA	0.419	<1 % of 132	Hungary	0.580	6.4 % of 9.13	Lithuania	0.350	14.3
			Italy	0.455	1.0 % of 44.4	Latvia	0.230	12
			Austria	0.380	5.7 % of 6.7	Finland	0.097	2.62
			CZ	0.325	2.1 % of 15.3	Estonia	0.050	4.44

Table 10.3 German radium companies: 1937–1945 [15, 16, 34]

Company	Site	Post-war zone
Auer Gesellschaft A.G.	Berlin	Divided
(a) Extraction plant	Oranienburg	USSR
(b) Chemical plant	Berlin	UK
Buchler Chininfabrik	Braunschweig	UK
Treibacher Chemische Werke	Treibach	Austria
Radium Chemie A.G.	Frankfurt am Main	USA

in ruins, getting the German technical and scientific experts and their families to immigrate to Russia was not a hard sell. Over 600 German scientists and technicians and over 7000 factory workers were relocated to Russia [15, 16, 30, 31]. The American Army had an Office of Strategic Services (OSS) program called Operation Paperclip, which was also "recruiting" German atomic/nuclear scientists to relocate them to the USA [32].

The Germans had never sought to investigate their own uranium deposits in Saxony. Though known since 1789, they had never been considered commercially viable for the small market in the ceramics and glass industry. This niche was filled by the Joachimsthal mines. The German research, on radon in the Erzgebirge confirmed the presence of uranium during 1935–1939, but it did not arouse any commercial interest. The discovery of nuclear fission in Berlin in 1938/1939 created a demand for research quantities and this was easily met from Belgium or Joachimsthal. The German research was confined to nuclear reactors and seeking an energy source from uranium. They did not have the industrial capabilities to process enough uranium for a bomb experiment and unlike the Soviets they did not know of the American effort [2, 16, 33].

German companies had taken over the Jachymov uranium mines in 1938/1939. Four German companies had been buying Jachymov uranium for radium production [16, 34–36]. They formed a German Radium Syndicate under the control of the *Reichstelle Chemie* at the State Administration Ministry (*Reichswirtschaftsministerium,* Table 10.3). Auer, Buchler, and Treibacher were directed to provide a 33.3 % investment in the confiscated uranium mines at Jachymov. They formed the *Joachimsthaler Bergwerke GmbH* (mining company), which doubled production between 1939/1940 and 1943/1944 from 1.65 to 3.3 g/year. The Syndicate made radium and mesothorium, which was being used for the production of luminous compounds (paint) for various military instruments. The Wehrmacht also supplied Italy and Japan. Auer was the primary producer for the paints and the work was done in their Berlin factory after the ores had first been processed at *Auer's* factory in *Oranienburg*, 21 miles north of Berlin [16, 35, 36]. At the war's end, the partition of Berlin gave the British control of the Auer factory in Berlin where they found 4120 mg of radium and 6042 mg of mesothorium [34].

From April of 1944 until October of 1945, the Manhattan Project sponsored the Alsos Mission [2, 33], which was to investigate the German effort in nuclear energy and to secure any German uranium to keep it from the Soviets and the French. In

March of 1945, Alsos learned that uranium and thorium research was being conducted at the Auer Company plant in *Oranienburg,* in the future Soviet zone. The Soviets also were aware that this was the major site for the German nuclear energy research. The American Army arranged for it to be bombed on 15 March to thwart Soviet efforts to acquire the uranium. After the Battle for Berlin, the Soviets quickly took control at *Oranienburg* and moved the scientists, the technical workers, the buildings, the remaining uranium, and all the processing equipment from Oranienburg to Russia [16].

On 8 May, Victory in Europe (V-E) Day, Saxony-Anhalt, Thuringia, and western Saxony were occupied by the American First Army, under General Hodges, who had reached Chemnitz and Leipzig in March (Fig. 10.1). The Erzgebirge remained essentially unoccupied by a military. On 12 May, the American Army was ordered to start withdrawing from east Germany so that the Red Army could move in during 21–28 June [21]. The American, British, and Soviet governments were unaware of the significance of the Saxon uranium deposits as were the Germans. Therefore, during June of 1945, the entire eastern 40 % of Germany, the future German Democratic Republic (GDR), was turned over to the Russians in accord with agreements signed at Yalta. If the USA did not adhere strictly to the Yalta partition/occupational zones, they feared that the Soviets would renege on the partition of Berlin and Vienna. The American Army pulled back from the Elbe and Mulde Rivers into the American Zone. The zones were confirmed again at Potsdam during 17 July to 2 August of 1945 [29, 30]. The Soviet Zone, Fig. 10.2, consisted of five states and their capitals [37]:

Saxony, Dresden
Thuringia, Erfurt
Saxony-Anhalt, Magdeburg
Brandenburg, Berlin
Mecklenburg-Western Pomerania, Schwerin

10.5 The Soviet Bonanza

The Trinity explosion in New Mexico on 16 July 1945, had confirmed for the Soviet spies and Stalin that the atomic bomb would work. Stalin was evidently as surprised as were the captured German physicists that the bomb actually worked [33, 36]. His spies had given him the recipe, now Stalin wanted the uranium to build it. The one place in Germany most famous for its radioactivity was the Oberschlema Radiumbad in Saxony, Chap. 9. In June of 1945, the Soviets wasted no time in rounding up the medical personal at *Oberschlema* and sending them to Moscow for interrogation about the local radioactivity. Stalin appointed Lavrentiy Beriya (1899–1953), the head of the Soviet secret police, the *Narodnyy Komissariat Vnutrennikh Del* (NKVD) to take over uranium procurement on 20 August 1945 [15, 16].

In the fall of 1945, the NKVD put 100 Saxon miners to work exploring several pits. By January of 1946 the Soviets had completed a preliminary survey of the Saxon Erzgebirge. The Soviet geologists reported they were stunned to find veins

Fig. 10.2 Soviet Zone [37]. (Public domain)

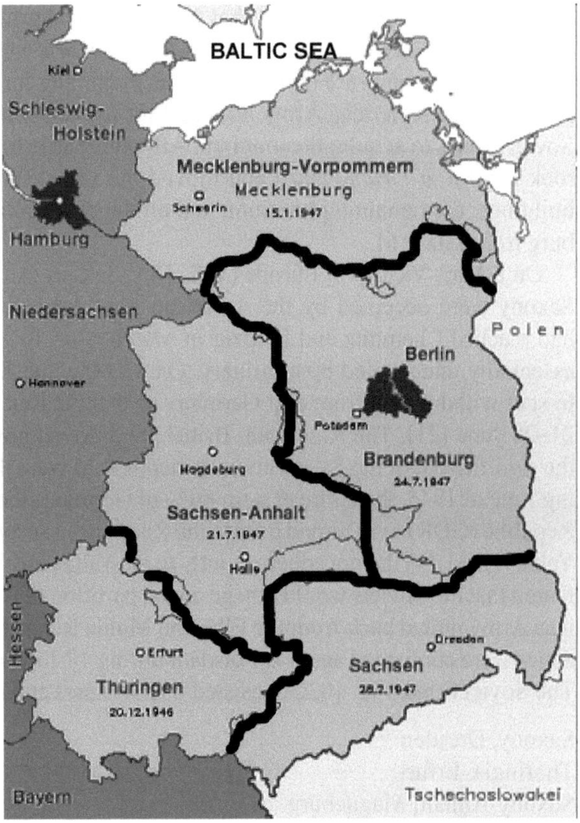

of uranium 11 yards, 33 ft., wide! Talk about "Glück Auf" (good luck)! The NKVD took over the mining operation and mine workers were being conscripted by July of 1946. By September, there were 4000–5000 miners underground reworking the old silver mine shafts. The first three locations investigated in 1946 were the old mining towns of (a) Schneeberg, (b) Oberschlema, and (c) John-George; all three well known for their radioactivity. These three were included within a triangle bordered by John-George, Annaberg, and Schneeberg, which was revealed to be one of the richest uranium-producing areas outside the Congo (Fig. 10.3).

In September, the mining areas were cordoned off with wire and wooden fences. In December, the Soviets called for the Germans to supply another 20,000 workers. The Soviets organized their mining business into a stock company they called *Wismut* (*German:* bismuth) in a feeble attempt at subterfuge. In 1947, the mining operation moved into Annaberg, Marienberg, and Niederschlag-Barenstein on the fringes of the original triangle. The Soviets needed large numbers of German miners but wanted to isolate them in secrecy while they supplied the Soviet physicists with uranium. However, there was already a chronic labor shortage in the Soviet Zone. They had to coerce and/or conscript mine workers to live in a confined and restricted forced labor camp very similar to a Russian Gulag or Nazi concentration

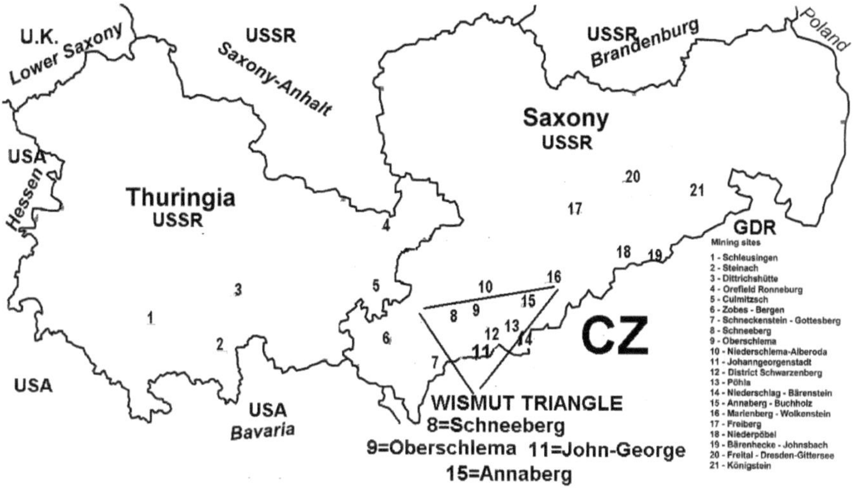

Fig. 10.3 Soviet uranium camps [15, 16, 38]. (Public domain)

camp. Beria appointed Major General Mikhail M. Maltsev (1904–n.d.), a Soviet Gulag specialist in forced labor, as the director general stationed in Chemnitz. An imprisoned Nazi, Johannes Schmidt, who was an experienced mine director, an *Obersteiger,* from Schneeberg, was resurrected from a Soviet prison camp with the understanding that his pardon depended on production quotas. Wismut was run by General Maltsev and the Soviet troops as a forced labor camp from 1947 to 1954 [12, 15, 16].

Wismut operated outside the jurisdiction of the Red Army and the East German civilian government. They took complete control of the mining area of 17 towns, 80 villages, and 250,000 people. The town populations increased precipitously and housing could not keep pace. From 1946 to 1951, (a) Oberschlema swelled from 2400 citizens to 9500, (b) John-George increased from 6000 to 38,300, and (c) the Schneeberg/Aue area, from 110,000 to 212,000. In 1949/1950, Wismut moved east of the Erzgebirge into the Vogtland Mountains and opened mines in Zobes/Bergen and Schneckenstein/Gottesberg. Then following the veins north into Thuringia, they hit the "mother lode" in the Ronneberg/Gera area where there was a small radon spring. This area alone furnished 100,000 t for the Soviet bombs. Saxon uranium furnished 80 % of the uranium for the first Soviet bomb, detonated in August of 1949. Eventually, the mining would extend to 400 mine shafts and 15 milling plants. Over 220,000 t of uranium would be mined by 150,000 workers, guarded by 5000 Soviet troops [12, 15, 16, 39].

When Stalin died in 1953, the Soviets had 120 atomic bombs. Conditions for the German miners began to improve as the (a) *Stalinist Era* (1947–1954) gave way to the (b) *German Democratic Republic Era* (1955–1969) and then the return of the (c) *Saxon Guilds* (1970–1990). During *the Stalinist Era, 1947–1954,* shaft ventilation and worker safety were nonexistent. The primitive mining techniques with dry drilling and only natural ventilation resulted in an enormously high dust burden. There are no records of radon measurements ever being taken. Accidents were rife and it was estimated that one out of seven workers became injured or died on the job. During this rapid growth of uranium mining, Wismut ran out of unemployed derelicts, criminals, and vagrants. Therefore, German prisoners of war were released from the Soviet Union and sent through Frankfort am Oder to Aue where they were immediately "recruited" for mine work. The Soviets also drafted totally unsuitable and inexperienced white collar workers, students, and the physically handicapped or disabled. Physicians who did not certify a worker as fit for duty were fined or worse [12, 15, 16].

The German labor unions were made responsible for finding workers and they had to go beyond Saxony and Thuringia into northern East Germany to conscript workers. All of this led to increased covert immigration into West Germany. In 1948, they tried enlisting transient or unattached women for mostly aboveground work. However, in what sounds almost like a Central Intelligence Agency plot, at least 50 % of the women had lues or gonorrhea, which led to more worker disability [12, 16]. A Communist puppet state, the German Democratic Republic was created in 1949. A Communist party, the *Sozialistische Einheitspartei Deutschlands,* had been formed in 1946 and in 1949 the Soviets turned nominal control of East Germany over to that Communist Party. The Soviets outwardly surrendered the functions of their military government to the *Sozialistische Einheitspartei Deutschlands* but the Soviet troops had to remain to keep the Communists in power. After Stalin died in 1953, things changed for the better for East Germany and for the Wismut miners [12, 15, 16].

Lavrentiy Beriya (1899–1953), the head of the NKVD, had done a masterful job overseeing the Soviet atomic bomb program and had delivered the bomb in less than 5 years. This feat was aided by his enormously successful recruitment of ideological spies in the New Deal State Department, the British secret service, Los Alamos and the Chalk River, Ontario, nuclear reactor. He had also organized the Communist takeover of the countries now in the Warsaw Pact. Naturally, he had some hopes of succeeding Stalin. Instead, he was shot during the coup d'état by Nikita Khrushchev and Marshall Georgy Zhukov in December of 1953. Khrushchev succeeded Stalin and in 1954 he replaced the NKVD with the KGB; Russian: *Komitet gosudarstvennoy bezopasnosti,* English: Committee for State Security [12, 15, 16, 29–31].

The *German Democratic Republic Era,* 1955–1969, followed the Stalinist Era for the miners. In 1954, Wismut joint ventured with the Republic to form the Soviet Deutsche Aktien Gesellschaft-Wismut; a 50/50 stock corporation. Up to this point, the uranium had all been considered as part of the US$10 billion war repatriations demanded by the Soviet Union. Now East Germany was a shareholder and would be paid something for their ore. The Red Army passed control of Wismut to the new company and dismissed M. M. Malstev. The East Germans now had more control over the daily operations of the mining. Prisoners were released. The Wismut workers became the best paid in East Germany; with access to the purchase of cars, refrigerators, and other home appliances. Safety features also improved with outside ventilation and wet drilling to reduce the dust [12, 15, 16].

Table 10.4 Soviet Zone uranium camps: 1947–1990 [38]

	UR mining camps	Tons	Operation	Map #
A	Ronnenburg Thuringia	100,000	1949–1990	4
B	Schneeberg; Oberschlema, Niederschlema	80,400	1946–1989	8, 9, 10
C	Konigstein	17,800	1967–1990	21
D	Culmitzsch Thuringia	11,000	1950–1967	5
E	Zobes/Bergen Vogtland	4862	1949–1961	6
F	Dresden/Freital	3700	1968–1989	20
G	John-George	3600	1946–1958	11
H	Schwartzenberg	1350	< 1959	12
I	Pohla/Fichtelberg	1200	1983–1990	13
J	Schneckenstein/Gottesberg	953	1950–1960	7
K	Annaberg/jBucholz	450	1947–1958	15
L	Thuringia Wald	166	1950s	1–3
M	Niederschlag/Barenstein	133	1947–1954	14
N	Marieberg	120	1947–1954	16
O	Freiberg	5	?	17
	Total	220,000		

The Saxon Guild Era, 1970–1990, began in 1970 when the miners were encouraged to resume the ancient and respected Saxon miner rituals and traditions that marked them apart as a fearless, indefatigable, and loyal brotherhood in the tenth century days of silver mining. The Communists saw this as a kind of a socialist super union. For the miners this was a return to the status of the venerable Saxon guilds. Their pay, benefits, and now health care, continued to exceed those of the average East German worker. During 1985–1990, a young Vladimir Putin served the KGB in their Dresden office (Table 10.4).

After the collapse of the Soviet Union in 1991 and subsequent German unification, the Wismut health records became available for the epidemiological study of the health hazards of uranium mining in Soviet Saxony and Thuringia. This constitutes the largest group of miners exposed to radon in the world; 60,000. From 1946 to 1990, 400,000 Wismut workers mined over 220,000 t of uranium for the Soviets. The records and tissue samples were studied by the German Cancer Research Center in Heidelberg and formed the basis for the German uranium miner's (GUM) studies. Autopsy data were available on workers who died between 1957 and 1989 and included 16,620 uranium workers and 9003 other Wismut employees who were not significantly exposed to uranium. The male uranium workers had a 32 % incidence of lung cancer compared to 12 % for other male workers and 2–2.6 % for female workers. The GUM studies also revealed a decrease in both lung cancer deaths and silicosis after the Stalinist period (Table 10.5). This appeared to be related to the improvements in the mine environment due to ventilation and wet drilling, etc. [40, 41].

Table 10.5 Cancer incidence in GUM workers [40, 41]. (From the archives of the Institute of Pathology of the Health Service of Wismut Co.)

Era	Stalinist	GDR	Guild
Years	1946–1954	1955–1969	1970–1090
Workforce #	100–140,000	45,000	20–27,000
Autopsies	10,303	2.284	227
% Lung cancer	34	23	17
% Other cancers	16	21	19
% Silicosis	35.8	12.8	20.7

GDR German Democratic Republic
Tissue from 28,975 autopsies transferred to the German Cancer Research Center in Heidelberg; 25,623 evaluated

Wismut left enough pollution to require decades of remedial effort by the German government. At least 15 uranium mills and 400 mine shafts were abandoned by the Soviets. The resort town of Oberschlema was destroyed and sank 40 ft. in altitude. In a lake just west of Chemnitz, 22,500 t of radioactive waste, arsenic, sulfuric acid and lead were dumped. South of Leipzig, 500 million t of radioactive and chemical wastes were scattered. Emitting radon into the wind were 3500 radioactive dumps. Saxony and Thuringia became the largest environmental cleanup project in Europe, with estimates of US$ 9.3 billion or 10–15 billion Deutsche Marks [12, 42].

Homes near the mines and nuclear waste dumps in Saxony developed elevated *indoor radon* levels. The 2009 World Health Organization (WHO) statistics for indoor radon found a worldwide average of 39 Bq/m^3 and the recommendations call for a maximum indoor level of 100–300 Bq/m^3 or less if possible [43]. After mining was stopped in 1990, Schneeberg buildings were averaging 300 Bq/m^3 and in 60 houses the median value was over 10,000 Bq/m^3 [44]. In 1993, measurements of *indoor radon* were taken from the Wismut mining centers at John-George, and from Schneeberg where half of the samples had radon concentrations four times those found in West Germany. Pooled European data revealed the risk of lung cancer death by age 75 for both *nonsmokers* and smokers exposed to the same indoor radon levels. Indoor levels of 0–100 Bq/m^3 resulted in 100–120 cases of lung cancer per 1000 smokers compared to only 4–5 cases in nonsmokers. With an indoor level of 800 Bq/m^3 there were 220 lung cancers per 1000 smokers and 10 per 1000 nonsmokers [45].

10.6 American Wismut

In 1946, the Atomic Energy Commission (AEC) was created to deal with the production of fissionable material and with the research and development of atomic weapons. No one else could buy or own uranium. The AEC was also specifically charged with regulating the safety of nuclear power without imposing excessive requirements that would inhibit the growth of the nuclear industry. The AEC had two divisions: (a) Military Applications and (b) Biology and Medicine. The later was to guarantee the *health and safety* of over 600,000 workers. The US Congress appointed nine senators and nine house representatives to act as a Committee on Atomic Energy to oversee the AEC. Additional safe guards were to be furnished by a Division of Mine Safety and Miner's Health, the Department of Labor, the Bureau of Indian Affairs within the Department of the Interior, the state mining inspectors, the US Geological Service, and the US Public Health Service [9, 46, 47].

Immediately prior to World War II, there was little mining activity on the Colorado Plateau. During the Manhattan Project, some of the old vanadium tailings were reworked for their uranium. However, most of the uranium for the American bombs came from Belgium and Canada. There were no more than 350 miners on the plateau until 1948/1949. In 1948, the AEC announced they would guarantee the price of uranium ore and this resulted in a rush of prospectors to the Colorado Plateau [12]. In August of 1949, America got a rude shock when the Soviets exploded their bomb, nicknamed Joe 1; an event that was not predicted to occur before 1965, if ever.

By 1950 there were about 550 miners on the plateau and the number almost doubled every 18 months. In 1950, the industry was comprised of numerous small mines, frequently with less than ten miners. Large deposits were discovered near Moab, Utah, in 1953 and near Grants, New Mexico, in 1954. Massive uranium mining and milling began in the effort to try and maintain nuclear superiority over the Soviet Union. After 1959, large mining corporations, frequently employing over 100 underground miners, came to dominate the business: Union Carbide, Anaconda, Kerr-McGee, United Nuclear, and Homesake Mining. By 1966, there were 600 uranium mines on the Colorado Plateau, employing over 5000 miners, who were predominantly Mormon but included hundreds of Navajo Indians. The work was 90 % underground. After 1973, several oil companies entered the uranium business: Standard of Ohio, Atlantic Richfield, Getty Oil, Conoco, Mobil, Gulf, and Exxon [12, 46–49].

In *1949,* the Colorado State Department of Public Health, the Colorado Bureau of Mines, and the US Vanadium Company requested the Public Health Service to determine *if* there was a potential hazard to uranium mining! The Public Health Service measured radon in 38 mines of the Colorado Plateau (Colorado, New Mexico, Arizona and Utah) during *1949/1950.* In order to gain access to the mines, the Public Health Service had to make an agreement with the mine owners that they would not inform the miners, "not alarm them," of the radiation hazards. The Public Health Service then found that radon concentrations in the unventilated mines of the Colorado Plateau were frequently above 3000–4000 pCi/L and as high as 10,000 pCi/L or 27.5 ME [48]. This was 10^3 times the level suggested in 1940 by Evans and Goodman, who had recommended 10 pCi/L as safe and 1000 pCi/L as dangerous [50]. A confidential memo by the AEC stated that the *1950/1951* samples showed a median level of 3100 pCi/L or 8.5 ME. The Public Health Service considered this a "serious problem, beyond all expectations and requiring a control program as soon as possible" [48, 49, 51].

However, the AEC was no more interested in the health hazards of uranium mining than was the NKVD and for the same reason; the arms race. The AEC informed the Public Health Service that the miner's health problems were the responsibility of the individual states where the mines were located. The AEC would only take responsibility for regulating exposure *after* the uranium had been removed from the mines. In addition, the AEC stated in *1951* that it was "unrealistic to set a definite level that mine operators should meet." Meanwhile, Dr. W. Hueper from the National Cancer Institute was to address the Colorado Medical Society in 1952. Shields Warren, from the AEC, requested that the National Cancer Institute silence him and/or fire him to prevent him from ever mentioning the German Erzgebirge lung cancer studies. Hueper was not fired but was forbidden to travel east of the Mississippi River [49]. Hueper must have felt like he was living in East Germany. In *1952,* the Public Health Service measured the radon concentrations in 157 mines on the Colorado Plateau. The *highest* concentrations were at Marysville, Utah, (48,700 pCi/L) and at Bull Canyon (59,000 pCi/L), south of the Paradox Valley in western Colorado. Varying local environments and mine sizes accounted for a wide range of concentrations but the American mines had radon concentrations that resembled the Erzgebirge mines *before* World War II [46–49, 51].

In 1953, the Public Health Service completed a series of ventilation studies and informed the mine owners in Colorado. They stressed that proper ventilation would significantly abate the hazards and could reduce radon from 26,900 to 500 pCi/L. The large mining companies gradually introduced ventilation. However, the Biology and Medicine Division of the AEC did not want to push for ventilation for fear of "getting the government in the Mining Business." A suggestion that the federal government ventilate the very uranium mines that they were subsidizing or increase the price they paid for uranium so that the mine owners could afford it, fell on deaf ears at the AEC [49].

A method for measuring the radon progeny in the mines became available by 1952. Until that time, only radon concentration had been measured. This was expressed as picocuries per liter of air in the USA and Canada while using becquerel per cubicmeter in Europe. In 1951, physicist William Bale from the University of Rochester, New York, and the AEC Health and Safety Branch reported that the radon progeny and their latent alpha energy could be measured [52]. The unstable progeny were found to release a total of 1.3×10^5 MeV of ionizing radiation energy during their decay, when the decay products were in equilibrium with *100 pCi/L of radon* [53]. This measurement was now used to measure the *exposure* hazard for the miners.

An exposure to the ionizing radiation of 1.3×10^5 MeV from alpha emission was called a working level (WL). *The WL was defined as the concentration of the short half-life daughters or radon that will release* 1.3×10^5 million eV of alpha energy per liter of air. The WL was adopted for compliance sampling in underground mines. By 1955, the Public Health Service recommended all states adopt a radon concentration of *100 pCi/L* as the standard equivalent of one WL, which represented exposure to 1.3×10^5 MeV of alpha irradiation. The American uranium mines were averaging 2.0–200 times the WL while non-uranium metal mines had 0.5–1.5 times the WL [51, 53].

The Public Health Service studied 678 mines from 1950 to 1959. By 1952, they found that 55% of the mines and 66% of the workers were being exposed to *1000 pCi/L*. However, by *1957–1959* only half of the mines exceeded the WL of *100 pCi/L and* only 20% exceeded 10 times the WL [48]. In 1960, the Public Health Service presented evidence to the governors of the Colorado Plateau states that linked uranium mining to lung cancer. In a study of 371 mines, lung cancer was observed 4.5 times more often than was expected. The Public Health Service stated that lung cancer was caused by uranium mining and that: "it was the most well documented occupational disease that (ever) existed" [9]. The states of New Mexico, Colorado, and Utah began to enforce some standards and the Public Health Service circulated a pamphlet that warned miners of a lung cancer hazard. By 1960, the AEC finally concluded that it could enforce safety measures in the mines that it leased. They had the Bureau of Mines inspect and follow up on the new standards and in 10 months the WL exposures dramatically decreased and continued to do so during 1961–1967 [49] (Table 10.6).

In 1967, Secretary of Labor Willard Wirtz announced the first federal standard for all the uranium mines which were leased by the federal government. He invoked the 1936 Walsh-Healy Act as his authority and set a limit for *exposure* to radon in the mines at one WL. To monitor compliance with the recommended monthly *exposure* limit for miners, the WL month (WLM) was introduced: the *exposure* to a concentration of radon progeny of one WL for one working month. One month was considered to have 170 working hours; 40 h/week and 4.25 weeks a month. The number of WLMs is the number of WLs times the number of months worked. Miner *exposure* was set at 1 WLM/month and 12 WLM/year [49, 53]. Shortly thereafter data was released on radon measurements from various mines during 1950–1968 which revealed a range of 120–3720 WLMs. Excess lung cancer was seen when the cumulative dose was $\geq$100 WLM. After 1971, the monthly *exposure* limit was reduced from 1 to 0.333 WLM and the yearly exposure reduced to only 4 WLM [54].

President Nixon created the Environmental Protection Agency (EPA) in 1970. In 1974, the American Congress responded to increasing criticism of the AEC by abolishing it. The Nuclear Regulatory Commission was created to properly manage the safety of workers and the public. The Department of Energy took over the other functions of the AEC The EPA and the Public Health Service collaborated in 1990 on establishing a "Toxicology Profile for Radon" by reviewing the various Public Health Service studies of 1973–1988 [55]. Their report confirmed the statistically significant excess in lung cancer for uranium miners in the USA, Canada, and Czechoslovakia. The deaths were linked to a total cumulative exposure greater than 100 WLM. Even non-uranium hard rock miners working in a low-exposure

Table 10.6 Improving WL exposures [51]

Year	# mines	%<1 WL	%>10 WL
1961	266	21	29
1962	327	32	4
1965	229	39	1
1967	217	70	1

environment developed lung cancer when their length of exposure led to a cumulative dose over 100 WLM. Nonsmoking uranium miners experienced no increased lung cancer deaths with less than 465 WLMs.

The American government began actions in May of 1989 to compensate the uranium miners and persons exposed to atmospheric nuclear testing during the Cold War. A Radiation Exposure Compensation Act was introduced in 1989, passed by the House and Senate during the summer of 1990 and signed into law in October by President G. H. W. Bush. Uranium miners who worked in Colorado, New Mexico, Arizona, Utah, and Wyoming during 1947–1971 and who developed cancer were to be compensated. It provided US$ 100,000 if the afflicted miner was exposed to from 200 to 500 WLM. The lower WLM total was for nonsmokers while smokers needed 300–500 to qualify [56, 57]. Finally, the USA had joined Czechoslovakia (1932) and Nazi Germany (1939/1940) in recognition of the occupational hazards of uranium mining.

As this embarrassing consequence of the Cold War was being addressed, the war ended.

The Soviet obsession with nuclear parity and Communist economic fallacies bankrupted the Soviet Union. In 1989/1990, the Communist puppet states of Poland, Czechoslovakia, and Hungary were overthrown. When they opened their borders, a mass exodus ensued from the German Democratic Republic and it collapsed in 1990. It was absorbed by a united Germany. Then in 1991 the three Baltic States and the Ukraine declared their independence as did seven more of the 15 Republics of the Union of Soviet Socialist Republics (USSR). The Communist Party was dissolved in August of 1991. The Soviet Union (USSR) was terminated on 26 December and was replaced by the Russian Federation [58]. Russia's western border reverted approximately to what it had been before the expansions of Tsar Peter, Empress Catherine, and Tsar Nicholas I. The border was also fundamentally what the Germans had negotiated in March of 1918 at the Treaty of Brest-Litovsk, which had also freed the Baltic states, Finland, Ukraine, and Poland from the Russian Bear [59, 60] (Fig. 10.4).

10.7 End Note

The Cold War and its uranium frenzy were over. And uranium mining in the USA was essentially over. The peacetime use of uranium fueled nuclear generators for electricity has never developed enough in the USA to support a domestic uranium mining industry. China and India are now the big customers for uranium, which is mined in Canada and Australia. Jachymov has recovered from the Soviet abuses and has returned as an international health spa with radioactive bath waters. The Katanga Province is still mining uranium but it is done by looters and scavengers who have made it even more dangerous than derelict Detroit. The old Carnotite Mineral District is now quiet. The Paradox Valley has reverted to open cattle range and the solitude is broken only by the rare tourists bicycling over the Rio Dolores Bridge at Bedrock or rafting north under the Bridge.

Fig. 10.4 Russian border 1991. (Authors collection)

It has been a century since the liquor-crazed miners in Bedrock would gallop off on a Saturday night, with guns blazing, to tattoo the schoolhouse door. However, when we enter the picturesque tranquil valley with its majestic high canyon walls and we scan the horizon for signs of life, it is easy for our imagination to conjure up visions of past inhabitants. Looking high up on the northwestern mesas, we may catch a glimpse of the rustlers cutting out brands on Sew-Em-Up Mesa while the McCarty brothers head out to rob a bank. On the southwest La Sal Creek, there is the Frenchman Poulot peering intensely at the canary yellow carnotite and ignoring his copper mine duties at the Cash-In to seek his fortune in uranium; while the ghost of headless Slim Hecox searches for his cranium. And there is Richard Pearce, Queen Victoria's Man in Denver, back from hobnobbing with the gentry at Briarhurst and looking for radium salts to take back to Cornwall to spike his tonic water. Over on the southeast Joe Flannery and a mule team are blazing a wagon road from Coke Ovens to the Rio Grande Southern railhead. In the north canyon's Long Park, Dr. Kelly and Professor Douglas are at the doctor's camp exalting the miners to avoid

Fig. 10.5 Paradox 2012 (Author)

"demon rum" and high above them on the Rio San Miguel is the bustling refinery at Joe Jr. Camp which proclaims Standard as the world's radium king. And then they are gone; these ghosts of the Paradox (Fig. 10.5). God rest their adventurous souls.

References

1. Segre E (1980) From x-rays to quarks. Freeman, San Francisco
2. Rhodes R (1986) The making of the atomic bomb. Simon & Schuster, New York
3. Gochfeld M (2005) Chronological history of occupational medicine. J Occup Environ Med 47(2):96–114
4. Hueper WC (2012) Biography of W.C. Hueper. University of California San Francisco. Legacy Tobacco Documents Library. http://www.legacy.library.ucsf.edu/tid/pmf89d00/pdf. Accessed 15 July 2012
5. Davis D (2007) The secret history of the war on cancer. Basic Books, New York
6. Sellers C (1997) Discovering environmental cancer: Wilhelm Hueper. Am J Public Health 87(11):1824–1835
7. Hueper WC (1942) Occupational tumors and allied diseases. Thomas, Springfield
8. Lorenz E (1944) Radioactivity and lung cancer. J Natl Cancer Inst 5(1):1–15
9. Caulfield C (1989) Multiple exposures. Harper and Row, New York

10. Groves LR (1962) Now it can be told. Harper and Row, New York
11. Brown AC, MacDonald CB (eds) (1997) The secret history of the atomic bomb. Dial Press/James Wade, New York
12. Zoellner T (2009) Uranium. Penquin Books, New York
13. Helmreich JE (1982) Gathering rare ores. Princeton University Press, Princeton
14. Lange E (1935) Krebser-krankungen und geologishe verhaltnisse im erzgebirge. Z Krebsforsh 42:306–310
15. Naimark NM (1995) The Russians in Germany. Belknap/Harvard Press, Cambridge
16. Zeman Z, Karksch R (2008) Uranium matters. CEU Press, Budapest
17. Gilbert M (1989) The second world war. Holt, New York
18. Ferrell RH (1998) The dying president, Franklin D. Roosevelt 1944–1945. University of Missouri, Columbia
19. Albright J, Kunstel M (1997) Bombshell. Times Books, New York
20. Romerstein H, Breindel E (2000) The venona secrets. Regnery History, Washington, DC
21. Bauer E (1979) The history of world war II. Galahad Books, New York
22. Hart BL (ed) (1980) History of the second world war. Exeter, New York
23. Ambrose SE, Sulzberger CL (2009) World war II. American heritage, tess press. Black Dog & Levanthal, New York
24. Ziemke EF (1975) U.S Army in the occupation of Germany map. http://www.en.wikipedia.org/wiki/Allied-occupied_Germany#mediaviewer/File:Germany_occupation_zones_with_border.jpg. Accessed 22 Aug 2014
25. Proctor RN (1999) The Nazi war on cancer. Princeton University Press, Princeton
26. Marshall JL, Marshall VR (2008) Jachymov (Joachimsthal), Czech Republic. Hexag Alpha Chi Sigma 99(4):68–71, 75
27. Marshall JL, Marshall VR (2010) Photographs of cities Jachymov CR. In: Rediscovery of the elements DVD. Jenny Marshall Computer Services, P.O. Box 6, Denton Texas. http://www.jennymarshall.xom/products.htm. Accessed 22 Aug 2014
28. Kunz A (2005) Occupation zones and states map 1947. http://www.germanhistorydocs.ghi-dc.org/map.cfm?map_id=525. Accessed 22 Aug 2014
29. Curtis GE (ed) (1992) Post war East Germany. http://www.shsu.edu/~his_ncp/EGermPW.html. Accessed 22 Aug 2014
30. US Dept. State (2014) East Germany (German Democratic Republic). http://www.history.state.gov/countries/german-democratic-republic
31. Applebaum A (2012) Iron curtain. Doubleday, New York
32. Callahan M (2014) Behind the secret plan to bring Nazi scientists to US. New York Post 1 Feb 2014. http://www.nypost.com/2014/02/01/behind-the-secret-plan-to-smuggle-nazi-scientists-to-america/. Accessed 22 Aug 2014
33. Goudsmit SA (1947) Alsos. H Schuman, New York
34. BIOS (2014) British intelligence objective sub-committee BIOS final report no. 303: production of luminous compounds at the works of Auer Gesellschaft A.G. circa 1945. http://www.cdvandt.org/fiat-cios-bios.htm. Accessed 22 Aug 2014
35. Hayes P (2004) From cooperation to complicity, degussa in the 3rd reich. Cambridge University Press, New York
36. Riehl N, Seitz F (1996) Stalin's captive: Nikolaus Riehl and the soviet race for the bomb. Chemical Heritage Foundation, Philadelphia
37. DDR (1947) Germany Laender map 1947 1990.png. http://www.commons.wikimedia.org/wiki/File:Germany_Laender_1947_1990_DDR.png. Accessed 3 June 2014
38. Geomartin (2013) WISMUT location map (Saxony and Thuringia). http://www.en.wikipedia.org/wiki/Wismut_%28mining_company%29#mediaviewer/File:Wismut_locati on_map.jpg. Accessed 14 Jan 2013
39. Otten H et al (2011) Case study: Wismut-a uranium exposure revisited. In: Stellman SD (ed) Encyclopedia of occupational health and safety. http://www.ilo.org/oshenc/part-iv/record-systems-and-surveillance/item/738-case-study-wismut-a-uranium-exposure-revisited. Accessed 29 Aug 2014

40. Wesch H, Wiethege T et al (1991) German uranium miner study—historical background and available histo-pathological material. Radiat Res 152:48–51
41. Wiethege T, Wesch H et al (1999) German uranium miner study—pathological and molecular genetic findings. Radiat Res 152:52–55
42. Schuttmann W (1993) Schneeberg lung disease and uranium mining in the Saxon ore mountains (erzgebirge). Am J Ind Med 2:355–368
43. WHO (2013) World health organization handbook on indoor radon. http://www.nrsb.org/pdf/WHORadonHandbook.pdf. Accessed 27 Feb 2013
44. Keller G (1993) Radiological aspects of former mining activities in the Saxon erzgebirge. Environ Int 19:449–454
45. National Research Council (1999) Health effects of exposure to radon: BEIR VI. The National Academies Press, Washington, DC. http://www.nap.edu/catalog.php?record_id=5499. Accessed 27 Feb 2013
46. Ball H (1993) Cancer factories. Greenwood Press, Westport
47. Proctor RN (1995) Cancer wars. Basic books, New York
48. Cooper WC (1962) Session III, Case Study No. 1, uranium milling and mining. J Occup Med 4(Suppl):614–627
49. National Security Archives (1995) ACHRE (The Advisory Committee on Human Radiation Experiments) Part II. Chapter 12: the uranium miners. In: The National Security Archive/ACHRE; ACHRE REPORT 3 Oct 1995. http://www.gwu.edu/~nsarchiv/radiation/. Accessed 27 Feb 2013
50. Evans RD, Goodman C (1940) Determinations of the thoron content of air. J Ind Hyg Toxicol 22(3):89–99
51. Holaday DH (1969) History of the exposure of miners to radon. Health Phys 16:547–552
52. Bale WF (1980) Memorandum to the files: hazards associated with radon and thoron. Health Phys 38(6):1061–1066
53. Evans RD (1980) Engineers' guide to the elementary behavior of radon daughters. Health Phys 38(6):1173–1197
54. MSHA (2013) Mine safety health administration—handbook series—PH06-IV-1(1) Metal And Nonmetal. Chapter 10 Ionizing radiation. http://www.msha.gov/Readroom/HANDBOOK/MNMInspChapters/Chapter10.pdf. Accessed 27 Feb 2013
55. US Public Health Service (2013) Toxicological profile for radon. http://www.bvsde.paho.org/bvstox/i/fulltext/toxprofiles/radon.pdf. Accessed 27 Feb 2013
56. George Washington University (2014) Radiation exposure compensation act. http://www2.gwu.edu/~nsarchiv/radiation/dir/mstreet/commeet/meet3/brief3.gfr/tab_g/br3g1e.txt. Accessed 2 June 2014
57. Dept. of Justice (2014) Radiation exposure compensation act. http://www.justice.gov/civil/common/reca.html. Accessed 2 June 2014
58. Clines FX (2010) Gorbachev resigns. NY Times 26 Dec 1991. http://www.nytimes.com/learning/general/onthisday/big/1225.html#article. Accessed 22 Aug 2014
59. Toland J (1980) No man's land. Doubleday, New York
60. Baldwin HW (1962) World war I. Grove, New York

Appendices

Appendix I: Periodic Table

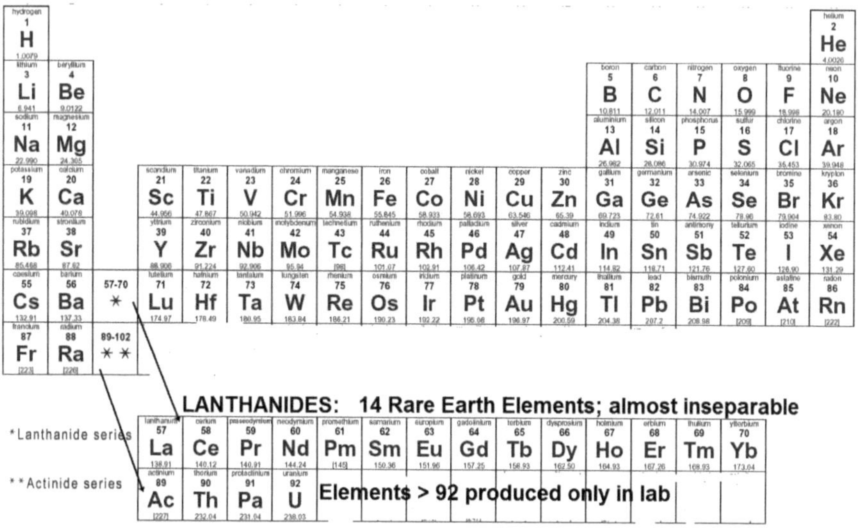

R. F. Robison, *Mining and Selling Radium and Uranium*,
DOI 10.1007/978-3-319-11830-7

Appendix II: Radiation Therapy Trials: 1900–1909

II A	X-Ray therapy	Trials 1896–1905	II B	Radium therapy	Trials 1900–1909
Year	Disease	Author—country/city		Disease	Author—country/city
1900	Lupus vulgaris; Lupus erythemateux	J Jutassy—Hungary	*1901*	Lupus erythemateux	H Danlos; P Block—Paris
	Alopecia areata	R Kienbeck; G Holtzknecht—Vienna	*1902*	Lupus vulgaris; psoriasis; SCC[a]; port wine stain[b]	H Danlos—Paris
	Vascular nevus; eczema	J Jutassy	*1903*	Lupus (2)	J MacIntyre—UK
	Psoriasis	Strator—German		Psoriasis; lupus: oral cavity ca; melanoma:	G Holtzknecht—Vienna A. Exner—Vienna
	Lupus vulgaris	R Hahn—Hamburg		Cured skin cancer	SW Goldberg and ES London—Russia
	Superficial sarcoma	BM Ricketts—USA		Cervix; sarcoma	M. Cleaves—NYC
	Acne rosacea	J Jutassy; R Hahn		Nasal; oral cavity; uterus	GH Stover—Denver
	Epithelioma[a] (SCC)	W Merrill; W Johnson—USA	*1904*	SCC nose, face; BCC; nevus	M. Davidson—UK
1901	Cured skin cancer: SCC	T Sjogren—Sweden		Fibroma; lipoma; neuroma	Bondet—Lyon
	Ekzem, acne rosacea, acne vulgaris; psoriasis	R Hahn and Albers—Schonberg		Keloids; acne; eczema; psoriasis; BCC(5); SCC(33); breast (4)	FH Williams—Boston
	Leprosy	JH Sequeira—UK		Breast cancer; eczema	O Lassar—Berlin
	Breast cancer	GC Hopkins—Philadelphia		Pigmented nevi; angioma	TJP Hartigan—UK

II A	X-Ray therapy	Trials 1896–1905	II B	Radium therapy	Trials 1900–1909
Year	Disease	Author—country/city		Disease	Author—country/city
	Systemic tuberculosis	J Rudis-Jucinsky—Iowa		Esophagus (9)	M Einhorn—USA
1902	Mycosis fungoides; lichen planis	W Scholtz—German		Breast cancer; warts (5); lupus; cervix; vagina	R Abbe—NYC
	SCC; esophagus	WA Pusey—Chicago	*1905*	Rodent ulcer = BCC^; lupus	M Metzenbaum—USA
	SCC eyelids; analgesia	WH Sweet—Philadelphia		SCC cervix; fibroids; goiter	R Abbe—NYC
	Neuralgia	Bondurant—New York		SCC; pigmented nevus	Wickham and Degrais
1903	Lupus vulgaris; SCC	EH Grubb—Chicago		Sycosis; eczema; acne roseacea; lichen ruber	A. Blaschko—France
	Lymphoma	WA Pusey		Sarcoma	W J Morton—USA
	Lymphoma, leukemia	N Senn—Chicago	*1906*	SCC nose	Wickham and Degrais
	Breast cancer	WA Pusey		SCC cervix-intracavitary	P Oudin and Verchere
	Keloids	HK Pancoast—USA		Eczema; angiomas	Wickham and Degrais
1904	Hyperthyroid goiter	CH Mayo—USA		Lupus (35)	R Abbe
1905	Leprosy (13)	HB Wilkinson—Manila	*1907*	SCC ear and eyelids	Wickham and Dominici
	Breast cancer (23)	RH Boggs—Pittsburgh	*1908*	SCC lips	Dominici
	Breast cancer	GH Stover—Denver	*1909*	SCC tongue, cheek	Abbe; Dominci

[a] *SCC* squamous cell carcinoma skin cancer, *BCC^* basal cell carcinoma skin cancer
[b] Birth mark, angioma, port wine stain, and vascular nevus

Appendix III: Uranium Minerals: Tested by Marie Curie and Used by European Factories

Name and Formulae	Chemical name:	Sites where found:	% Uranium
Autunite Ca $(UO_2)_2$ $(Po_4)_2 \times 10–12\ H_2O$	A hydrated uranyl calcium phosphate (1852); yellow-green fluorescent; or calcium uranium phosphate	1 = Portugal, Cornwall, Madagascar 2 = France	48–50
Betafite $(CaU)_2(Ti, Nb, Ta)_2O_6(OH)$	A niobo-titanate of uranium, calcium, REE[a] (1906)	Madagascar 1908	17–20
Carnotite $K_2(UO_2)_2$ $(VO_4)_2 \times 3\ H_2O$	A vanadium-uranate (1899); canary yellow and green yellow	1 = Colorado-Utah border 2 = Australia, Portugal	<5, rarely 50
Clevite $UO_2 + 10\%$ $(REE)_2O_3$	A urinate with thorium and REE, (1895)—black	1 = Sweden, Norway 2 = USA, Saxony	48–61
Fergusonite: REE (NbO_4); most common = $YNbO_4$	A niobate of yttrium, eribium and/or other REE—brown black	1 = Ytterby Sweden, Norway 2 = Russia, USA, Ceylon	2.1–8.0
Gummite (Pb, Ca, Ba) U_3, $SiO_{12} \times 6\ H_2O$	A uraninite variant—red, yellow, orange, brown	S axony, Bohemia	55–60
Pechblende UO_2, or U_2O_3, or U_3O_8	Uranite or uranium oxide (1789)—black/brown	1 = Bohemia, Saxony 2 = Cornwall, E. Colorado	50; 50–80 range
Samarskite (Y, Fe, U)(Nb, Ta) O_4	An oxide of Y, Fe, Nb, Ta, U. (1847)—black	1 = Russia, Sweden, Norway	3–14
Tobernite (chalcolite) $Cu(UO_2)_2(PO_4)_2 \times 8–12\ H_2O$	Copper urinate (1793)—bright green copper uranium phosphate—green	1 = Cornwall, Portugal 2 = French Indo-China	50–60
Monazite (Th)(PO4)(REE) or (Ce-La-Nd-Y) $ThPO_4$	A thorium phosphate and REE mineral with three different species; Ce, La, Nd, or Y	1 = Brazil, Norway 2 = German factories	0.12–0.46 Th=2.7–15.2
Thorite (Th, U)SiO_4	A neosilicate of thorium and uranium. (1828)—brown-black to yellow-orange	Norway	6.26 % Th=46.6

Name and Formulae	Chemical name:	Sites where found:	% Uranium
Orangite	A less common orange colored variety of thorite		7.76 % Th = 72–82
Thorianite ThO_2 and REE (mostly Ce)	A rare thorium oxide mineral (1904)—dark gray, brown black	Sri Lanka, India, the USA Madagascar, Russia, Congo, Canada, South Africa	4–10

1 primary, *2* secondary
[a] *REE* rare earth elements: Ce, La, Nd, Ca, Pr, Sm

Appendix IV: Radioactive Decay Tables: Unstable to Stable

Uranium		Thorium		Actinium	
Isotope-particle	Half-life	Isotope-particle	Half-life	Isotope-particle	Half-life
$^{238}U_{92}$-alpha	4.58×10^9 year	–	–	–	–
$^{234}Th_{90}$-beta	27 days	–	–	–	–
$^{234}Pa_{91}$-beta	1.17 min	–	–	–	–
$^{234}U_{92}$-alpha	2.5×10^5 year	$^{232}Th_{90}$-alpha	1.4×10^{10} year	$^{235}U_{92}$-alpha	7.1×10^8 year
$^{230}Th_{90}$-alpha	7.5×10^4 year	$^{228}Ra_{88}$-beta	6.7 years	$^{231}Th_{90}$-beta	25.5 hours
$^{226}Ra_{88}$-alpha	1602 years	$^{228}Ac_{89}$-beta	6.13 h	$^{231}Pa_{91}$-alpha	3.2×10^4 year
$^{222}Rn_{86}$-alpha	3.80 days	$^{228}Th_{90}$-alpha	1.91 years	$^{227}Ac_{89}$-beta	21.6 years
$^{218}Po_{84}$-alpha	3.1 min	$^{224}Ra_{88}$-alpha	3.64 days	$^{227}Th_{90}$-alpha	18.2 days
$^{214}Pb_{82}$-beta	26.8 min	$^{220}Rn_{86}$-alpha	55 s	$^{223}Ra_{88}$-alpha	11.4 days
$^{214}Bi_{83}$-beta	19.7 min	$^{216}Po_{84}$-alpha	0.16 s	$^{219}Rn_{86}$-alpha	4.0 s
$^{214}Po_{84}$-alpha	164.3 µs	$^{212}Pb_{82}$-beta	10.6 h	$^{215}Po_{84}$-alpha	1.78 ms
$^{210}Pb_{82}$-beta	22.3 years	$^{212}Bi_{83}$-beta	60.6 min	$^{211}Pb_{82}$-beta	36.1 min
$^{210}Bi_{83}$-beta	5 days	$^{212}Po_{84}$-alpha	300 ns	$^{211}Bi_{83}$-alpha	2.15 min
$^{210}Po_{84}$-alpha	138 days	$^{208}Tl_{81}$-beta	3.1 min	$^{207}Th_{81}$-beta	4.79 min
$^{206}Pb_{82}$	*Stable*	$^{208}Pb_{82}$	*Stable*	$^{207}Pb_{82}$	*Stable*

A# mass number = atomic weight = total number of nuclear particles = protons and neutrons
Z# atomic number = number of protons = number of electrons
A alpha particle; loss causes Z# minus 4 and A# minus 2
B beta particle; loss converts neutron to proton = A# plus one
ms = milliseconds; *ns* = nanoseconds

Appendix V: Uranium Decay Emissions

Stage of decay	Element Z#	Isotope and A#	Half-life	MeV of alpha decay	MeV of beta decay	% of gamma occurrence	MeV energy of gamma rays
1	# 92	U-238	4.58×10^9 year			23 %	<0.1
6	# 88	Ra-226	1600 years	4.79		4 %	<0.2
7	# 86	Rn-222	3.83 days	5.5	None	<1 %	0.5
8	*# 84*	*Po-218*	*3 min*	*6.0*	*None*	–	–
9	# 82	Pb-214	27 min	None	0.65–1.02	4.0 %	0.24, 0.29, 0.35
10	# 83	Bi-214	20 min	none	3.17–3.3	47 %	0.61, 0.77, 0.93, 1.12, 1.24, 1.38, 1.76, 2.2
11	*# 84*	*Po-214*	*0.164 ms*	*7.7*		<1 %	0.8
12	# 82	Pb-210	22.3 years				

Appendix VI: Beta and Gamma Decay Emissions

Uranium decay series				
Parent	Half-life	Daughter	Emitted	Energy in MeV
$^{214}Pb_{82}$	26.8 min	$^{214}Bi_{83}$	A beta particle of energy	0.65–1.02
			+3 gamma rays of	0.24, 0.29, 0.35
$^{214}Bi_{83}$	20 min	$^{214}Po_{84}$	A beta particle of energy	3.17–3.3
			+4 gamma rays of	0.61, 0.77, 0.93, 1.12
			+4 gamma rays of	1.24, 1.38, 1.76, 2.2
Thorium decay series				
$^{228}Ac_{89}$	6.13 h	$^{228}Th_{90}$	One beta of energy	1.11
			Gamma ray energy of	0.90
$^{208}Tl_{81}$	3.1 min	$^{208}Pb_{82}$	One beta of energy	1.79
			Gamma ray energy of	2.62

Actinium-228 ($^{228}Ac_{89}$) and thalium-208 ($^{208}Tl_{81}$)

Appendix VII: Price of Radium, in US Dollars

	Lubinau: 2008	Hahne: 2006	Landa: 1982
1902	3000	–	–
1903–1904	10,000	10,000	–
1904–1905	15,000–18,600	15,000	–
1905	25,000–50,000	25,000	20,000
1905–1906	50,000	50,000	–
1906	50,000–90,000	60,000	40,000
1907	–	–	60,000
1908	–	–	80,000
1909–1910	75,000–135,000	75,000–135,000	100,000
1910	135,000	135,000	120,000
1911	120,000	–	130,000
1911–1912	150,000	150,000	140,000
1912–1913	180,000	180,000	160,000
1914	120,000–170,000	180,000	–
1915–1916	120,000–160,000	160,000	160,000
1916–1918	90,000–160,000	120,000	140,000
World War I	135,000	–	120,000–170,000
1920	–	–	110,000–115,000
1921–1922	120,000	100,000–105,000	–
1923–1933	70,000	70,000	70,000
1934	50,000	–	50,000
1938	25,000	–	25,000
1958	16,000–21,000	–	20,000

References

1. Abbe T (1904) Notes on the physiological and the therapeutic action of radium. Wash Med Ann 2(6):363–367
2. Adloff JP, Kauffman GB (2007) The troubled story of polonium. Chem Educ 12:94–101
3. Bonewitz RL et al (2005) Smithsonian rock & gems. DK Publishing, New York
4. Cleaves MA (1904) Light energy. Rebman, New York
5. Curie M, Curie I (1926) Radium. Encyclopedia Britannica 13th edition. http://www.britannica.com/nobelprize/article-9435287. Accessed 9 April 2013
6. Hahne F (1989) Early uranium mining in the United States. World Nuclear Association web site. http://www.world-nuclear.org/usumin.htm. Accessed 10 Feb 2006
7. Kassabian MK (1907) Electro-therapeutics and roentgen rays. Lippincott, Philadelphia
8. Landa ER (1982) Radium—the first nuclear industry. Sci Am 247(5):180–193
9. Lubinau JO, Mould RF (2008) The roller coaster price of radium. Nowotw J Oncol 58(4):9e–14e
10. MacKee GM (1938) X-rays and radium in the treatment of skin diseases, 3rd edn. Lea & Febiger, Philadelphia
11. Moore RB, Kithil KL (1913) A preliminary report on uranium, radium, and vanadium bulletin 70. Dept. of Interior, Bureau of Mines, Washington Govt. Printing Office
12. Mould RF (2007) Radium history mosaic. Nowotw J Oncol 57(Suppl 4):99–127 (Warsaw)
13. Viol CH (1921) History and development of radium-therapy. J Radiol 2:29–34
14. Wickham L, Degrais P (1910) Radiumtherapy. Funk & Wagnalls, New York

Index

R. F. Robison, *Mining and Selling Radium and Uranium*,
DOI 10.1007/978-3-319-11830-7

MIX
Papier aus verantwortungsvollen Quellen
Paper from responsible sources
FSC® C105338

If you have any concerns about our products,
you can contact us on
ProductSafety@springernature.com

In case Publisher is established outside the EU,
the EU authorized representative is:
Springer Nature Customer Service Center GmbH
Europaplatz 3, 69115 Heidelberg, Germany

Printed by Libri Plureos GmbH
in Hamburg, Germany